STAR CLUSTERS AND BLACK HOLES
IN GALAXIES ACROSS COSMIC TIME

IAU SYMPOSIUM 312

COVER ILLUSTRATION:

The background for the symposium poster is comprised of several images relating to the event. From bottom to top, the logos of the Silk Road Project, which is funded by the Chinese Government's Thousand Talent (Qiān Rén) Programme, and contributed many person-hours to the organisation of the conference, of the two hosting institutions, the Kavli Institute for Astronomy and Astrophysics and the National Astronomical Observatories of China, as well as their parent institutions of Peking University and the Chinese Academy of Sciences. Further up is a silhouette of the Gate of Heavenly Peace (Tiān'ān Mén) in central Beijing. The sinusoidal patters in the middle of the picture represent gravitational waves while the image on the right is an artist's impression of tidal disruption of a star around the supermassive black hole binary in the galaxy SDSS J120136.02+300305.5, discovered by Fukun Liu, Shuo Li, and Stefanie Komossa. Finally, at the top, NGC 869 and NGC 884 in the Perseus Constellation, also known as the Double Cluster and Caldwell 14. This pair of open clusters lies approximately two kiloparsecs from the Earth and is visible to the naked eye under some conditions (rarely in Beijing) or with small binoculars.

INTERNATIONAL ASTRONOMICAL UNION

UNION ASTRONOMIQUE INTERNATIONALE

STAR CLUSTERS AND BLACK HOLES IN GALAXIES ACROSS COSMIC TIME

PROCEEDINGS OF THE 312th SYMPOSIUM OF THE INTERNATIONAL ASTRONOMICAL UNION HELD IN BEIJING, CHINA AUGUST 25–29, 2014

Edited by

YOHAI MEIRON

Kavli Institute for Astronomy and Astrophysics at Peking University

SHUO LI

National Astronomical Observatories, Chinese Academy of Sciences

FUKUN LIU

Department of Astronomy, Peking University

and

RAINER SPURZEM

National Astronomical Observatories, Chinese Academy of Sciences

CAMBRIDGE UNIVERSITY PRESS
University Printing House, Cambridge CB2 8BS, United Kingdom
40 West 20th Street, New York, NY 10011–4211, USA
10 Stamford Road, Oakleigh, Melbourne 3166, Australia

First published 2016

Printed in the UK by Bell & Bain, Glasgow, UK

Typeset in System LaTeX 2_ε

A catalogue record for this book is available from the British Library Library of Congress Cataloguing in Publication data

ISBN 9781107078727 hardback
ISSN 1743-9213

Table of Contents

PART ONE

Black Holes in Galactic Nuclei

Session 1: Galaxy Mergers, AGN Feedback, Binary Black Holes, Tidal Disruption

Session 2: Dynamics of Stars and Gas around Black Hole

Session 3: Accretion Disks around Supermassive Black Holes

PART TWO
Galactic and Extragalactic Globular Clusters

Session 1: Extragalactic Globular Cluster Systems

Session 2: Globular Star Clusters in the Local Group

Session 3: Galactic Globular Clusters

Session 4: Dwarf Galaxies, Nuclear Star Clusters

Session 5: Nuclear Star Clusters

PART THREE
Gravitational wave emission, observations, and the link to astrophysics

Preface

Star clusters and black holes are moving into the focus of high resolution astrophysics, computationally as well as observationally. For the first time, Observations in many regions of the electromagnetic spectrum are converging with theoretical modelling and computer simulations. These cosmological and galaxy formation models reach down to the supermassive black hole level and follow their formation and growth in the centres of galaxies, by gas and star accretion. High star formation activity in the early universe leads to the formation of dense and very compact clusters of stars around these black holes, where stars can diffuse into the depth of the potential well and finally get disrupted by tidal forces of the central black hole, or by direct stellar collisions. Gas which settles in nuclear discs around supermassive black holes feeds the central engine of active galactic nuclei, where relativistic dynamics of stars and gas is coupled to the larger galactic scales through energy and momentum feedback.

IAU Symposium (IAUS) 312 brought together experts on high resolution observations as well as theoretical modelling and computational simulations, who presented their research on star clusters, black holes and their interrelation. IAUS 312 continues a tradition of former IAU symposia on stellar dynamics and related fields, beginning with IAUS 69 in Besançon, France (1969; *Dynamics of Stellar Systems*) and IAUS 113 in Princeton, New Jersey, United States (1984; *Dynamics of Star Clusters*), continued by IAUS 174 in Tokyo, Japan (1995; *Dynamical Evolution of Star Clusters – Confrontation of Theory and Observations*), and IAUS 246 in Capri, Italy (2007; *Dynamical Evolution of Dense Stellar Systems*). The interested reader could nicely follow the evolution of the field and its main scientific actors by browsing through these historic volumes. One may also realise how modern astrophysical techniques (computational as well as observational) have widened the field, even though one can still find methods and physical concepts today, active and well used in science, which were introduced as early as *e.g.* in IAUS 69.

IAUS 312 was divided into three main parts:

(*a*) black holes in galactic nuclei

(*b*) galactic and extragalactic star clusters

(*c*) gravitational waves and multi-messenger astrophysics

Compared to the past symposia mentioned above, the connection with gravitational wave astrophysics was new and could prepare the ground for future collaborations between astrophysicists, instrument developers and data processing experts for ground and space based gravitational wave instruments. Just like the modelling of star cluster dynamics was inspired many decades ago by optical star counts and observations of stars using electromagnetic waves, we may see in the coming decades scientific inspirations coming from gravitational wave signals originating from compact objects (*e.g.* black holes) in star clusters. Certainly, the immense progress in astronomy in the optical, as well as in other regions of the electromagnetic spectrum, is and will be a main driver in the field in the near future.

During IAU Symposium 312, about 130 astronomers from 22 countries in all continents (except Antarctica) assembled to exchange their new results on black holes, star clusters, and gravitational wave research. This was the first IAU Symposium in the field to take place in China, and while many domestic researchers and students attended, it also attracted a large number of high level international scientists to the country, some of them for the first time.

The chairs and members of the scientific and local organising committees and the editors would like to cordially thank all colleagues, administrative staff and students who have contributed to make this conference a success, for their help and support. We thank the International Astronomical Union for supporting the proposal for this Symposium and also for providing significant support for young scientists. We also thank the National Astronomical Observatories of Chinese Academy of Sciences, the Thousand Talent Programme of the Government of China, and the National Science Foundation of China for significant support.

Yohai Meiron, Shuo Li,
Fukun Liu & Rainer Spurzem
Beijing, China, February 9, 2015

THE ORGANIZING COMMITTEE

Scientific

Rainer Spurzem (China; co-chair)
Fukun Liu (China; co-chair)
Joss Bland-Hawthorn (Australia)
Joan Centrella (USA)
Ron Ekers (Australia)
Martin Gaskell (USA)
Douglas Heggie (UK)

Stefanie Komossa (Germany/China)
Hyung-Mok Lee (Korea)
Jufu Lu (China)
Steve McMillan (USA)
Giampaolo Piotto (Italy)
Alison Sills (Canada)
Kim Venn (Canada)

Local

Rainer Spurzem (NAOC[1]/KIAA[2]; co-chair)
Fukun Liu (DoA[3]/KIAA; co-chair)
Peter Anders (NAOC)
Peter Berczik (NAOC)
Licai Deng (NAOC)
Jose Fiestas (NAOC)
Lijun Gou (NAOC)
Gareth Kennedy (NAOC)
Thijs Kouwenhoven (KIAA)

Lixin Li (KIAA)
Shuo Li (NAOC)
Jifeng Liu (NAOC)
Youjun Lu (NAOC)
Yohai Meiron (KIAA)
Eric Peng (KIAA/DoA)
Maxwell Xu Tsai (NAOC/KIAA)
Qingjuan Yu (KIAA)

[1] National Astronomical Observatories, Chinese Academy of Sciences

[2] Kavli Institute for Astronomy and Astrophysics at Peking University

[3] Department of Astronomy, Peking University

Acknowledgements

The symposium was sponsored and supported by the coordinating IAU division H (Interstellar Matter and Local Universe), and further supported by the IAU divisions D (High Energy Phenomena and Fundamental Physics), G (Stars and Stellar Physics), and J (Galaxies and Cosmology); and by the IAU Commissions No. 28 (Galaxies), 33 (Structure and Dynamics of the Galactic System), 37 (Star Clusters and Associations), 44 (Space and High Energy Astrophysics), and 47 (Cosmology).

Funding by the International Astronomical Union, Thousand Talent (Qiān Rén) Programme of the Government of China, National Astronomical Observatories of Chinese Academy of Sciences and its Silk Road Project, National Natural Science Foundation of China (NSFC grant No. 11410301047), Sugon, National Science Library of Chinese Academy of Sciences, is gratefully acknowledged.

CONFERENCE PHOTOGRAPH

Participants

Marek **Abramowicz**, Copernicus Center — marek.abramowicz@physics.gu.se
Danor **Aharon**, Technion – Israel Institute of Technology — danor@tx.technion.ac.il
Karla **Alamo-Martínez**, Peking University — k.alamo@crya.unam.mx
Tal **Alexander**, Weizmann Institute of Science — tal.alexander@weizmann.ac.il
Pau **Amaro Seoane**, Albert Einstein Institute — Pau.Amaro-Seoane@aei.mpg.de
Peter **Anders**, National Astronomical Observatories, Chinese Academy of Sciences — anders@bao.ac.cn
Farruh **Atamurotov**, Institute of Nuclear Physics — farruh@astrin.uz
Giacomo **Beccari**, ESO — gbeccari@eso.org
Matthew **Benacquista**, University of Texas at Brownsville — benacquista@phys.utb.edu
Peter **Berczik**, National Astronomical Observatories of China, Chinese Academy of Sciences — berczik@nao.cas.cn
Paolo **Bianchini**, Max-Planck Institute for Astronomy — bianchini@mpia.de
Omer **Blaes**, University of California, Santa Barbara — blaes@physics.ucsb.edu
Patrick **Brem**, Albert-Einstein-Institute — pbrem@aei.mpg.de
Maxwell Xu **Cai**, National Astronomical Observatories, Chinese Academy of Sciences — maxwell@nao.cas.cn
Raymond **Carlberg**, University of Toronto — raymond.carlberg@utoronto.ca
Joan **Centrella**, Astrophysics Science Division — Joan.Centrella@nasa.gov
Xian **Chen**, Max-Planck Institute for Gravitational Physics — xian.chen@aei.mpg.de
Yuguang **Chen**, Peking University — yuguang.chen.1@gmail.com
Zihan **Chen**, National Astronomical Observatories, Chinese Academy of Sciences — zhchen@nao.cas.cn
Huaqing **Cheng**, National Astronomical Observatories, Chinese Academy of Sciences — hqcheng@nao.cas.cn
Kyungwon **Chun**, School of Space Research, Kyung Hee University — kwchun@khu.ac.kr
Jorge **Cuadra**, PUC — jcuadra@astro.puc.cl
Lixin **Dai**, University of Maryland / Yale University / University of Chile — phycosimo@gmail.com
Ashkbiz **Danehkar**, Macquarie University — ashkbiz.danehkar@students.mq.edu.au
Richard **de Grijs**, Kavli Institute for Astronomy and Astrophysics, Peking University — grijs@pku.edu.cn
Bililign T. **Dullo**, Swinburne University — bdullo@astro.swin.edu.au
Ron **Ekers**, CSIRO — ron.ekers@csiro.au
Zhou **Fan**, National Astronomical Observatories, Chinese Academy of Sciences — zfan@bao.ac.cn
Francesco R. **Ferraro**, Dipartimento di Fisica e astronomia — francesco.ferraro3@unibo.it
Jose **Fiestas**, National Astronomical Observatories, Chinese Academy of Sciences — fiestas@nao.cas.cn
Zhaoming **Gan**, Shanghai Astronomical Observatory — zmgan@shao.ac.cn
Qing **Gao**, National Astronomical Observatories, Chinese Academy of Sciences — gaoqing10@mails.gucas.ac.cn
Felipe **Garrido**, Instituto de Astrofísica, Universidad Católica de Chile — fagarri1@uc.cl
Mirek **Giersz**, Nicolaus Copernicus Astronomical Center, Polish Academy of Sciences — mig@camk.edu.pl
Xuefei **Gong**, Academy of Mathematics and Systems Science, Chinese Academy of Sciences — xfgong@amss.ac.cn
Hang **Gong**, National Astronomical Observatories, Chinese Academy of Sciences — ghang.naoc@gmail.com
Rosa A. **González-Lópezlira**, UNAM — r.gonzalez@crya.unam.mx
Lijun **Gou**, National Astronomical Observatories, Chinese Academy of Sciences — lgou@nao.cas.cn
Alister **Graham**, Swinburne University of Technology — agraham@astro.swin.edu.au
Eva **Grebel**, Heidelberg University — grebel@ari.uni-heidelberg.de
Difeng **Guo**, University of Heidelberg — difengguo@gmail.com
Wei **Hao**, Max Plank Institute for Astrophysics — elvis@mpa-garching.mpg.de
Luis **Ho**, Kavli Institute for Astronomy and Astrophysics (KIAA) — lho.pku@gmail.com
Jongsuk **Hong**, Indiana University — hongjong@indiana.edu
Siyi **Huang**, National Astronomical Observatories of China, Chinese Academy of Sciences — huang41@nao.cas.cn
Jarrod **Hurley**, Swinburne University of Technology — jhurley@swin.edu.au
Natalia **Ivanova**, University of Alberta — nata.ivanova@ualberta.ca
Siyao **Jia**, Peking University — luckyjsy@126.com
Dongming **Jin**, University of Texas at Brownsville — domi.kingdom@gmail.com
Kenneth **Kellermannk**, NRAO — kkellerm@nrao.edu
Simon **Kemp**, Instituto de Astronomia, Universidad de Guadalajara — snk@astro.iam.udg.mx
Gareth **Kennedy**, National Astronomical Observatories, Chinese Academy of Sciences — gareth.f.kennedy@gmail.com
Stefanie **Komossa**, Max-Planck-Institut fuer Radioastronomie — skomossa@mpifr.de
M.B.N. (Thijs) **Kouwenhoven**, Kavli Institute for Astronomy and Astrophysics, Peking University — thijskouwenhoven@gmail.com
Pavel **Kroupa**, University of Bonn / Helmholtz-Institut fuer Strahlen- und Kernphysik — pavel@astro.uni-bonn.de
Diederik **Kruijssen**, Max-Planck Institut für Astrophysik — kruijssen@mpa-garching.mpg.de
Cheng-Yu **Kuo**, ASIAA — cykuo@asiaa.sinica.edu.tw
Kazuaki **Kuroda**, ICRR, UTokyo — kuroda@icrr.u-tokyo.ac.jp
Barbara **Lanzoni**, Department of Physics and Astronomy, University of Bologna — barbara.lanzoni3@unibo.it
Yun-Kau **Lau**, Institute of Appl. Maths, Chinese Academy of Sciences — lau@amss.ac.cn
Joowon **Lee**, Kyung Hee University — jwlee9033@khu.ac.kr
Kejia **Lee**, Peking University — kjlee@pku.edu.cn
Shuo **Li**, National Astronomical Observatories, Chinese Academy of Sciences — lishuo@nao.cas.cn
Biao **Li**, Kavli Institute for Astronomy and Astrophysics at Peking University — hslibiao@163.com
Jinzhong **Liu**, Xinjiang astronomical Observatory — liujinzh@xao.ac.cn
Chengze **Liu**, Shanghai Jiao Tong University — czliu@sjtu.edu.cn

Fukun **Liu**, Peking University   fkliu@pku.edu.cn
Bifang **Liu**, National Astronomical Observatories, Chinese Academy of Sciences   bfliu@nao.cas.cn
Heyang **Liu**, National Astronomical Observatories, Chinese Academy of Sciences   liuheyang@nao.cas.cn
Yiqing **Liu**, Peking University   yiqing.liu@pku.edu.cn
Xiang **Liu**, Xinjiang Astronomical Observatory, Chinese Academy of Sciences   liux@xao.ac.cn
Luis-Fernando **Lomeli-Nuñez**, Universidad Nacional Autónoma de México, Centro de Radioastronomía y Astrofísica   l.lomeli@crya.unam.mx
Nora **Lützgendorf**, European Space Agency (ESA/ESTEC)   nluetzge@cosmos.esa.int
Alessandra **Mastrobuono Battisti**, Technion – Israel Institute of Technology   amastrobuono@ph.technion.ac.il
Cristián **Maureira**, Albert-Einstein-Institute   cristian.maureira.fredes@aei.mpg.de
Yohai **Meiron**, Kavli Institute for Astronomy and Astrophysics at Peking University   sahmes@gmail.com
Chris **Messenger**, University of Glasgow   christopher.messenger@glasgow.ac.uk
Amin **Mosallanezhad**, Shanghai Astronomical Observatory (SHAO)   amin@shao.ac.cn
Guobin **Mou**, Shanghai Astronomical Observatory, Chinese Academy of Sciences   gbmou@shao.ac.cn
Naohito **Nakasato**, Univerisy of Aizu   nakasato@u-aizu.ac.jp
Sakurako **Okamoto**, Shanghai Astronomical Observatory   sakurako.okamoto@gmail.com
Haiwu **Pan**, National Astronomical Observatories, Chinese Academy of Sciences   panhaiwu@bao.ac.cn
Xiaoying **Pang**, Shanghai Institute of Technology   xypang@bao.ac.cn
Dawoo **Park**, Seoul National University   dawoo@astro.snu.ac.kr
Eric **Peng**, Peking University   peng@pku.edu.cn
Giampaolo **Piotto**, Università di Padova   giampaolo.piotto@unipd.it
Erlin **Qiao**, National Astronomical Observatories, Chinese Academy of Sciences   qiaoel@nao.cas.cn
Yanli **Qiu**, National Astronomical Observatories, Chinese Academy of Sciences   ylqiu@bao.ac.cn
Zara **Randriamanakoto**, South African Astronomical Observatory   zara@saao.ac.za
Carl **Rodriguez**, Northwestern University   carllouisrodriguez@gmail.com
Rainer **Schödel**, IAA (CSIC)   rainer@iaa.es
Nathan **Secrest**, George Mason University   nathansecrest@msn.com
Alberto **Sesana**, Albert Einstein Institute   alberto.sesana@aei.mpg.de
Jinyi **Shangguan**, Department of astronomy, Peking University   shangguan@pku.edu.cn
Jihye **Shin**, Kavli Institute for Astronomy and Astrophysics   jhshin.jhshin@gmail.com
Bekdaulet **Shukirgaliyev**, Fesenkov Astrophysical Institute   bekdaulet.s@gmail.com
Margarita **Sobolenko**, Main Astronomical Observatory NAS of Ukraine   sobolenko@mao.kiev.ua
Roberto **Soria**, Curtin Institute of Radio Astronomy   roberto.soria@curtin.edu.au
Mario **Spera**, INAF – Astronomical Observatory of Padova   mario.spera@oapd.inaf.it
Rainer **Spurzem**, National Astronomical Observatories, Chinese Academy of Sciences   spurzem@nao.cas.cn
Smitha **Subramanian**, Indian Institute of Astrophysics   smitha@iiap.res.in
Edwin **van der Helm**, Leiden observatory   vdhelm@strw.leidenuniv.nl
Eugene **Vasiliev**, Lebedev Physical Institute   eugvas@lpi.ru
Oleksandr **Veles**, Astronomisches Rechen-Institut   veles@ari.uni-heidelberg.de
Enrico **Vesperini**, Indiana University   evesperi@indiana.edu
Jingbo **Wang**, Xinjiang Astronomical Observatory, Chinese Academy of Sciences   wangjingbo@xao.ac.cn
Long **Wang**, Department of Astronomy at Peking University   longwang.astro@gmail.com
Song **Wang**, National Astronomical Observatories, Chinese Academy of Sciences   songw@bao.ac.cn
Junfeng **Wang**, Xiamen University   jfwang@xmu.edu.cn
Maciej **Wielgus**, Nicolaus Copernicus Astronomical Center   maciek.wielgus@gmail.com
Qingwen **Wu**, Huazhong University of Science and Technology   qwwu@mail.hust.edu.cn
Fu-Guo **Xie**, Shanghai Astronomical Observatory, China   fgxie@shao.ac.cn
Dawei **Xu**, National Astronomical Observatories, Chinese Academy of Sciences   dwxu@nao.cas.cn
Weiwei **Xu**, National Astronomical Observatories, Chinese Academy of Sciences   weiweixu@bao.ac.cn
Li **Xue**, Department of Astronomy, Xiamen University   lixue@xmu.edu.cn
Changshuo **Yan**, NAOC   yancs@nao.cas.cn
Xiaolong **Yang**, Xinjiang Astronomical Observatory, Chinese Academy of Sciences   yangxiaolong@xao.ac.cn
Su **Yao**, National Astronomical Observatories, Chinese Academy of Sciences   yaosu@nao.cas.cn
Bei **You**, Shanghai Astronomical Observatory   youbeiyb@gmail.com
Feng **Yuan**, Shanghai Astronomical Observatory   fyuan@shao.ac.cn
Weimin **Yuan**, National Astronomical Observatories, Chinese Academy of Sciences   wmy@nao.cas.cn
Yu **Zhang**, Xinjiang Observatory   zhy@xao.ac.cn
Hongxin **Zhang**, Peking University   hongxin@pku.edu.cn
Shiyan **Zhong**, National Astronomical Observatories, Chinese Academy of Sciences   zhongshiyan09@mails.gucas.ac.cn
Zhiqin **Zhou**, Astronomy Department, Peking University   zhiqinzhou@qq.com
Alice **Zocchi**, University of Surrey   a.zocchi@surrey.ac.uk

PART ONE
Black holes in galactic nuclei

Star clusters and black holes in galaxies across cosmic time
Proceedings IAU Symposium No. 312, 2014
Y. Meiron, S. Li, F.-K. Liu & R. Spurzem, eds.

© International Astronomical Union 2016
doi:10.1017/S1743921315007383

Radio evidence for AGN activity: relativistic jets as tracers of SMBHs

Kenneth I. Kellermann

National Radio Astronomy Observatory, 520 Edgemont Rd., Charlottesville, VA, USA
email: `kkellerm@nrao.edu`

Abstract. Although the radio emission from most quasars appears to be associated with star forming activity in the host galaxy, about ten percent of optically selected quasars have very luminous relativistic jets apparently powered by a SMBH which is located at the base of the jet. When these jets are pointed close to the line of sight their apparent luminosity is enhanced by Doppler boosting and appears highly variable. High resolution radio interferometry shows directly the outflow of relativistic plasma jets from the SMBH. Apparent transverse velocities in these so-called "blazars" are typically about 7c but reach as much as 50c indicating true velocities within one percent of the speed of light. The jets appear to be collimated and accelerated in regions as much as a hundred parsecs downstream from the SMBH. Measurements made with Earth to space interferometers indicate apparent brightness temperatures of $\sim 10^{14}$ K or more. This is well in excess of the limits imposed by inverse Compton cooling. The modest Doppler factors deduced from the observed ejection speeds appear to be inadequate to explain the high observed brightness temperatures in terms of relativistic boosting.

Keywords. AGN, quasars, radio galaxies, jets, SMBHs

1. Why radio?

Historically the first speculations about the existence of active galactic nuclei (AGN) and supermassive black holes (SMBHs) came from the huge energy requirements implied by the discovery of distant powerful radio galaxies and quasars. Today, radio observations remain crucial to understanding the role of SMBHs in astrophysics. Only at radio wavelengths is it possible to image the region immediately surrounding the SMBH central engine and the relativistic jets which apparently originate with the SMBH. Typical resolution obtained with Very Long Baseline Interferometer (VLBI) observations at centimeter wavelengths is of the order of 0.001 arcsecond (1 milliarcsec). Thus, for nearby sources such as those located in the Virgo cluster, a linear resolution of 1 milliarcsec corresponds to only about 0.1 parsec or about 100 Schwartzchild radii for the SMBH located in the nucleus of M87. For $z \sim 1$ the resolution for Earth-based VLBI is ~ 10 parsecs. However, using an Earth to space interferometry at ~ 1 cm or Earth-based systems at millimeter wavelengths, the resolution is improved by more than another order of magnitude. Finally, we note that at radio wavelengths, there is no obscuration, even close to the SMBH.

However, not all radio sources are due to AGN; and not all AGN and SMBHs are radio sources. Nearly all observed extragalactic radio emission is probably due to synchrotron radiation from ultra relativistic electron with energies ~ 10 GeV moving in weak magnetic fields with $B \sim 10^{-5}$ to 10^{-4} Gauss. The high energy electrons are thought to be accelerated in one of two ways; either by a central engine associated with accretion onto a SMBH in elliptical galaxies or in quasars, or by supernovae following massive star formation (starbursts) in the nucleus of early type galaxies. Unfortunately, both processes are often referred to as AGN, and this has led to considerable confusion in the literature.

Above ~ 1 mJy, the radio source number-flux density relation is dominated by sources driven by SMBHs. These more powerful sources are characterized by extended radio lobes, and by highly beamed relativistic jets extending from a few parsecs to hundreds of kiloparsecs from the SMBH. At microJy levels there is an increasing contribution from star formation related activity rather than from a SMBH. However, there is an uncomfortably large spread in the observed microJy source count even among different observers using the same instrument, the VLA, in the same field. Most likely these discrepancies are the result of systematic errors in the reported flux densities due to uncertainties in corrections for the effects of resolution (Condon *et al.* 2012).

A recent complexity comes from the ARCADE 2 balloon measurement of a 3 GHz sky brightness of 54 ± 6 K which is about 5 sigma above that expected from known radio sources, suggesting a possible population of previously unrecognized weak sub microJansky sources (Fixen *et al.* 2011). However, deep 3 GHz VLA observations showed no evidence for any source population greater than about 30 nanoJy. Any population of weaker sources that could produce the excess sky brightness would need to have a sky density greater than 6×10^4 per ster, or 60 times greater than the density of the faintest (mag 29) galaxies in the Hubble Ultra Deep Field (Condon *et al.* 2012). Thus if the excess background temperature observed by ARCADE is real and due to discrete sources, these sources cannot be associated with any known galaxy population. It will be important to verify that the ARCADE 2 results were not contaminated by unrecognized Galactic radio emission.

Radio emission due to AGN can usually be distinguished from that due to star formation in a variety of ways.

• **Morphology:** Star formation sources may have dimensions of the order of a few tenths of an arcsecond or a few kiloparsecs at cosmological distances while SMBH driven AGN sources are typically very small, of the order of 0.001 arcsec (10 pc) or less, and are coincident with the galaxy nucleus or QSO. Quasars and AGN powered by SMBHs are often variable on time scales as short as days with corresponding changes in their morphology indicating highly collimated outflows with apparent superluminal velocity. SMBH driven AGN may also contain extended lobes tens or hundreds of kiloparsecs distant from the compact nucleus, and sometimes show optical and radio jets joining the nucleus and radio lobes.

• **Radio Spectra:** Star forming sources and the extended radio lobes of AGN generally have steep radio spectra. Due to synchrotron self absorption, the compact sources generally have flat or even inverted spectra.

• **Brightness Temperature:** Star forming sources mostly have measured brightness temperature up to $\sim 10^6$ K while the compact flat spectrum sources associated with AGN have brightness temperature 10^{11-12} K or more.

• **Radio Luminosity:** Star forming regions typically have a radio luminosity close to 10^{22-23} W/Hz and follow the well known correlation between radio and FIR luminosity Condon(1992). Radio galaxies and quasars driven by SMBHs may be 10^{4-5} times more luminous so their radio luminosity greatly exceeds that expected from the radio/FIR relation characteristic of star forming regions.

• **X and γ-ray emission:** Star forming regions are only weak x-ray sources with typical luminosity $\sim 10^{42}$ ergs/sec while SMBH driven AGN can be strong x-ray, γ-ray, and TeV sources.

• **Host Galaxies:** SMBH driven radio sources are located in the nuclei of elliptical galaxies or are associated with quasars which themselves are thought to be the bright nuclei of elliptical galaxies that greatly outshine their host galaxy. Low (optical) luminosity AGN are typically found in early type spiral (often classified as Seyfert) galaxies.

Radio emission from star forming regions is typically associated with spiral galaxies but may also be found in the host galaxies of radio quiet quasars (see Section 4).

2. Early evidence for AGN and SMBHs

Perhaps the first suggestions that the nuclei of galaxies may contain more than just stars came from Sir James Jeans in 1929 who remarked in his book on Astronomy and Cosmogony (Jeans 1929),

> The centres of the nebulae are of the nature of singular points at which matter is poured into our universe from some other and entirely spatial dimension so that to a denizen of our universe, they appear as points at which matter is being continuously created.

The modern understanding of the important role of galactic nuclei probably began with the famous paper by Karl Seyfert (1943) who reported on his study of broad strong emission lines in the nucleus of seven spiral nebulae. Interestingly, although Seyfert's name ultimately became attached to the broad category of spiral galaxies with active nuclei, his 1943 paper received no citations until 1951, and apparently went unnoticed until Baade and Minkowski (1954) drew attention to the similarity of the Cygnus A radio source spectrum with that of the galaxies studied by Seyfert.

Not until the 1949 Nature paper by Bolton, Stanley, and Slee (1949) did astronomers finally recognize the vast energy requirements of radio galaxies. Bolton *et al.* had identified three of the strongest discrete radio sources with the Crab Nebula, M87, and NGC 5128, Until that time the discrete radio sources were widely thought to be associated with galactic stars. This was understandable, as Karl Jansky and Grote Reber had observed radio emission from the Milky Way. The Milky Way is composed of stars, so it was natural to assume that the discrete radio sources had a stellar origin. Bolton *et al.* understood the importance of their identification of the Taurus A radio source with the Crab Nebula which was widely recognized as the remnant of the 1054 supernova reported by Chinese observers. BSS correctly identified two other strong sources with M87 and NGC 5128, but realizing that if they were extragalactic, their absolute radio luminosity would need to be a million times more luminous than that of the Crab Nebula, they argued that "NGC 5128 and NGC 4486 (M87) have not been resolved into stars, so there is little direct evidence that they are true galaxies." So they concluded that they are within our own Galaxy. Indeed their paper carried the title "Positions of Three Discrete Radio Sources of Galactic Radio Frequency Radiation." John Bolton later argued that he really did understand that M87 and NGC 5128 were very luminous radio sources, but that he was concerned that that in view of their apparent extraordinary radio luminosity, Nature might not publish their paper.

The following years saw the identification of more radio galaxies, and the changed paradigm which had previously considered all discrete radio sources to be stellar to one with most high latitude sources were assumed to be extragalactic. The energy requirements were exacerbated in 1951 with the identification of Cygnus A, the second strongest radio source in the sky with a magnitude 18 galaxy at what was then considered a high redshift of 0.056 and a corresponding radio luminosity about 10^3 times more luminous than M87 and NGC 5128 (Baade and Minkowski 1954). The total energy contained in relativistic particles and magnetic fields in the radio lobes of Cygnus A and other powerful radio galaxies was estimated to be at least 10^{60-61} ergs (Burbidge 1959).

Hoyle, Fowler, Burbidge and Burbidge (1964) were apparently the first to call attention to gravitational collapse as a possible energy source to power radio galaxies. By the middle of 1960, many radio sources had been identified with galaxies having red shifts up to 0.24 (Bolton 1960). Typically the optical counterpart of strong radio sources was identified with an elliptical galaxy that was the brightest member of a cluster. In 1960, Rudolph Minkowski (1960) identified 3C 295 with a mag 20 galaxy at $z = 0.46$. 3C 295 is about ten times smaller than Cygnus A and ten times more distant consistent with the idea that the smallest radio sources might be path finders to finding very distant galaxies. But a few months later Caltech radio astronomers identified the first of several very small sources with what appeared to be galactic stars, thus raising questions about the extragalactic nature of other small diameter radio sources.

3. The first quasars

While searching for ever more distant radio galaxies, Caltech radio astronomers John Bolton and Tom Matthews identified 3C 48 with an apparent stellar object. At the 107th meting of the American Astronomical Society held in New York in December 1960, Allan Sandage (1960) reported the discovery of "The First True Radio Star." Before he left to return to Australia, John Bolton (1990) speculated that 3C 48 had a high redshift of 0.37, but was apparently dissuaded by Jesse Greenstein and Ira Bowen on the grounds that there was a 3 or 4 Angstrom discrepancy among the corresponding rest wavelengths. In a subsequent analysis of the complex emission line spectrum, Jesse Greenstein (1962) interpreted the 3C 48 spectrum in terms of emission lines from highly ionized states of rare earth elements. He briefly considered a possible redshift of 0.37, but quickly dismissed the possibility that 3C 48 was extragalactic. Nearly two years would pass, and other compact radio sources would be identified as galactic stars before a series of lunar occultations would lead to the identification of 3C 273 with a star like object at a redshift of 0.16 and the immediate realization that 3C 48 was also extragalactic with a redshift of 0.37 leading to the recognition of quasi stellar radio sources or "quasars" as the extremely bright nuclei of galaxies. The apparent high radio as well as optical luminosity of quasars, coupled with their very small dimensions presented a further challenge to understanding the source of energy and how this energy is converted to relativistic particles and magnetic fields.

4. Radio loud and radio quiet quasars

The following years led to the identification of more quasars at ever larger redshifts and the suggestion that quasars are powered by accretion onto supermassive black holes (SMBH) with masses up to 10^9 solar masses or more (Lynden-Bell 1969). Generally, the identified quasars had a significant UV excess compared with stars, so due to the redshift of their spectrum, they appeared blue on photographic plates facilitating their identification with radio sources with even modest position accuracy.

In 1965, Sandage noted that the density of blue stellar objects on the sky was some thousand time greater than that of 3C radio sources. Sandage argued that what he called "quasi stellar galaxies" are related to quasars, except that they are not strong radio sources. But, his paper was widely attacked, perhaps in part because of the perceived irregular treatment by the Astrophsyical Journal. Sandage's paper was received on May 15, 1965 at the Astrophysical Journal, but S. Chandrasekar, the ApJ editor was apparently so impressed by Sandage's claim for a "New Constituent of Universe" that he held up publication of the Journal, and Sandage's paper appeared in the May 15 issue. Tom Kinman (1965) along with Lynds and Villere (1965) argued that most of Sandage's

Blue Stellar Objects were only blue galactic stars, while Fritz Zwicky (1965) pointed out that he had previously called attention to this phenomena, and he later accused Sandage of "one of the most astounding feats of plagiarism" (Zwicky & Zwicky 1971).

As it turned out, most of Sandage's Blue Stellar Objects were just that, "blue stellar objects," and only some ten percent of optically identified quasars are strong radio sources. But, it has now been more than half a century since we have divided quasars into the two classes of radio loud and radio quiet quasars, and it has still not been clear if there are two distinct populations or rather whether the radio loud population is merely the extreme end of a continuous distribution of radio luminosity. Proponents of each interpretation claim that the other interpretation is due to selection effects.

Many of the previous investigations designed to distinguish between radio loud and radio quiet quasars were limited by contamination from low luminosity AGN with absolute optical magnitudes greater than -23, biased samples based on radio rather than optical selection criteria, and inadequate sensitivity to detect radio emission from most of the radio quiet population.

In an attempt to overcome these limitations, Kimball *et al.* (2011) observed 179 quasars selected from the SDSS. All of the these quasars were within the redshift range 0.2 to 0.3 and were brighter than $M_i = -23$ so were genuine quasars that presumably contained a SMBH. The observations were made with the Jansky Very Large Array at 6 GHz reaching an rms noise of 6 μJy. All but about 6 quasars were detected as radio sources with an observed radio luminosity sharply peaked between 10^{22} and 10^{23} Watts/Hz characteristic of the radio luminosity typically observed from star forming galaxies. About ten percent of the SDSS sample are strong radio sources with radio luminosities ranging up to 10^{27} W/Hz. Kimball *et al.* concluded that the radio emission from radio quiet quasars is due to star formation in the host galaxy. Similar conclusions were reached by Padovani *et al.* (2011, 2014) based on the identification and classification of the microJy radio sources found in a deep VLA survey of the Extended Chandra Deep Field South. Based on radio, optical, IR, and X-ray data, Padovani *et al.* concluded that the microJy radio emission from AGN, like that of galaxies, is powered primarily by starbursts, and not the SMBHs which powers the AGN. Condon *et al.* (2013) argue that these starbursts are fueled by the same gas that flows into the SMBH that powers the quasar and thus accounts for the co-evolution of star formation and SMBHs.

5. Jet kinematics and relativistic beaming

Shortly after the recognition of quasars, radio source observations in both the Soviet Union (Sholomitskii 1965) and the U.S. (Dent 1965) demonstrated variability on time scales of months or less. This presented a problem. Causality arguments suggested linear dimensions, $d \leqslant c\tau$ where c is the speed of light and τ the characteristic time scale of the observed variability. Knowing the quasar redshift and corresponding distance puts a limit to the angular size which for many variable sources was only $\sim 10^{-5}$ arcseconds and the corresponding lower limit to the brightness temperature which appeared to be significantly in excess of the inverse Compton limit of $\sim 10^{11.5}$ K.

For most variable sources, the apparent violation of the inverse Compton limit is now understood in terms of relativistic beaming. Due to relativistic effects, we observe apparent jet speeds, luminosities, and brightness temperatures which are related to the corresponding intrinsic quantities in the AGN rest frame through the Doppler factor, δ, the Lorentz factor, γ, and the jet orientation, θ, with respect to the line of sight (Cohen *et al.* 2007.

The apparent velocity transverse to the line of sight, β_{app}, the apparent luminosity, L, the apparent brightness temperature, T_{app} and the Doppler factor, δ, can be calculated from the Lorentz factor, γ, θ, and the intrinsic luminosity, L_o. The apparent transverse velocity β_{app} is given by

$$\beta_{\mathrm{app}} = \frac{\beta \sin \theta}{1 - \beta \cos \theta}, \tag{5.1}$$

where $\beta = v/c$.

For small values of θ, because the radiating source is almost catching up with its own radiation, equation 5.1 shows that the apparent transverse can exceed the speed of light, which is commonly referred to as "superluminal motion." The apparent luminosity, L, is given by

$$L = L_o \delta^n, \tag{5.2}$$

where the Doppler factor, δ, is

$$\delta = \gamma^{-1}(1 - \beta \cos \theta)^{-1}, \tag{5.3}$$

and where L_o is the luminosity that would be measured by an observer in the AGN frame. The quantity n depends on the geometry and spectral index and is typically in the range between 2 and 3.

The Lorentz factor, γ, is given by

$$\gamma = (1 - \beta^2)^{-1/2}. \tag{5.4}$$

Quasars or AGN with highly Doppler boosted relativistic jets pointed nearly along the line-of-sight are often referred to as "blazars." Blazars are characterized by rapid flux density variability, apparent superluminal motion, and strong x-ray and γ-ray emission. High resolution observations of blazars provide unique insight to the process by which relativistic jets are accelerated and collimated in the region close to the SMBH. We want to understand:

- How and where is the relativistic beam accelerated and collimated into narrow jets? Are there accelerations or decelerations? Do all parts of the jet move at the same speed?
- What causes the curvature of jets? Does the flow follow a curved trajectory or is the motion ballistic and characteristic of a rotating nozzle?
- Does the observed apparent velocity reflect the true bulk velocity of motion? What determines the jet velocity? Is the velocity related to other properties such as radio, optical, x, or γ-ray luminosity?
- What is the maximum observed brightness temperature? Does it exceed the inverse Compton limit?
- What is the energy production mechanism?
- What can we learn from radio observations about the nature of the SMBH?

Very Long Baseline observations made since 1971 have confirmed the apparent superluminal motion expected from highly relativistic bulk motion. Since 1995, the NRAO Very Long Baseline Array (VLBA) has been used to study the motions of a large sample of quasars and AGN at 7mm by a group from Boston University (Marscher 2012) and by the international MOJAVE group (Kellermann *et al.* 2004, Lister *et al.* 2009, Homan *et al.* 2009, Lister *et al.* 2013, Homan *et al.* 2014). More detailed information may be found on the respective web sites:

Boston University 7 mm program: **http://www.bu.edu/blazars/VLBAproject.html**

MOJAVE 2 cm program: **: //www.physics.purdue.edu/ mlister/MOJAVE/**

The results of these programs may be summarized as follows.

• Radio loud quasars and AGN show highly relativistic bulk motion with a broad distribution of apparent velocities. In general, the jets appear one sided, probably due to differential Doppler boosting so that the approaching jet appears much brighter than the receding one. Each jet appears to have its own characteristic velocity but there is an appreciable spread in the apparent velocity of the different features within a given jet. The typical apparent velocity, $\beta_{app} \sim 8$ corresponding to an intrinsic value of $\beta \sim 0.99$. The maximum observed apparent velocity, $\beta \sim 50$ corresponds to an intrinsic value of $\beta \sim 0.999$. The parent jet population is mostly only mildly relativistic, but is under represented in flux density limited samples due to the effect of Doppler boosting.

• The jets with the fasted apparent velocities have the highest apparent luminosity, likely reflecting a correlation between intrinsic speed and intrinsic luminosity rather than simply being the result of Doppler boosting (Cohen *et al.* 2007).

• Apparent inward motions are uncommon and are likely the result of a feature moving outward along a curved trajectory approaches the line-of-sight so that the apparent separation from the jet base transverse to the line-of-sight appears to decrease with time.

• Individual jet features may show both apparent accelerations and decelerations. Both the apparent speed and direction of motion may change with time, but changes in speed are more common than changes in direction, indicating real changes in the Lorentz factor as features propagate down the jet. In general the apparent speed is greater further down the jet, so that acceleration must take place at distances at least up to ~ 100 pc from the base of the jet (Homan *et al.* 2009, 2014).

• Many jets show a curved structure and in a few cases there is evidence of an oscillatory behavior. Sometimes the flow appears to follow pre-existing channels; other times the flow appears ballistic as from a rotating nozzle, perhaps due to precession possibly resulting from a binary black hole pair (Lister *et al.* 2013).

• In some cases the direction of ejection appears to vary within a well defined cone forming what appears to be an edge brightened jet (Lister *et al.* 2013) such as shown by the jet in the nearby radio galaxy M87 where there is sufficient linear resolution to resolve the jet transverse to its structure (Kovalev *et al.* 2007).

• There appears to be a relation between radio and gamma ray emission. There is statistical evidence that radio outbursts follow a γ-ray event by ~ 1 month, but it has been difficult to convincingly establish a one to one correlation between individual radio and γ-ray events (Pushkarev *et al.* 2010).

6. Brightness temperature issues

As described above, inverse Compton scattering limits the maximum observed brightness temperature, $T <\sim 10^{11.5}$. At the inverse Compton limit, the energy contained in relativistic particles greatly exceeds that in the magnetic field which is perhaps not unreasonable in a very young source. If the particle and magnetic energies are in equilibrium, then the corresponding brightness temperature is only $\sim 10^{10.5}$.

The observed brightness temperature may be calculated from (Kovalev *et al.* 2005)

$$T_b = \frac{2\ln 2}{k\pi} \frac{S\lambda^2}{\theta^2} \quad K = 1.4 \times 10^9 S \frac{\lambda^2}{\theta^2}(1+z)\,K\,, \tag{6.1}$$

where k is the Boltzman constant, θ the angular size in milliarcsec, S is the flux density in Janskys, and λ the wavelength in cm. The resolution of a radio interferometer, is given by the ratio of the observing wavelength, to the interferometer baseline, D; or $\theta = \lambda/D$. Putting this back into eqn. 6.1 gives

$$T_b = 80SD^2(1+z)\ K\,, \tag{6.2}$$

so the maximum brightness temperature that may be measured depends only on the flux density and baseline length, and is independent of wavelength. For ground based observations with a maximum baseline of $\sim 8,000$ km, the highest brightness temperatures which can be reached are $\sim 10^{13}$ K. Recent observations with the Russian RadioAstron space VLBI satellite have suggested lower limits to brightness temperatures of 3C 273 and other sources $\sim 10^{14}$ K (Kellermann $et\ al.$ 2014, Kovalev 2014), or at least two to three orders of magnitude greater than the limit set by inverse Compton cooling. Several explanations are possible (Kellermann $et\ al.$ 2014).

1. For a relativistically beamed source the apparent brightness temperature is boosted by a factor δ. To explain the high observed brightness temperatures in this way would require Doppler factors, $\delta \sim 10^2$ to 10^3. But typical observed values of $\delta \sim \gamma \sim 10$ with maximum observed values ~ 50, and for 3C 273, $\gamma \sim 15$ (Lister $et\ al.$ 2013). Possibly the bulk flow which is related to the Doppler boosting might be much greater than the pattern flow observed by the VLBA, but Cohen $et\ al.$ (2007) have shown that this is unlikely.

2. The observed emission may be coherent such as observed in pulsars or the Sun including possible stimulated synchrotron emission.

3) The radio emission might be the result of synchrotron emission from protons rather than electrons which would enhance the upper limit to the brightness temperature by about the ratio of the proton to electron mass or more than a factor of 1000. However, proton synchrotron radiation would require a magnetic field strength more than 10^6 times stronger than needed for electron synchrotron radiation of the same strength at the same wavelength.

4) There may be a continuous acceleration of relativistic particles which balances the energy losses due to inverse Compton cooling.

7. Summary and issues

About ten percent of quasars and bright elliptical galaxies are strong radio sources with radio luminosity, $P_r > 10^{23}\ W/Hz$ and are thought to be driven by accretion onto a SMBH. The weaker radio sources, $P_r < 10^{23}\ W/Hz$ are mostly due to star formation in the host galaxy. The observed properties of radio jets can be interpreted in terms of a highly relativistic outflow from a central engine driven by a SMBH of up to 10^9 solar masses. But, many questions remain.

$\bullet$ Most quasars and AGN are not strong radio sources. Why are only $\sim 10\%$ of quasars strong radio sources, although all quasars presumably contain a SMBH to account for their extraordinary optical luminosity.

$\bullet$ How do SMBHs generate relativistic jets?

$\bullet$ How are the jets confined and shaped as they propagate away from the SMBH? What are the relative roles of velocity shear, hydrodynamic turbulence, shocks, and plasma instabilities in shaping the form and kinematics of relativistic jets?

$\bullet$ Why do only some jets produce γ-rays? How and where are the γ-rays produced? What is the relation between radio and γ-ray emission?

$\bullet$ Is there evidence for binary black hole pairs? See the paper by Ekers (2015) in this volume.

- If confirmed, observations of 3.3 GHz sky brightness combined with the density of faint sources detected in deep VLA observations suggest the possible existence of new population of faint radio sources not due to star formation or to AGN and unrelated to any known galaxy population?

- Are there other emission processes which play a role beside incoherent synchrotron radiation?

Acknowledgements

The National Radio Astronomy Observatory is operated by Associated Universities, Inc. under cooperative agreement with the National Science Foundation. I am indebted to many colleagues, especially Ron Ekers and members of the MOJAVE team for numerous discussions that have contributed to this paper.

References

Baade, W. & Minkowski, R. 1954, *ApJ*, 119, 206

Bolton, J. G. 1960, paper presented to the URSI General Assembly, London, September 1960, *Observations of the Owens Valley Radio Observatory*, 1960, No. 5

Bolton, J. G. 1990, *Proceedings of the Astronomical Society of Australia*, 8, 381

Bolton, J. G., Stanley, G. J., & Slee, O. B. 1949, *Nature*, 164, 101

Burbidge, G. R. 1959, *ApJ*, 129, 849

Cohen, M. H., Lister, M. L., Homan, D. C., *et al.* 2007, *ApJ*, 658, 232

Condon, J. J. 1992, *ARAA*, 30, pg. 575

Condon, J. J., Cotton, W. D., Fomalont, E. B., *et al.* 2012, *ApJ*, 758, 23

Condon, J. J., Kellermann, K. I., Kimball, A. E., Ivezić, Ž., & Perley, R. A. 2013, *ApJ*, 768, 37

Dent, W. A. 1965, *Science*, 148, 1458

Ekers, R. D. 2015, *IAU Symposium 312, Star Clusters and Black Holes in Galaxies and Across Cosmic Time*

Fixsen, D. J., Kogut, A., Levin, S., *et al.* 2011, *ApJ*, 734, 5

Greenstein, J. L., 1962, manuscript submitted to the Astrophysical Journal, but subsequently withdrawn

Homan, D. C., Kadler, M., Kellermann, K. I., *et al.* 2009, *ApJ*, 706, 1253

Homan, D. C., Lister, M. L., Kovalev, Y. Y., *et al.* 2014, arXiv:1410.8502

Hoyle, F., Fowler, W. A., Burbidge, G. R., & Burbidge, E. M. 1964, *ApJ*, 139, 909

Jeans, J. H. 1961, *Astronomy and Cosmogony*, New York: Dover, 1961, p. 360

Kellermann, K. I., Lister, M. L., Homan, D. C., *et al.* 2004, *ApJ*, 609, 539

Kellermann, K. I., RadioAstron AGN Early Science Team 2014, *American Astronomical Society Meeting Abstracts* 223, #421.04

Kimball, A. E., Kellermann, K. I., Condon, J. J., Ivezić, Ž., & Perley, R. A. 2011, *ApJL*, 739, L29

Kinman, T. D. 1965, *ApJ*, 142, 1241

Kovalev, Y. Y., Kellermann, K. I., Lister, M. L., *et al.* 2005, *AJ*, 130, 2473

Kovalev, Y. Y., Lister, M. L., Homan, D. C., & Kellermann, K. I. 2007, *ApJL*, 668, L27

Kovalev, Y. 2014, *IAU Symposium 304*, pg. 78

Lister, M. L., Cohen, M. H., Homan, D. C., *et al.* 2009, *ApJ*, 138, 1874

Lister, M. L., Aller, M. F., Aller, H. D., *et al.* 2013, *AJ*, 146, 120

Lynden-Bell, D. 1969, *Nature*, 223, 690

Lynds, C. R. & Villere, G. 1965, *ApJ*, 142, 1296

Marscher, A. P. 2012, *International Journal of Modern Physics Conference Series*, 8, 151

Matthews, T. A., Bolton, J. G., Greenstein, J. G., Munch, G., and Sandage, A. H., 107th Meeting of the American Astronomical Society

Minkowski, R. 1960, *ApJ*, 132, 908

Padovani, P., Miller, N., Kellermann, K. I., *et al.* 2011, *ApJ*, 740, 20

Padovani, P., Bonzini, M., Miller, N., *et al.* 2014, *IAU Symposium*, 304, pg. 79
Pushkarev, A. B., Kovalev, Y. Y., & Lister, M. L. 2010, *ApJL*, 722, L7
Seyfert, C. K. 1943, *ApJ*, 97, 28
Sholomitskii, G. B. 1965, *Soviet Astronomy*, 9, 516
Zwicky, F. 1965, *ApJ*, 142, 1293
Zwicky, F. & Zwicky, M. A. 1971, Guemligen: Zwicky, *Catalogue of Selected Compact Galaxies and of Post-Eruptive Galaxies* pg. *xix*

Star clusters and black holes in galaxies across cosmic time
Proceedings IAU Symposium No. 312, 2014
Y. Meiron, S. Li, F.-K. Liu & R. Spurzem, eds.

© International Astronomical Union 2016
doi:10.1017/S1743921315007395

Compact object mergers: observations of supermassive binary black holes and stellar tidal disruption events

S. Komossa and J. A. Zensus

Max-Planck-Institut für Radioastronomie, Auf dem Hügel 69, 53121 Bonn, Germany
email: skomossa@mpifr.de

Abstract. The capture and disruption of stars by supermassive black holes (SMBHs), and the formation and coalescence of binaries, are inevitable consequences of the presence of SMBHs at the cores of galaxies. Pairs of active galactic nuclei (AGN) and binary SMBHs are important stages in the evolution of galaxy mergers, and an intense search for these systems is currently ongoing. In the early and advanced stages of galaxy merging, observations of the triggering of accretion onto one or both BHs inform us about feedback processes and BH growth. Identification of the compact binary SMBHs at parsec and sub-parsec scales provides us with important constraints on the interaction processes that govern the shrinkage of the binary beyond the "final parsec". Coalescing binary SMBHs are among the most powerful sources of gravitational waves (GWs) in the universe. Stellar tidal disruption events (TDEs) appear as luminous, transient, accretion flares when part of the stellar material is accreted by the SMBH. About 30 events have been identified by multi-wavelength observations by now, and they will be detected in the thousands in future ground-based or space-based transient surveys. The study of TDEs provides us with a variety of new astrophysical tools and applications, related to fundamental physics or astrophysics. Here, we provide a review of the current status of observations of SMBH pairs and binaries, and TDEs, and discuss astrophysical implications.

Keywords. Black holes, galaxies, mergers, accretion, jets

---

1. Introduction: galaxy mergers and supermassive binary black holes

Galaxies have merged frequently with each other throughout the history of the universe. Galaxy mergers trigger quasars, are the sites of major black hole growth, and are believed to fix the scaling relations between SMBH mass and host properties – either by merging repeatedly with each other, and/or by feedback processes following the onset of accretion onto one or both black holes. If both galaxies harbor SMBHs at their centers, these two will ultimately form a bound pair. Coalescing supermassive binary black holes (SMBBHs) are among the most powerful emitters of gravitational waves in the universe. The subsequent gravitational wave recoil of the newly formed single black hole will, in rare cases, lead to kick velocities exceeding the host's escape velocity, and astrophysical consequences of this phenomenon are now being explored.

Galaxy and SMBH mergers evolve in several stages (e.g. Begelman *et al.* 1980; Roos 1981; Merritt & Milosavljević 2005; Colpi 2014; our Fig. 1). During the first stage, merging of the two galaxies is dominated by dynamical friction. At separations on the order of parsecs, the two BHs form a bound pair. That binary then hardens by interactions with stars and gas. The efficiency of these processes in shrinking the binary orbit has been much discussed in the literature, and is known as the "final-parsec problem", reflecting early results and concerns that the binary may stall at parsec-scale separations for more than a Hubble time, rarely reaching a regime where efficient GW emission leads to

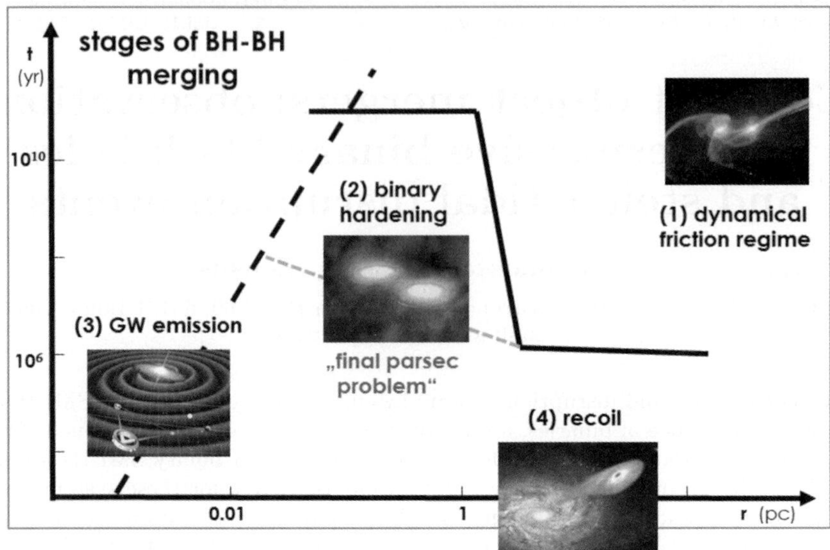

Figure 1. Stages of the evolution of SMBH pairs in the course of galaxy merging (following Begelman *et al.* 1980).

rapid coalescence of the system. Recent results indicate that non-axisymmetric galaxy potentials, the abundance of stars with centrophilic orbits, and/or the presence of large amounts of gas are, in many cases, sufficient to shrink the orbit in less than a Hubble time (e.g. Perets & Alexander 2008; Preto *et al.* 2011; Khan *et al.* 2013; Mayer 2013; Chapon *et al.* 2013; Ivanov *et al.* 2015; Aly *et al.* 2015; Vasiliev 2015; review by Colpi 2014). At coalescence, the emitted GWs carry away linear momentum, so that the SMBH receives a kick velocity, and then oscillates about the core of its host galaxy, or in rare cases escapes (e.g. Campanelli *et al.* 2007; Lousto & Zlochower 2011; review by Centrella *et al.* 2010).

Key questions related to all stages of galaxy merging include the following: (a) when does the accretion process start, (b) how long does it last, (c) how much matter is accreted before and after binary coalescence, (d) how much do the SMBHs grow in each phase, (e) how often are both SMBHs active, (f) how efficient are feedback processes, (g) how efficient is the loss of angular momentum due to interactions with gas and stars, (h) how much do the SMBHs' spins change during accretion, (i) how quick does the SMBBH coalesce, (j) how frequent are coalescences in the universe, and (k) what is the amplitude of GW recoil?

The answers to these questions are central to our understanding of the assembly history and demography of black holes, and of galaxy formation and evolution across cosmic times. Identifying SMBBHs in all stages of their evolution is therefore of great interest, and an intense search is currently ongoing.

A variety of signatures have been used to search for and identify pairs, SMBBHs, and candidates (Fig. 2). Detection of the wide systems, when the two black holes are spatially resolved from each other, is observationally most easy and robust. More indirect methods are in use to search for the closest SMBBH systems, no longer spatially resolved. Semi-periodicities in lightcurves or spatial structures in radio jets, double-peaked emission lines, and other features, have all been used to identify candidates. Most methods require that both, or at least one SMBH, is active. Most difficult to recognize are SMBBHs at the cores of non-active galaxies. They could be widely present, and with current methods (and with the single exception of our own Galactic Center), we would have almost no way of detecting them. A recent suggestion has therefore been to use the lightcurves of flares from tidally disrupted stars to search for the tell-tale signatures of binaries in otherwise quiescent galaxies (Sect. 3.3).

This review provides a short overview of observations of, and search strategies for, pairs and binary SMBHs at wide and close orbital separations. We will not cover the widest systems of AGN pairs, in early stages of interaction, or multiple AGN in clusters of galaxies, due to lack of space. An accompanying review (Liu 2015, these proceedings) will elaborate in much greater depth on theoretical aspects, and theoretical predictions of signatures of SMBBHs which have not yet been observed, but can be used for future searches. Further, this contribution will focus on main principles and detection methods, and a few prime representative systems. There is not enough space to reference all publications that have contributed to this exciting and rapidly growing field. Our apologies in advance.

2. Spatially resolved systems in single galaxies and advanced mergers

Wide pairs of accreting SMBHs, spatially resolved, can be identified by the characteristic signatures of AGN activity from both BHs, in form of luminous (hard) X-ray emission, compact radio cores, typical optical emission-line ratios, or IR colours. Only a few systems have been identified at projected separations of $r \sim 1$ kpc or less. In X-rays, these are NGC 6240 (at $r = 1$ kpc; Komossa *et al.* 2003) and NGC 3393 (at $r = 150$ pc; Fabbiano *et al.* 2011), both based on high-resolution *Chandra* imaging spectroscopy. In the radio regime, two compact, variable, flat-spectrum cores were found in 0402+379 (at $r = 7$ pc; Rodriguez *et al.* 2006; Burke-Spolaor 2011). In the optical band, two candidate AGN cores exist in SDSSJ132323.33−15941.9 (at $r = 0.8$ kpc; Woo *et al.* 2014).

The galaxy pair SDSSJ1502+1115 (Sect. 3.1) is remarkable for its overall radio structure. It consists of two bright radio cores at 7.4 kpc separation (Fu *et al.* 2011b). One of the two is further resolved into two knots of about equal brightness at 140 pc projected separation. These have been interpreted as representing either two separate SMBHs, or else double hot spots around a single SMBH (Deane *et al.* 2014; Wrobel *et al.* 2014).

3. Candidate spatially unresolved systems

3.1. *Double-peaked emission lines*

Optical spectra of AGN are characterized by narrow and broad emission lines. If these appear double, they may indicate the presence of two accreting SMBHs (Gaskell 1983; 1996; Zhou *et al.* 2004; review by Popović 2012). Further, single-peaked emission lines, which are kinematically shifted with respect to their host galaxy, may imply the presence of a merger (Comerford *et al.* 2009).

In recent years, larger samples of AGN with doubled-peaked narrow lines ("narrow-line double-peakers") have been identified thanks to large spectroscopic surveys like SDSS, AGES and LAMOST (e.g. Wang *et al.* 2009; Komossa & Xu 2009; Liu *et al.* 2010; Smith *et al.* 2010; Ge *et al.* 2012; Comerford *et al.* 2013; Barrows *et al.* 2013; Shi *et al.* 2014). A challenge when identifying the true binary AGN among them arises from the fact, that several other mechanisms do exist, which also produce double-peaked lines, but only involve a single AGN. These include the presence of two-sided jets or outflows, rotating disks, or a *single* AGN which ionizes the interstellar media of *two* host galaxies (e.g. Xu & Komossa 2009). Further, double-peakers are only expected for a short fraction of

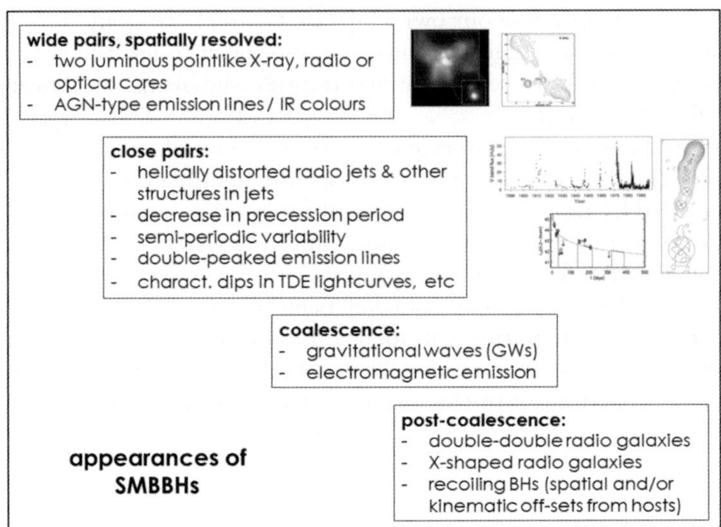

Figure 2. Signatures and detection methods of SMBH pairs and binaries.

the total merger time (Yu *et al.* 2011, van Wassenhove *et al.* 2012; Blecha *et al.* 2013). Therefore, multi-wavelength follow-up observations are required, in order to select the binaries among the large numbers of double-peakers. Such follow-ups have shown that only a small fraction of them, $\sim 2\% - 10\%$, harbor AGN pairs (e.g. Fu *et al.* 2011a; Fu *et al.* 2012; Shen *et al.* 2011; Smith *et al.* 2012; Comerford *et al.* 2012). One of the confirmed systems is SDSSJ1502+1115, with two luminous radio cores at a projected spatial separation of 7.4 kpc (Fu *et al.* 2011b).†

A fraction of all quasar spectra exhibits double-peaked *broad* emission lines. If these are due to two broad-line regions bound to two SMBHs orbiting each other, we should see the characteristic Doppler-shifts of the emission lines reflecting the orbital motion (Gaskell 1983; Shen & Loeb 2010). Broad-line double-peakers carefully monitored in the 1980s and 90s (e.g. Halpern & Filippenko 1988; Halpern & Eracleous 2000) did not reveal the expected orbital motions, and have been interpreted as systems with warped accretion disks around single SMBHs instead. New large samples of broad-line double-peakers, or of systems with single, kinematically shifted broad lines, have now been selected from SDSS (e.g. Tsalmantza *et al.* 2011; Eracleous *et al.* 2012; Decarli *et al.* 2013; Shen *et al.* 2013; Ju *et al.* 2013), and some binaries may hide among them.

Recently, Bon *et al.* (2012) presented a SMBBH model for the well-known, nearby, broad-line Seyfert galaxy NGC 4151, based on evidence for periodic variations of the observed Hα emission line, in many years of spectroscopic monitoring. The observations have been explained with a sub-parsec binary with an orbital period of ~ 16 yr.

3.2. *Semi-periodic variability*

. A number of blazars show evidence for semi-periodic optical variability, which might be linked to the presence of a second SMBH (e.g. Sillanpää *et al.* 1988; Raiteri *et al.* 2001, Fan *et al.* 2002; de Paolis *et al.* 2002; Rieger 2004; Ostorero *et al.* 2004; Liu *et al.* 2006;

† Note that we list narrow double-peakers under the Section of "unresolved sources", because their initial selection criterion is (in most cases) based on spatially unresolved emission. Follow-up imaging, when available, then often did resolve the sources. However, they usually consist of wider pairs of galaxies with core separations above a kpc, not fitting in the category discussed in Section 2.

Qian *et al.* 2007; Volvach *et al.* 2007; Xie *et al.* 2008; Karouzos *et al.* 2010; Kudryavtseva *et al.* 2011; Graham *et al.* 2015). The best studied such case is the blazar OJ287. Its optical lightcurve has been observed for more than a century (thanks to photographic plate archives all around the world, and intense dedicated monitoring during the last few decades) and shows repeat outbursts every $\sim$ 12 yr (e.g. Sillanpää *et al.* 1988; Valtaoja *et al.* 2000, Valtonen *et al.* 2012), each of which is composed of two peaks separated by $\sim$1 yr. The best explored SMBBH model consists of a secondary BH in a precessing orbit, which impacts a warped accretion disk around the primary twice each orbit. Precise timing of the past optical peaks then allows to derive the orbital parameters of the system. Calculating Keplerian orbits with post-Newtonian corrections, Valtonen (2007) presented an orbital solution with a primary mass of $\sim$ 2 $\times$ 10^{10} $M_\odot$, a mass ratio $\sim$0.01, an eccentricity of $\epsilon = 0.66$, and a semi-major axis of 0.045 pc. They also reported tentative evidence for an orbital shrinkage due to emission of GWs, of order $\Delta T_{\rm GW} \approx 0.01$ yr/period (Valtonen *et al.* 2008). The next optical maximum is expected in a few years, allowing to test new predictions of all recent models for OJ287.

3.3. *Dips in TDE lightcurves*

MBBHs imprint their presence on the outburst lightcurves of stellar tidal disruption events (TDEs). The secondary temporary interrupts the accretion stream on the primary, causing characteristic deep dips in the decline lightcurves (Liu *et al.* 2009). This signature has been observed in the lightcurve of the TDE from SDSSJ120136.02+300305.5, which is well modelled with a binary of mass ratio $q \sim 0.1$ at $\sim$0.6 mpc spatial separation (Liu *et al.* 2014). In the future, SMBBHs with TDEs may also be recognized by reprocessed emission lines which will show a tilted response function due to the off-centre location of one of the SMBHs (Brem *et al.* 2014).

3.4. *Structures in radio jets*

Several blazar radio jets show semi-periodic deviations from a straight line, and/or some other unusual structures. One way to explain these observations is involving the presence of a binary SMBH, which causes either (1) a modulation due to orbital motion of the jet-emitting BH around the primary BH, or (2) jet precession (e.g. Begelman *et al.* 1980; Roos 1988; Hardee *et al.* 1994, Britzen *et al.* 2001)[†]. If the jet precession is caused by a binary, then a prediction of this scenario is the acceleration of jet precession, observable on long timescales (Liu & Chen 2007).

Radio interferometry has provided us with the highest-resolution observations of jets over decades. Here, we would like to mention three representative candidate SMBBH systems. These are among the well-studied systems, but there is a number of others which would deserve mentioning, and are not due to lack of space. The quasar S5 1928+738 has long been suspected to harbor a SMBBH (Hummel *et al.* 1992; Roos *et al.* 1993; Murphy *et al.* 2003; see also Roland *et al.* 2015). Kun *et al.* (2014), analyzing 20 yr of VLBI data, presented evidence that the jet-emitting SMBH is actually spinning. Their orbital modelling implies a binary separation of $\sim$ 10 mpc, and an orbital period of $\sim$ 5 yr. The helical distortions of the jet of the BL Lac object Mrk 501 have been interpreted with a SMBBH model by Conway & Wrobel (1995) and Villata & Raiteri (1999). A SMBBH scenario was also involved in order to explain evidence for semi-periodic variability of this source (Rieger & Manheim 2000; see also de Paolis *et al.* 2002; Rödig *et al.* 2009). Lobanov & Roland (2005) presented a SMBBH model at $\sim$ 0.3 pc separation for the quasar 3C345,

---

† See, e.g. Britzen *et al.* 2010; Lobanov & Roland 2005, and Godfrey *et al.* 2012 (and references therein), for a discussion including alternative scenarios such as disk oscillations or Kelvin-Helmholtz instabilities.

which can reproduce both, its optical and radio variability, and the morphology and kinematics of the parsec-scale jet.

The most powerful method to date of *spatially* resolving the orbit of a jet-emitting BH in a compact binary is phase-referencing of VLBI radio data. Using that technique, Sudou *et al.* (2003) reported evidence for systematic changes in radio position, which they interpreted as orbital motion of the radio core of 3C66B with a period of 1.05 yr. Part of the possible orbital solutions could be excluded based on current pulsar timing constraints (Jenet *et al.* 2004), while the rest remains a possibility (Iguchi *et al.* 2010). Future phase-reference measurements of this and other systems, along with simulations of the jet base, core-shift measurements, and studies of transverse motions will provide us with strong tests of the SMBBH model.

4. Post-coalescence candidates

Certain signatures of compact and coalescing binaries remain imprinted on their large-scale environment, and can therefore be recovered from multi-wavelength observations long after the actual coalescence. For instance, accretion temporarily interrupts in compact binaries, because of the fast orbital shrinkage due to GW emission, dominating over viscous processes, so that the inner disk no longer catches up (e.g. Liu *et al.* 2003; Milosavljević & Phinney 2005; Farris *et al.* 2015). If these systems launch radio jets, jet formation will be temporarily interrupted, too, and this may explain the presence of double-double radio galaxies (Liu *et al.* 2003). If the hole's spin direction changes after coalescence, the jet will be launched in a new direction, and this is one possibility to account for the structure of X-shaped radio galaxies (Merritt & Ekers 2002; see Gopal-Krishna *et al.* 2012 for a recent overview; see also Mezcua *et al.* 2012). If the newly formed single SMBH receives a significant kick velocity after coalescence, it will appear spatially or kinematically off-set from its host galaxy, and several candidate recoiling SMBHs have emerged in recent years (review by Komossa 2012a). Further, it has been suggested that the central stellar light deficits observed in some ellipticals and bulges were created by SMBBHs which had shrunk their orbits by slingshot ejection of stars, consistent with recent observations (e.g. Dullo & Graham 2014).

5. Future missions and searches

A number of ongoing and future missions and surveys will be sensitive to SMBBHs in all stages of evolution. For instance, space VLBI and mm VLBI at the shortest wavelengths feasible will provide us with the highest spatial resolution (e.g. Fish *et al.* 2013; Tilanus *et al.* 2014), while the *Square Kilometer Array* (SKA) will provide high sensitivity (e.g. Deane *et al.* 2015). SKA and other current and upcoming PTA (pulsar timing array) experiments will detect the gravitational wave signatures of the most massive coalescing SMBBHs using pulsar timing (e.g. Lazio 2013; Hobbs 2013; Kramer & Champion 2013; Sesana 2015). Future high-sensitivity X-ray observatories may kinematically resolve binary effects on the iron line profile from one or two disks (Yu & Lu 2001; McKernan *et al.* 2013; Jovanović *et al.* 2014), while optical integral field spectroscopy may reveal the kinematic signature of the inspiral phase (Meiron & Laor 2013).

Further breakthroughs in the field are expected once space-based gravitational-wave interferometers are in operation, providing measurements of coalescence rates, SMBH masses and spins (e.g. Babak *et al.* 2011; review by Barausse *et al.* 2015). eLISA is currently scheduled for launch around 2030. Electromagnetic counterparts to GWs (e.g.

Schutz 1986) from coalescing SMBBHs, and characteristic signals before or after coalescence, may appear as transients in current or future transient surveys (reviews by Schnittman 2011; Haiman 2012).

6. Tidal disruption of stars by supermassive black holes

The tidal disruption, and subsequent accretion, of a star by a supermassive black hole produces a luminous flare of electromagnetic radiation (e.g. Rees 1990; Luminet 1985). A star is disrupted, once the tidal forces of the hole exceed the self-gravity of the star (Hills 1975). The distance at which this happens, the tidal radius, is given by

$$r_{\rm t} \simeq 7 \times 10^{12} \left(\frac{M_{\rm BH}}{10^6\,{\rm M_\odot}} \right)^{\frac{1}{3}} \left(\frac{M_*}{{\rm M_\odot}} \right)^{-\frac{1}{3}} \left(\frac{r_*}{{\rm R_\odot}} \right) \ {\rm cm} \qquad (6.1)$$

A fraction of the stellar material will be on unbound orbits and escape, while the rest will eventually be accreted (Fig. 3). The events appear as luminous transients with peak in the UV or soft X-rays, declining on the timescale of months to years (e.g. Rees 1990; Evans & Kochanek 1989). Recent state of the art modelling has addressed the different stages of TDE evolution under various conditions (e.g. Lodato *et al.* 2009; Brassart & Luminet 2010; Strubbe & Quataert 2011; Lodato & Rossi 2011; Cheng *et al.* 2012; Kesden 2012; Guillochon & Ramirez-Ruiz 2013; Hayasaki *et al.* 2013, Dai & Blandford 2013; Cheng & Bogdanovic 2014; Shiokawa *et al.* 2015; and references therein). If the doomed star is compact (e.g. a white dwarf), partial disruption will produce an electromagnetic *and* a GW signal (review by Amaro-Seoane *et al.* 2007).

Observing TDEs, out to large cosmic distances, provides us with a variety of new astrophysical tools and applications, related to fundamental physics or astrophysics, including studying precession effects in the Kerr metric, measuring BH spin, observing other relativistic effects (at 10^8 M$_\odot$, the tidal radius is on the order of the Schwarzschild radius), probing accretion physics under extreme conditions and near $L_{\rm edd}$, understanding the physics of jet formation and early evolution, reverberation-mapping the gaseous core environment via its emission-line response, searching for a population of (so far elusive) intermediate mass BHs, detecting supermassive binary BHs at the cores of quiescent galaxies (from TDE lightcurves; Sect. 3.3), probing stellar kinematics on spatial scales which cannot be resolved directly (via disruption rates in different types of galaxies), or spotting recoiling BHs by off-nuclear TDEs.

7. Multi-wavelength observations

A key signpost of TDEs is their luminous, transient high-energy emission, peaking in the UV or soft X-rays, arising from the accretion of the stellar material. First events from quiescent galaxies have been identified in the course of the ROSAT all sky survey, which was ideal for detection because of its repeat coverage of almost the whole sky, and its high sensitivity in the soft X-ray band (0.1-2.4 keV). Events appeared as luminous transients, reaching peak luminosities up to $> 10^{43-44}$ erg s^{-1} just in soft X-rays (e.g. Bade *et al.* 1996; Komossa & Bade 1999; Grupe *et al.* 1999). They then faded away by factors larger than a few thousand (Halpern *et al.* 2004; Komossa *et al.* 2004), their initially supersoft X-ray spectra ($kT \sim 0.04-0.1$ keV) showed a hardening with time, and optical spectroscopy of the host galaxies revealed little or no activity at all (review by Komossa 2002). Two of these events, NGC 5905 and RXJ1242–1119, continue to be the best-monitored events in terms of their *long-term* X-ray lightcurves, spanning time intervals of more than a decade

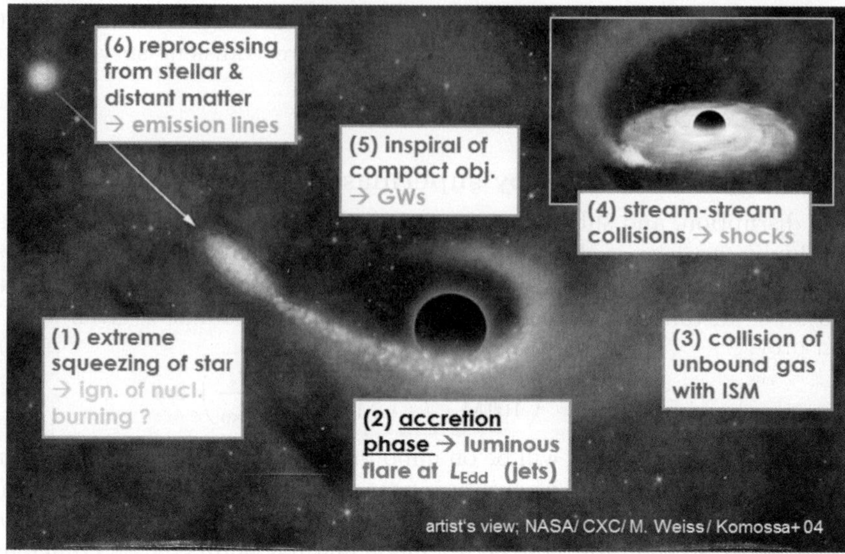

Figure 3. Evolution of stellar tidal disruption events and sources of radiation. In most cases, the accretion phase is the most luminous electromagnetic phase.

(Komossa *et al.* 2004, Halpern *et al.* 2004). All the event properties agree very well with order-of-magnitude predictions from tidal disruption theory (e.g. Rees 1988; 1990). More recently, in X-rays similar events have been found with *Chandra* and *XMM-Newton* (e.g. Esquej *et al.* 2008; Maksym *et al.* 2010; Lin *et al.* 2011; Saxton *et al.* 2012; Nikolajuk & Walter 2013; Maksym *et al.* 2013; Donato *et al.* 2014), some of them with well-covered lightcurves during the first few years. Events were also found at longer wavelengths, in the UV and optical (e.g. Gezari *et al.* 2006; Komossa *et al.* 2008; van Velzen *et al.* 2011; Cenko *et al.* 2012a; Gezari *et al.* 2012; Chornock *et al.* 2014), some of them caught before their peak (see Komossa 2012b for a more extended review of multi-λ observations). Several estimates of TDE rates are all on the order of $10^{-4} - 10^{-5}$ yr^{-1} galaxy^{-1} (e.g. Donley *et al.* 2002; Esquej *et al.* 2008; Maksym *et al.* 2010; Wang *et al.* 2012) and agree well with theoretical predictions (e.g. Brockamp *et al.* 2011).

8. Emission-line transients

TDEs which occur in gas-rich galaxies will provide us with a powerful new tool of performing reverberation mapping of the cores of these galaxies. As the luminous electromagnetic radiation travels across the galaxy core, it will photoionize any circum-nuclear material (including the tidal debris itself) and is reprocessed into line radiation. Recently, SDSS and other surveys have enabled the discovery of several well-observed cases of transient optical emission lines, of a kind not observed before, and arising from otherwise quiescent galaxies†: All of them exhibit bright, broad, fading emission from Helium and/or Hydrogen (Komossa *et al.* 2008; 2009, Wang *et al.* 2011; 2012; Gezari *et al.* 2012;

† These emission-line transients are markedly different from the mild line variability seen in AGN, with two exceptions: (1) The AGN IC 3599; which underwent a high-amplitude X-ray outburst accompanied by a strong increase in its optical emission lines (Brandt *et al.* 1995; Grupe *et al.* 1995; Komossa & Bade 1999), and (2) the AGN NGC 1097, which shows strong, broad, double-peaked Balmer lines which emerged abruptly (e.g. Storchi-Bergmann *et al.* 1995). The underlying mechanism remains unknown, but high-amplitude Narrow-line Seyfert 1 variability (only IC 3599), variants of accretion-disk instabilities, or a TDE have all been considered.

Gaskell & Rojas Lobos 2014; Holoien *et al.* 2014, Arcavi *et al.* 2014), while some of them show transient super-strong iron coronal lines in addition, up to ionization stages of Fe^{13+} (Komossa *et al.* 2008; 2009; Wang *et al.* 2011; 2012).

9. Jetted TDEs

The possibility that TDEs launch radio jets, came up with the detection of the first few X-ray TDEs with ROSAT. Dedicated follow-ups of NGC 5905 did not detect any radio emission from a jet, however (Komossa 2002).

Two events recently discovered with *Swift*, Swift J1644+57 and Swift J2058.4+0516, differ from previous TDEs, in the sense that they had much harder X-ray spectra, were accompanied by strong (beamed) radio emission, and exhibit some other remarkable properties (e.g. Burrows *et al.* 2011; Bloom *et al.* 2011; Zauderer *et al.* 2011; 2013; Levan *et al.* 2011; Cenko *et al.* 2012b). Swift J1644+57 was detected with *Swift* BAT in 2011. Its (isotropic) peak luminosity exceeded 10^{48} erg s^{-1}. The X-ray lightcurve shows a general downward trend, on which rapid, high-amplitude variability is superposed, as fast as 100s. After ~ 1.5 yr, the X-rays suddenly dropped by a large factor, and have remained faint so far. The host galaxy at redshift $z = 0.35$ does not show signs of permanent optical AGN activity. The event is accompanied by unresolved and variable radio emission, which has been interpreted as the rapid onset of a powerful jet after stellar tidal disruption. The event has motivated a large number of follow-ups and theoretical studies (review by Komossa 2015; in prep.), with an emphasis on the question of jet launching under TDE conditions, and the role of magnetic fields (e.g. Tchekhovskoy *et al.* 2014).

These and future observations of jetted TDEs provide us with a completely new probe of the early phases of jet formation and evolution in an otherwise quiescent environment without past radio-AGN activity.

10. Future missions and surveys

TDEs will be detected in large numbers with future sky surveys, including in the radio with SKA (Donnarumma *et al.* 2015), in the optical with LSST (Gezari *et al.* 2009), in hard X-rays with LOFT (Rossi *et al.* 2015), and in soft X-rays with the proposed mission *Einstein Probe* (Yuan *et al.* 2015). Well-covered lightcurves will enable a wealth of new science, and X-rays will be sensitive to relativistic effects (Sect. 6).

Acknowledgements

SK would like to thank ISSI/Bern for supporting and hosting two workshops on "Unveiling multiple AGN activity in galaxy mergers", and the participants for many stimulating discussions. SK would also like to thank NAOC Beijing for their great hospitality and support over many years. Many thanks to S. Britzen, T. Krichbaum, A. Lobanov, and E. Ros for a critical reading of the manuscript and very useful comments.

References

Aly, H., Dehnen, W., Nixon, C., & King, A. 2015, *MNRAS*, 449, 65
Amaro-Seoane, P., Gair, J. R., Freitag, M., *et al.* 2007, *Classical and Quantum Gravity*, 24, 113
Arcavi, I., Gal-Yam, A., Sullivan, M., *et al.* 2014, *ApJ*, 793, 38
Babak, S., Gair, J. R., Petiteau, A., & Sesana, A. 2011, *Classical and Quantum Gravity*, 28, 114001
Bade, N., Komossa, S., & Dahlem, M. 1996, *A&A*, 309, L35

Barausse, E., Bellovary, J., Berti, E., *et al.* 2015, *Journal of Physics Conference Series*, 610, 012001

Barrows, R. S., Sandberg Lacy, C. H., Kennefick, J., *et al.* 2013, *ApJ*, 769, 95

Begelman, M. C., Blandford, R. D., & Rees, M. J. 1980, *Nature*, 287, 307

Blecha, L., Loeb, A., & Narayan, R. 2013, *MNRAS*, 429, 2594

Bloom, J. S., Giannios, D., Metzger, B. D., *et al.* 2011, *Science*, 333, 203

Bon, E., Jovanović, P., Marziani, P., *et al.* 2012, *ApJ*, 759, 118

Brandt, W. N., Pounds, K. A., & Fink, H. 1995, *MNRAS*, 273, L47

Brassart, M. & Luminet, J.-P. 2010, *A&A*, 511, A80

Brem, P., Cuadra, J., Amaro-Seoane, P., & Komossa, S. 2014, *ApJ*, 792, 100

Britzen, S., Roland, J., Laskar, J., *et al.* 2001, *A&A*, 374, 784

Britzen, S., Kudryavtseva, N. A., Witzel, A., *et al.* 2010, *A&A*, 511, A57

Brockamp, M., Baumgardt, H., & Kroupa, P. 2011, *MNRAS*, 418, 1308

Burke-Spolaor, S. 2011, *MNRAS*, 410, 2113

Burrows, D. N., Kennea, J. A., Ghisellini, G., *et al.* 2011, *Nature*, 476, 421

Campanelli, M., Lousto, C., Zlochower, Y., & Merritt, D. 2007, *ApJ*, 659, L5

Cenko, S. B., Bloom, J. S., Kulkarni, S. R., *et al.* 2012a, *MNRAS*, 420, 2684

Cenko, S. B., Krimm, H. A., Horesh, A., *et al.* 2012b, *ApJ*, 753, 77

Centrella, J., Baker, J. G., Kelly, B. J., & van Meter, J. R. 2010, *Reviews of Modern Physics*, 82, 3069

Chapon, D., Mayer, L., & Teyssier, R. 2013, *MNRAS*, 429, 3114

Cheng, K.-S., Chernyshov, D. O., Dogiel, V. A., *et al.* 2012, *ApJ*, 746, 116

Cheng, R. M. & Bogdanović, T. 2014, *Phys. Rev. D*, 90, 064020

Chornock, R., Berger, E., Gezari, S., *et al.* 2014, *ApJ*, 780, 44

Colpi, M. 2014, *Space Sci. Rev.*, 183, 189

Comerford, J. M., Gerke, B. F., Stern, D., *et al.* 2012, *ApJ*, 753, 42

Comerford, J. M., Schluns, K., Greene, J. E., & Cool, R. J. 2013, *ApJ*, 777, 64

Comerford, J. M., Gerke, B. F., Newman, J. A., *et al.* 2009, *ApJ*, 698, 956

Conway, J. E. & Wrobel, J. M. 1995, *ApJ*, 439, 98

Dai, L. & Blandford, R. 2013, *MNRAS*, 434, 2948

De Paolis, F., Ingrosso, G., & Nucita, A. A. 2002, *A&A*, 388, 470

Deane, R., Paragi, Z., Jarvis, M., *et al.* 2015, *Advancing Astrophysics with the Square Kilometre Array (AASKA14)*, 151

Deane, R. P., Paragi, Z., Jarvis, M. J., *et al.* 2014, *Nature*, 511, 57

Decarli, R., Dotti, M., Fumagalli, M., *et al.* 2013, *MNRAS*, 433, 1492

Donato, D., Cenko, S. B., Covino, S., *et al.* 2014, *ApJ*, 781, 59

Donley, J. L., Brandt, W. N., Eracleous, M., & Boller, T. 2002, *AJ*, 124, 1308

Donnarumma, I., Rossi, E. M., Fender, R., *et al.* 2015, *Advancing Astrophysics with the Square Kilometre Array (AASKA14)*, 54

Dullo, B. T. & Graham, A. W. 2014, *MNRAS*, 444, 2700

Eracleous, M., Boroson, T. A., Halpern, J. P., & Liu, J. 2012, *ApJS*, 201, 23

Esquej, P., Saxton, R. D., Komossa, S., *et al.* 2008, *A&A*, 489, 543

Evans, C. R. & Kochanek, C. S. 1989, *ApJ*, 346, L13

Fabbiano, G., Wang, J., Elvis, M., & Risaliti, G. 2011, *Nature*, 477, 431

Fan, J. H., Lin, R. G., Xie, G. Z., *et al.* 2002, *A&A*, 381, 1

Farris, B. D., Duffell, P., MacFadyen, A. I., & Haiman, Z. 2015, *MNRAS*, 447, L80

Fish, V., Alef, W., Anderson, J., *et al.* 2013, *ArXiv e-prints*, arXiv:1309.3519

Fu, H., Myers, A. D., Djorgovski, S. G., & Yan, L. 2011a, *ApJ*, 733, 103

Fu, H., Yan, L., Myers, A. D., *et al.* 2012, *ApJ*, 745, 67

Fu, H., Zhang, Z.-Y., Assef, R. J., *et al.* 2011b, *ApJ*, 740, L44

Gaskell, C. M. 1996, *ApJ*, 464, L107

Gaskell, C. M. & Rojas Lobos, P. A. 2014, *MNRAS*, 438, L36

Gaskell, M. 1983, in Proceedings of the 24th Liege Int. Astrophys. Coll., 473

Ge, J.-Q., Hu, C., Wang, J.-M., Bai, J.-M., & Zhang, S. 2012, *ApJS*, 201, 31

Gezari, S., Martin, D. C., Milliard, B., *et al.* 2006, *ApJ*, 653, L25

Gezari, S., Strubbe, L., Bloom, J. S., *et al.* 2009, in ArXiv Astrophysics e-prints, Vol. 2010, astro2010: The Astronomy and Astrophysics Decadal Survey, 88

Gezari, S., Chornock, R., Rest, A., *et al.* 2012, *Nature*, 485, 217

Godfrey, L. E. H., Lovell, J. E. J., Burke-Spolaor, S., *et al.* 2012, *ApJ*, 758, L27

Gopal-Krishna, Biermann, P. L., Gergely, L. Á., & Wiita, P. J. 2012, *Research in Astronomy and Astrophysics*, 12, 127

Graham, M. J., Djorgovski, S. G., Stern, D., *et al.* 2015, *Nature*, 518, 74

Grupe, D., Beuermann, K., Mannheim, K., *et al.* 1995, *A&A*, 299, L5

Grupe, D., Thomas, H.-C., & Leighly, K. M. 1999, *A&A*, 350, L31

Guillochon, J. & Ramirez-Ruiz, E. 2013, *ApJ*, 767, 25

Haiman, Z. 2012, in American Astronomical Society Meeting Abstracts, Vol. 220, American Astronomical Society Meeting Abstracts #220, #502.04

Halpern, J. P. & Eracleous, M. 2000, *ApJ*, 531, 647

Halpern, J. P. & Filippenko, A. V. 1988, *Nature*, 331, 46

Halpern, J. P., Gezari, S., & Komossa, S. 2004, *ApJ*, 604, 572

Hardee, P. E., Cooper, M. A., & Clarke, D. A. 1994, *ApJ*, 424, 126

Hayasaki, K., Stone, N., & Loeb, A. 2013, *MNRAS*, 434, 909

Hills, J. G. 1975, *Nature*, 254, 295

Hobbs, G. 2013, in IAU Symposium, Vol. 291, IAU Symposium, ed. J. van Leeuwen, 165–170

Holoien, T. W.-S., Prieto, J. L., Bersier, D., *et al.* 2014, *MNRAS*, 445, 3263

Hummel, C. A., Schalinski, C. J., Krichbaum, T. P., *et al.* 1992, *A&A*, 257, 489

Iguchi, S., Okuda, T., & Sudou, H. 2010, *ApJ*, 724, L166

Ivanov, P. B., Papaloizou, J. C. B., Paardekooper, S.-J., & Polnarev, A. G. 2015, *A&A*, 576, A29

Jenet, F. A., Lommen, A., Larson, S. L., & Wen, L. 2004, *ApJ*, 606, 799

Jovanović, P., Borka Jovanović, V., Borka, D., & Bogdanović, T. 2014, *Advances in Space Research*, 54, 1448

Ju, W., Greene, J. E., Rafikov, R. R., Bickerton, S. J., & Badenes, C. 2013, *ApJ*, 777, 44

Karouzos, M., Britzen, S., Eckart, A., Witzel, A., & Zensus, A. 2010, *A&A*, 519, A62

Kesden, M. 2012, *Phys. Rev. D*, 86, 064026

Khan, F. M., Holley-Bockelmann, K., Berczik, P., & Just, A. 2013, *ApJ*, 773, 100

Komossa, S. 2002, in Reviews in Modern Astronomy, Vol. 15, Reviews in Modern Astronomy, ed. R. E. Schielicke, 27

Komossa, S. 2012 a, *Advances in Astronomy*, 2012, 14, (id. 364973)

Komossa, S. 2012b, in European Physical Journal Web of Conferences, Vol. 39, European Physical Journal Web of Conferences, 2001

Komossa, S. & Bade, N. 1999, *A&A*, 343, 775

Komossa, S., Burwitz, V., Hasinger, G., *et al.* 2003, *ApJ*, 582, L15

Komossa, S., Halpern, J., Schartel, N., *et al.* 2004, *ApJ*, 603, L17

Komossa, S., Zhou, H., Wang, T., *et al.* 2008, *ApJ*, 678, L13

Komossa, S., Zhou, H., Rau, A., *et al.* 2009, *ApJ*, 701, 105

Kramer, M. & Champion, D. J. 2013, *Classical and Quantum Gravity*, 30, 224009

Kudryavtseva, N. A., Britzen, S., Witzel, A., *et al.* 2011, *A&A*, 526, A51

Kun, E., Gabányi, K. É., Karouzos, M., Britzen, S., & Gergely, L. Á. 2014, *MNRAS*, 445, 1370

Lazio, T. J. W. 2013, *Classical and Quantum Gravity*, 30, 224011

Levan, A. J., Tanvir, N. R., Cenko, S. B., *et al.* 2011, *Science*, 333, 199

Lin, D., Carrasco, E. R., Grupe, D., *et al.* 2011, *ApJ*, 738, 52

Liu, F. K. & Chen, X. 2007, *ApJ*, 671, 1272

Liu, F. K., Li, S., & Chen, X. 2009, *ApJ*, 706, L133

Liu, F. K., Li, S., & Komossa, S. 2014, *ApJ*, 786, 103

Liu, F. K., Wu, X.-B., & Cao, S. L. 2003, *MNRAS*, 340, 411

Liu, F. K., Zhao, G., & Wu, X.-B. 2006, *ApJ*, 650, 749

Liu, X., Shen, Y., Strauss, M. A., & Greene, J. E. 2010, *ApJ*, 708, 427

Lobanov, A. P. & Roland, J. 2005, *A&A*, 431, 831

Lodato, G., King, A. R., & Pringle, J. E. 2009, *MNRAS*, 392, 332

Lodato, G. & Rossi, E. M. 2011, *MNRAS*, 410, 359

Lousto, C. O. & Zlochower, Y. 2011, *Physical Review Letters*, 107, 231102

Luminet, J.-P. 1985, *Annales de Physique*, 10, 101

Maksym, W. P., Ulmer, M. P., & Eracleous, M. 2010, *ApJ*, 722, 1035

Maksym, W. P., Ulmer, M. P., Eracleous, M. C., Guennou, L., & Ho, L. C. 2013, *MNRAS*, 435, 1904

Mayer, L. 2013, *Classical and Quantum Gravity*, 30, 244008

McKernan, B., Ford, K. E. S., Kocsis, B., & Haiman, Z. 2013, *MNRAS*, 432, 1468

Meiron, Y. & Laor, A. 2013, *MNRAS*, 433, 2502

Merritt, D. & Ekers, R. D. 2002, *Science*, 297, 1310

Merritt, D. & Milosavljević, M. 2005, *Living Reviews in Relativity*, 8, 8

Mezcua, M., Chavushyan, V. H., Lobanov, A. P., & León-Tavares, J. 2012, *A&A*, 544, A36

Milosavljević, M. & Phinney, E. S. 2005, *ApJ*, 622, L93

Murphy, D. W., Preston, R. A., & Hirabayashi, H. 2003, *New A Rev.*, 47, 633

Nikołajuk, M. & Walter, R. 2013, *A&A*, 552, A75

Ostorero, L., Villata, M., & Raiteri, C. M. 2004, *A&A*, 419, 913

Perets, H. B. & Alexander, T. 2008, *ApJ*, 677, 146

Popović, L. Č. 2012, *New A Rev.*, 56, 74

Preto, M., Berentzen, I., Berczik, P., & Spurzem, R. 2011, *ApJ*, 732, L26

Qian, S.-J., Kudryavtseva, N. A., Britzen, S., *et al.* 2007, *Chinese J. Astron. Astrophys.*, 7, 364

Raiteri, C. M., Villata, M., Aller, H. D., *et al.* 2001, *A&A*, 377, 396

Rees, M. J. 1988, *Nature*, 333, 523

—. 1990, *Science*, 247, 817

Rieger, F. M. 2004, *ApJ*, 615, L5

Rieger, F. M. & Mannheim, K. 2000, *A&A*, 353, 473

Rödig, C., Burkart, T., Elbracht, O., & Spanier, F. 2009, *A&A*, 501, 925

Rodriguez, C., Taylor, G. B., Zavala, R. T., *et al.* 2006, *ApJ*, 646, 49

Roland, J., Britzen, S., Kun, E., *et al.* 2015, *A&A*, 578, A86

Roos, N. 1981, *A&A*, 104, 218

—. 1988, *ApJ*, 334, 95

Roos, N., Kaastra, J. S., & Hummel, C. A. 1993, *ApJ*, 409, 130

Rossi, E. M., Donnarumma, I., Fender, R., *et al.* 2015, *ArXiv e-prints*

Saxton, R. D., Read, A. M., Esquej, P., *et al.* 2012, *A&A*, 541, A106

Schnittman, J. D. 2011, *Classical and Quantum Gravity*, 28, 094021

Schutz, B. F. 1986, *Nature*, 323, 310

Sesana, A. 2015, *Astrophysics and Space Science Proceedings*, 40, 147

Shen, Y., Liu, X., Greene, J. E., & Strauss, M. A. 2011, *ApJ*, 735, 48

Shen, Y., Liu, X., Loeb, A., & Tremaine, S. 2013, *ApJ*, 775, 49

Shen, Y. & Loeb, A. 2010, *ApJ*, 725, 249

Shi, Z.-X., Luo, A.-L., Comte, G., *et al.* 2014, *Research in Astronomy and Astrophysics*, 14, 1234

Shiokawa, H., Krolik, J. H., Cheng, R. M., Piran, T., & Noble, S. C. 2015, *ApJ*, 804, 85

Sillanpää, A., Haarala, S., Valtonen, M. J., Sundelius, B., & Byrd, G. G. 1988, *ApJ*, 325, 628

Smith, K. L., Shields, G. A., Bonning, E. W., *et al.* 2010, *ApJ*, 716, 866

Smith, K. L., Shields, G. A., Salviander, S., Stevens, A. C., & Rosario, D. J. 2012, *ApJ*, 752, 63

Storchi-Bergmann, T., Eracleous, M., Livio, M., *et al.* 1995, *ApJ*, 443, 617

Strubbe, L. E. & Quataert, E. 2011, *MNRAS*, 415, 168

Sudou, H., Iguchi, S., Murata, Y., & Taniguchi, Y. 2003, *Science*, 300, 1263

Tchekhovskoy, A., Metzger, B. D., Giannios, D., & Kelley, L. Z. 2014, *MNRAS*, 437, 2744

Tilanus, R. P. J., Krichbaum, T. P., Zensus, J. A., *et al.* 2014, *ArXiv e-prints*, arXiv:1406.4650

Tsalmantza, P., Decarli, R., Dotti, M., & Hogg, D. W. 2011, *ApJ*, 738, 20

Valtaoja, E., Teräsranta, H., Tornikoski, M., *et al.* 2000, *ApJ*, 531, 744

Valtonen, M. J. 2007, *ApJ*, 659, 1074

Valtonen, M. J., Ciprini, S., & Lehto, H. J. 2012, *MNRAS*, 427, 77
Valtonen, M. J., Lehto, H. J., Nilsson, K., *et al.* 2008, *Nature*, 452, 851
van Velzen, S., Farrar, G. R., Gezari, S., *et al.* 2011, *ApJ*, 741, 73
van Wassenhove, S., Volonteri, M., Mayer, L., *et al.* 2012, *ApJ*, 748, L7
Vasiliev, E. 2015, *in these proceedings*
Villata, M. & Raiteri, C. M. 1999, *A&A*, 347, 30
Volvach, A. E., Volvach, L. N., Larionov, M. G., Aller, H. D., & Aller, M. F. 2007, *Astronomy Reports*, 51, 450
Wang, J.-M., Chen, Y.-M., Hu, C., *et al.* 2009, *ApJ*, 705, L76
Wang, T.-G., Zhou, H.-Y., Komossa, S., *et al.* 2012, *ApJ*, 749, 115
Wang, T.-G., Zhou, H.-Y., Wang, L.-F., Lu, H.-L., & Xu, D. 2011, *ApJ*, 740, 85
Woo, J.-H., Cho, H., Husemann, B., *et al.* 2014, *MNRAS*, 437, 32
Wrobel, J. M., Walker, R. C., & Fu, H. 2014, *ApJ*, 792, L8
Xie, G. Z., Yi, T. F., Li, H. Z., Zhou, S. B., & Chen, L. E. 2008, *AJ*, 135, 2212
Xu, D. & Komossa, S. 2009, *ApJ*, 705, L20
Yu, Q. & Lu, Y. 2001, *A&A*, 377, 17
Yu, Q., Lu, Y., Mohayaee, R., & Colin, J. 2011, *ApJ*, 738, 92
Yuan, W. 2015, *in these proceedings*
Zauderer, B. A., Berger, E., Margutti, R., *et al.* 2013, *ApJ*, 767, 152
Zauderer, B. A., Berger, E., Soderberg, A. M., *et al.* 2011, *Nature*, 476, 425
Zhou, H., Wang, T., Zhang, X., Dong, X., & Li, C. 2004, *ApJ*, 604, L33

Star clusters and black holes in galaxies across cosmic time
Proceedings IAU Symposium No. 312, 2014
Y. Meiron, S. Li, F.-K. Liu & R. Spurzem, eds.

© International Astronomical Union 2016
doi:10.1017/S1743921315007401

Radio evidence for
binary super massive black holes

R. D. Ekers

CSIRO Astronomy and Space Vimera and Pembroke Roads, Marsfield, NSW 2122, Australia
email: `ron.ekers@csiro.au`

Abstract. I present examples of radio AGN with binary nuclei which provide the direct radio evidence for binary Super Massive Black Holes (SMBH) driving the AGN activity. There is also other evidence for distorted radio morphology and periodic variability which may indicate the presence of a second (inactive) SMBH. Finally I enumerate a number of possible radio tracers for the binary SMBH merger events.

Keywords. Super Massive Black Holes (SMBH), binary, AGN

1. Introduction

The previous talk (Kellermann 2015) has summarized the evidence associating Super Massive Black Holes (SMBH) with AGN activity. When the AGN is radio loud we have an excellent SMBH tracer observable throughout the universe and VLBI observations have sufficient angular resolution to separate close (pc scale) SMBHs and to observe changes in the jet direction. All methods to investigate binary SMBHs need a radio loud AGN on at least one SMBH component to provide a tracer visible at radio wavelengths.

Radio images need to have sufficient resolution and quality to distinguish multiple AGN from multiple hot spots in a single AGN or to identify multiple images of a gravitational lensed AGN.

2. Multiple AGN

Spatially resolved binary AGN. 3C75, in Abel 400, (Owen *et al.* 1985) is one of the best examples of a binary AGN with two clearly separated AGN each with a nuclear source and very similar classical radio galaxy jet and lobe structure (Fig. 1). The separation of the two nuclei in 3C75 is 7.6 kpc. There are now many other examples of other similar double AGN e.g. PKS2359-139, 3C442, PKS2149-15, 3C338. In all cases the nuclei are separated by tens of kpc. Resolution at radio wavelengths is easily sufficient to see multiple AGN at smaller separation but these are much rarer. 0402+379 is the most compact known binary AGN with two compact radio components separated by 7 pc (Rodriguez *et al.* 2006).

On much smaller scales we can ask whether the evolving core-jet structures seen in the multi-epoch VLBI observations, described by Kellermann (2015) in the previous talk, would be affected by a second SMBH. For example, in 3C279, described by Homan *et al.* (2003), we see an abrupt change in jet direction at 3 milliarcseconds from the nucleus (1 kpc deprojected). There is a tendency for component trajectories to follow "pre-existing channels" and we see components deflected at this point in 1998 and again in 2004 - could this be a deflection by a second SMBH?

Other evidence for multiple AGN. Another option which can now be explored uses the new wideband spectra-polarimetry capabilities in JVLA (Perley *et al.* 2011) and ATCA

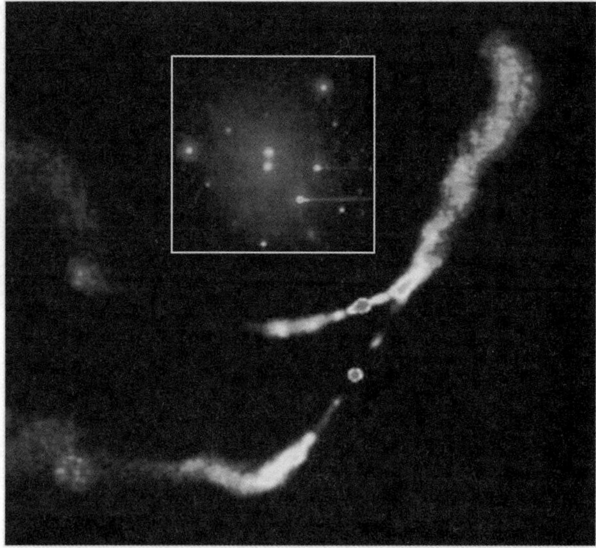

Figure 1. VLA observations of the radio galaxy 3C75, inset optical double nuclei

(Wilson *et al.* 2011) is to search for double rotation measures. It is expected that the line of sight to the two nuclei traverse regions of different density or magnetic field, and this will produce a double peak in the rotation measure synthesis even if the individual components are not spatially resolved.

Small diameter radio sources scintillate as a consequence of the propagation of the radio waves through irregularities in the interstellar medium. This is an analogue of the optical twinkling of stars due to irregularities in the troposphere. If a radio source scintillates this could be used to find double AGN. The sub pc scale of interest is easily accessible by looking for multiple peaks in scintillation structure function at sub-day time scales.

Periodic flux variability provides indirect evidence for the presence of a second SMBH. For example, OJ287 has shown quasi periodic variability with a 12 year cycle which was modelled by Valtonen *et al.* (2008) as a precessing binary black hole. Further observations, including observations of the changes in the resolved jet structure (Valtonen & Pihajoki 2013), have not directly confirmed initial predictions but have resulted in more complex models involving black hole spin and helical jets.

3. A systematic radio census of binary SMBH

Burke-Spolaor (2011) used archival Very Long Baseline Interferometry (VLBI) data for 3114 radio loud AGN, to search for binary supermassive black holes on pc scales. The only example found, was the 7 pc binary AGN, 0402+379, already mentioned.

Probability of detecting a binary SMBH. Note that the probability of detecting radio emission from an AGN is only 1% at Jy flux density levels, hence the probability of detecting both components in a binary SMBH is only 0.01%. At lower (mJy) flux levels this probability could increase to 10%, hence a 1% probability for two active SMBHs. Future studies will need much larger samples to find the close SMBH binaries. This requires more sensitivity to go to fainter sources but, sadly, SKA1 no longer has the long baseline needed to do this so we will have to wait for SKA2. A further complication which arises is the possible effect of one AGN on a second nearby SMBH. This could either quench the second AGN by blowing its accretion material away or enhance activity by

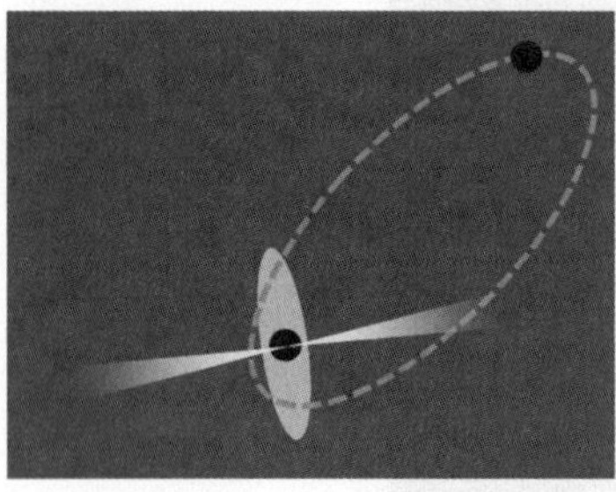 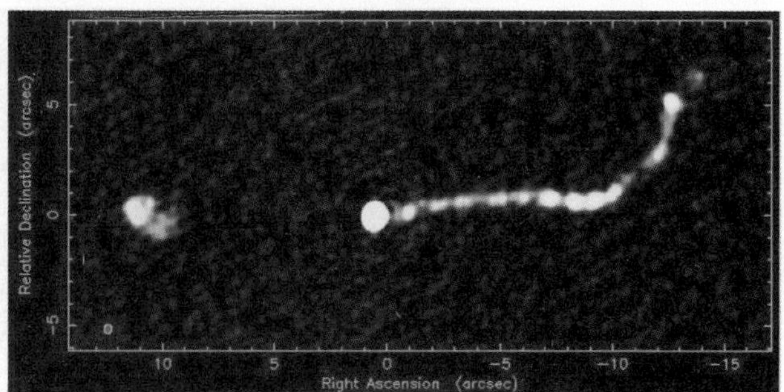

Figure 2. ATCA observations of radio galaxy 0637-752 with binary SMBH model

disturbing its accretion disk and increasing the amount of material accreted. The most likely outcome is that any collimated jet of flux from the first AGN will miss the second AGN and have little effect.

Binary scale sizes. AGN are good for probing the high-mass end of the SMBH distribution: $> 10^8 M_\odot$. The inspiral time scale due to gravitational radiation is 10^8 years (1% of age of universe) while the AGN phase lasts 10^9 years i.e. about 10 times longer. The dynamical stalling radius where the well known dynamical effects may stop working is about 3 pc and this is just the range of angular scale sizes accessible to VLBI techniques. Unfortunately, with earth scale baselines it is hard to reach the < 0.2 pc scale sizes for which gravitational radiation is detectable by pulsar timing. The Burke-Spolaor (2011) analysis suggests that binary pair evolution of supermassive black holes (masses $> 10^8 M_\odot$) spend less than 500 Myr in progression from the merging of galactic stellar cores to within the purported stalling radius for supermassive black hole pairs. The data show no evidence for an excess of stalled binary systems at small separations.

Triple SMBH System! Deane *et al.* (2014) have recently published a claim for a triple SMBH. This is SDSS J1502+1115, a $z = 0.39$ QSO with double peaked OIII. It contains one dust obscured AGN with a double flat spectrum radio core, separation 138 pc, and this is claimed to be a double SMBH. There is a second unobscured AGN with a third compact radio component 7.4 kpc away from the double core. Wrobel, Walker, & Fu (2014) obtained higher quality VLA data which indicates that the obscured AGN is actually a double sided AGN with very compact lobes and a single central SMBH. This source then becomes another example of a large separation binary SMBH.

4. Dynamical effects

Binary modulation of AGN activity. This spectacular "chain of beads" AGN, PKS 0637-752 (Fig. 2) could be explained by a bound second SMBH periodically plunging through the disk of the active SMBH which maintains its spin axis but has a burst of activity as the accretion disk is perturbed on each plunge, (Godfrey *et al.* 2012). The bend ($12''$) 80 kpc to the west may be an axis re-orientation event?

Astrometric measurement of orbit: 3C66. Sudou *et al.* (2003) suggested that 3C66B was in a 1 year orbit , based on astrometric VLBI measurements. This has been disproved by Jenet *et al.* (2004) because the gravitational wave emission would have already been detectable in the pulsar timing data.

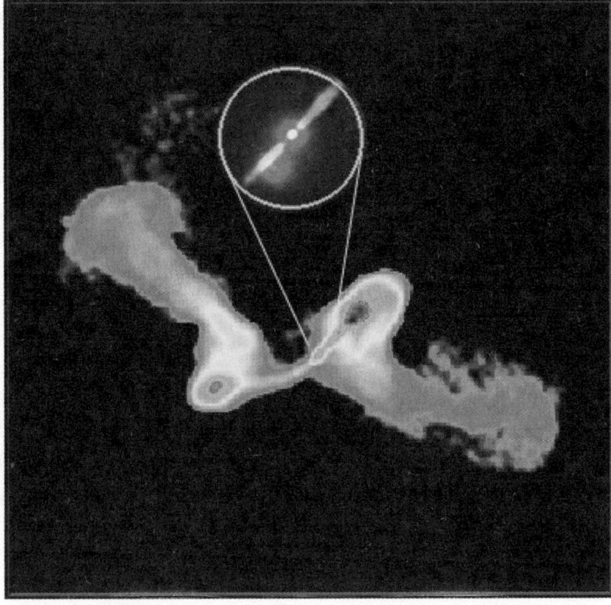

Figure 3. VLA observations of NGC326

5. Alignment of radio axes

In a giant radio galaxy, such as NGC 6251, the axis direction has been maintained for the life of the source, i.e. $> 10^9$ years. This is the normal situation seen in radio galaxies and this extraordinary long term stability of the jet axis is strong evidence that jet orientations are regulated by black hole spins (e.g. Begelman *et al.* 1984). More rarely we see examples where there is a change in axis direction. These are rare but about a dozen good examples are now known. They are referred to as the X sources. NGC326 (Ekers *et al.* 1978, Murgia *et al.* 2001) is the poster child example of an X-shaped radio galaxy in which the jet axis changed direction (Fig. 3). Rees (1978) summarised a range of possible models for changing direction of the spin axis and Merritt & Ekers (2002) explored the specific implications of merging binary SMBHs. These X structures occur in a few % of all radio galaxies but there is no consensus on whether this is an axis change due to a binary SMBH merger or whether it is due to a hydrodynamical effect in the external medium (e.g. Saripalli *et al.* 2013). Another proposed alternative is that the X-shaped sources may result from two pairs of jets that are associated with a pair of unresolved AGNs tracing a binary SMBH that have not yet merged (Lal & Rao 2007).

6. Observable consequences of SMBH mergers

In addition to the direct evidence for the presence of binary SMBH there are a number of potentially observable consequences of a binary SMBH merger.

- Spin axis changes generate X sources (see previous section)
- Dynamical kicks during the merger process will generate misaligned core jet structures.
- AGN is displaced from the centre of the galaxy.
- AGN activity is disrupted by the kick and emission is quenched.
- One sided jets may be a consequence of relative velocity or density asymmetries resulting from a merger event.

- Electromagnetic pulses may be generated in the surrounding plasma by the intense gravity wave generated by the merger.

Any future detection of gravitational waves (by LIGO, LISA etc.) from SMBH mergers or other gravitational events will need to be identified with their host galaxies to measure redshift but such identification will need more accurate positions than the many degree error boxes possible with the gravitational wave detectors. Any simultaneous transient radio emission could be used to obtain a sufficiently accurate position to make an identification. The wide field of view and possible rapid follow up of the future radio instruments will provide these opportunities.

7. Summary

Radio continuum emission can identify double AGNs and hence binary SMBH. There are many examples of binary activity on scales of a few kpc separation but these are very rare at pc separation. We have some evidence for periodic modulation of the radio properties which could be caused by a binary SMBH. The morphology of the X shaped radio sources shows some good examples of possible jet re-orientation or precession but interpretation of these observations is not unambiguous.

References

Begelman, M. C., Blandford, R. D., & Rees, M. J. 1984, *Rev. Mod. Phys.*, 56, 255

Burke-Spolaor, S. 2011, *MNRAS*, 410, 2113-2122

Deane, R. P., Paragi, Z., Jarvis, M. J., Coriat, M., Bernardi, G., Fender, R. P., Frey, S., Heywood, I., Klckner, H.-R., Grainge, K., & Rumsey, C. 2014, *Nature*, 511, 5760

Ekers, R. D., Fanti, R., Lari, C., & Parma, P. 1978, *Nature*, 276, 588-590

Godfrey, L. E. H.., Lovell, J. E. J.., Burke-Spolaor, S., Ekers, R. D., Bicknell, G. V., Birkinshaw, M., Worrall, D. M., Jauncey, D. L., Schwartz, D. A., Marshall, H. L., Gelbord, J., Perlman, E. S., & Georganopoulos, M. 2012, *ApJL*, 758, L27

Homan, D. C., Lister, M. L., Kellermann, K. I., Cohen, M. H., Ros, E., Zensus, J. A., Kadler, M.,& Vermeulen, R. C. 2003, *ApJL*, 589, L9-L12

Jenet, F. A., Lommen, A., Larson, S. L., & Wen, L. 2004, *ApJ*, 606 799-803

Kellermann, K. I. 2015, *IAU Symposium 312, Star Clusters and Black Holes in Galaxies and Across Cosmic Time*, ed. Fukun Liu

Lal, D. V. & Rao, A. P. 2007, *MNRAS*, 374, 1085

Merritt, D. & Ekers, R. D. 2002, *Science*, 297, 1310-1313

Murgia, M., Parma, P., De Ruiter, H. R., Bondi, M., Ekers, R. D., & Fanti, R., . Fomalont, E. B. 2001, *A&A*, 380, 102-116

Owen, F. N., O'Dea, C. P., Inoue, M., & Eilek, J. A. 1985, *ApJL*, 294, L85-L88

Perley, R. A., Chandler, C. J., Butler, B. J., & Wrobel, J. M. 2011, *ApJL*, 739, L1

Rees, M. J. 1978, *Nature*, 275, 516

Rodriguez, C., Taylor, G. B., Zavala, R. T., Peck, A. V., Pollock, L. K., & Romani, R. W. 2006, *ApJ*, 646, 49

Saripalli, L., Malarecki, J. M., Subrahmanyan, R., Jones, D. H., & Staveley-Smith, L. 2013, *MNRAS*, 436, 690-696

Sudou, H., Iguchi, S., Murata Y., & Taniguchi, Y. 2003, *Science*, 300, 1263-1265

Valtonen, M., Kidger, M., Lehto, H., & Poyner, G. 2008, *Astronomy & Astrophysics*, 477, 407-412

Valtonen, M. & Pihajoki, P. 2013, *A&A*, 557, 28-32

Wilson, W.E., Ferris, R.H., Axtens, P., Brown, A., Davis E., Hampson, G., Leach, M., Roberts, P., Saunders, S. Koribalski, B.S., & 23 coauthors 2011, *MNRAS*, 416, 832-56

Wrobel, J. M., Walker, R. C., & Fu, J. H. 2014, *ApJL*, 792, L8

Star clusters and black holes in galaxies across cosmic time
Proceedings IAU Symposium No. 312, 2014
Y. Meiron, S. Li, F.-K. Liu & R. Spurzem, eds.

© International Astronomical Union 2016
doi:10.1017/S1743921315007413

Investigating the AGN activity and black hole masses in low surface brightness galaxies

Smitha Subramanian[1], Ramya Sethuram[2], Mousumi Das[1],
Koshy George[1], Sivarani Thirupathi[1] and Tushar P. Prabhu[1]

[1]Indian Institute of Astrophysics, 2nd Block, Koramangala, Bangalore - 560034, India
email: `smitha@iiap.res.in`

[2]Shanghai Astronomical Observatory, Shanghai, China

Abstract. We present an analysis of the optical nuclear spectra from the active galactic nuclei (AGN) in a sample of giant low surface brightness (GLSB) galaxies. GLSB galaxies are extreme late type spirals that are large, isolated and poorly evolved compared to regular spiral galaxies. Earlier studies have indicated that their nuclei have relatively low mass black holes. Using data from the Sloan Digital Sky Survey (SDSS), we selected a sample of 30 GLSB galaxies that showed broad Hα emission lines in their AGN spectra. In some galaxies such as UGC 6284, the broad component of Hα is more related to outflows rather than the black hole. One galaxy (UGC 6614) showed two broad components in Hα, one associated with the black hole and the other associated with an outflow event. We derived the nuclear black hole (BH) masses of 29 galaxies from their broad Hα parameters. We find that the nuclear BH masses lie in the range $10^5 - 10^7$ M$_\odot$. The bulge stellar velocity dispersion σ_e was determined from the underlying stellar spectra. We compared our results with the existing BH mass - velocity dispersion ($M_{\rm BH}$–σ_e) correlations and found that the majority of our sample lie in the low BH mass regime and below the $M_{\rm BH}$–σ_e correlation. The effects of galaxy orientation in the measurement of σ_e and the increase of σ_e due to the effects of bar are probable reasons for the observed offset for some galaxies, but in many galaxies the offset is real. A possible explanation for the $M_{\rm BH}$–σ_e offset could be lack of mergers and accretion events in the history of these galaxies which leads to a lack of BH-bulge co-evolution.

Keywords. galaxies: active, galaxies: bulges, galaxies: nuclei

1. Introduction

Low surface brightness (LSB) galaxies are late type spiral galaxies (Sc or Sd) that have a central disk surface brightness of $\mu_{B(0)} \geqslant 22$ to 23 mag arcsec^{-2} (Impey & Bothun 1997; Impey *et al.* 2001). Disk LSBs are more rare and usually found in isolated environments (Bothun *et al.* 1993) and vary over a range of sizes. But the really large LSB galaxies, which are referred to as giant LSB (GLSB) galaxies, are generally found to lie close to the edges of voids (Rosenbaum *et al.* 2009). Nuclear activity is not common in LSB galaxies. This is in marked contrast to high surface brightness disk galaxies where the percentage having active galactic nuclei (AGN) can be as high as 50%, depending on the mean luminosity of the sample. The most probable explanation for this low fraction is that LSB disks generally lack two structural features that facilitate gas flows and the formation of a compact object in the nucleus - bars and strong spiral arms (Bothun *et al.* 1997).

However, a significant fraction of bulge dominated GLSB galaxies show AGN activity (Sprayberry *et al.* 1995). AGN activity in bulge dominated GLSBs is not surprising as studies indicate that the growth of nuclear black holes (BH) in galaxies is intimately linked to the growth of their bulges (Silk & Rees 1998; Heckman *et al.* 2004). The strong correlation of black hole mass ($M_{\rm BH}$) with bulge stellar velocity dispersion in galaxies ($M_{\rm BH}$–σ_e) is due to this supermassive black hole (SMBH) - bugle co-evolution (Gültekin *et al.* 2009 and McConnell & Ma 2013). But in LSB galaxies, their bulge velocity dispersion and disc rotation speeds suggest that they lie below the $M_{\rm BH}$–σ_e correlation for bright galaxies (Pizzella *et al.* 2005). Mei *et al.* (2009) estimated the black hole masses of three GLSB galaxies and found them to lie close to the $M_{\rm BH}$–σ_e correlation and the BH masses are found to be of the order of $\sim 10^7$ M$_\odot$. Later estimates from the study of another three GLSBs by Ramya *et al.* (2011) showed that their sample galaxies lie offset from the $M_{\rm BH}$–σ_e correlation. They suggested that the bulges of their sample might be well evolved, but the BH masses (3–9×10^5 M$_\odot$) are lower than those found in bright galaxies. Morelli *et al.* (2012) found that the bulges of LSB galaxies have similar properties as that of the bulges of normal galaxies. It is not clearly understood where the LSB galaxies in general lie on the $M_{\rm BH}$–σ_e correlation and also what are the possible reasons for the observed offset in a few sample galaxies. Low mass black holes in isolated LSB galaxies are are also very interesting candidates for the study of seed black holes in galaxies (Volonteri 2010).

2. Data and analysis

As an initial sample we selected 558 LSB galaxies from the literature (Sprayberry *et al.* 1995, Impey *et al.* 1996, Schombert *et al.* 1992 and Schombert *et al.* 1988) for which the Sloan Digital Sky Survey (SDSS) DR10 nuclear spectra are available. The emission line fluxes of these 558 galaxies are available for public in the SDSS DR10 database. The emission line fluxes are used to construct the [N$_{\rm II}$]$_{6583}$/Hα vs [O$_{\rm III}$]$_{5007}$/Hβ Baldwin-Phillips-Terlevich (BPT) (Baldwin *et al.* 1981) diagram. Out of 558 galaxies, 160 galaxies ($\sim$29%, with almost equal contribution from the composite and purely AGN candidates) are classified as AGN and the remaining 398 ($\sim$71%) galaxies as purely star forming systems. From the sample of 160 galaxies, we selected only those galaxies which have median S/N > 15 for further analysis. Thus, finally we used SDSS DR10 spectra of 115 LSB galaxies to identify sources with broad Balmer lines and hence estimate their black hole masses.

We used the pPXF (Penalized Pixel-Fitting stellar kinematics extraction) code by Cappellari & Emsellem (2004) to obtain the best fit model for the underlying stellar population. pPXF also provides the stellar velocity dispersion, σ. The Hα + [N$_{\rm II}$] doublet region of all the 115 galaxies are first fitted with three Gaussians, representing the three narrow lines of Hα and [N$_{\rm II}$] doublet. Then this region is again fitted including a fourth Gaussian component for the broad Hα line. The final sample with broad Hα component is selected based on the two criteria, that the the inclusion of the broad component should improve the reduced χ^2 value by at least 5% and also the broad Hα peak flux should be at least three times larger than the residue of the fit. Thus based on the these criteria, 52 LSB galaxies were selected as galaxies with broad Hα component. After the careful examination of the fits of each of these galaxies, 28 of them showed the presence of broad Hα at significant level and 2 galaxies showed an extra broad components, which are significantly blue/red shifted from the Hα central wavelength and are hence more likely to be associated with an outflow rather than with the active black hole. The 28 galaxies, with genuine broad Hα component, are of our interest as their black hole masses can be estimated and hence check their M-σ_e correlation. UGC 6614, which showed signatures of an outflow, has a clear broad Hα component along with the blue shifted outflow component. The other emission lines present in the spectra of these 30 galaxies (28 with broad Hα component and 2 with outflow signatures) were also analysed and fitted with

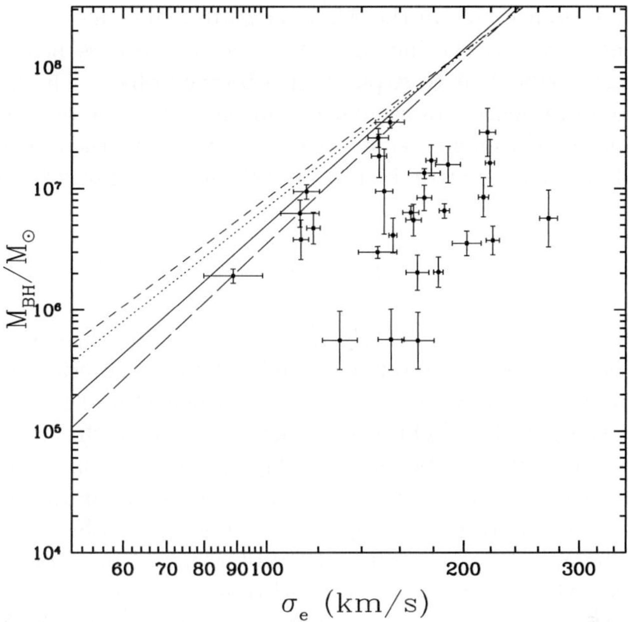

Figure 1. The M-σ_e plot with broad line AGN candidates. The linear regression lines given by Tremaine *et al.* (2002), Ferrarese & Merritt (2000), Gültekin *et al.* (2009) and McConnell & Ma (2013) relation for late type galaxies (dashed, solid, dotted and long-dashed lines, respectively) for $M_{\rm BH}$ against σ_e are also shown.

Gaussian profiles. IDL programs are used to fit the emission line profiles with Gaussian functions and the fluxes of all the emission lines are estimated.

3. Black hole masses and the $M-\sigma_e$ correlation

The virial black hole masses of 29 broad line AGN candidates are calculated using the equation given in Reines *et al.* (2013), using the luminosity and FWHM of broad Hα. The equation is given below. Here, the scale factor that depends on the broad line region geometry, ϵ is assumed to be 1.

$$\log \frac{M_{\rm BH}}{\rm M_\odot} = 6.57 + 0.47 \log \frac{L_{\rm H\alpha}}{10^{42}\,\rm erg\,s^{-1}} + 2.06 \log \frac{\rm FWHM_{H\alpha}}{10^3\,\rm km\,s^{-1}} \tag{3.1}$$

The masses estimated are in the range 5.5×10^5 M$_\odot$ to 3.5×10^7 M$_\odot$. The median mass is found to be 5.6×10^6 M$_\odot$. The median mass suggests that the LSB galaxies have black hole masses which are slightly higher than intermediate mass black holes.

The stellar velocity dispersion values obtained from pPXF were transformed to the equivalent velocity dispersion σ_e at a radius of $r_e/8$ in the galaxy bulge, where r_e is the bulge exponential scale length, using the transformation equation given by Jorgensen *et al.* (1995). The BH mass estimates for the 29 galaxies in our sample plotted against the equivalent velocity dispersion values, σ_e are shown in the M -σ_e plot in Fig. 1. Also plotted in the figure are the linear regression lines given by Tremaine *et al.* (2002), Ferrarese & Merritt (2000), Gültekin *et al.* (2009) and McConnell & Ma (2013) for late type galaxies (dashed, solid, dotted and long-dashed lines, respectively) for $M_{\rm BH}$ against σ_e. Most of our sample lie below the M–σ_e correlation. The probable reasons for the observed offset can be due the the effects of galaxy orientation in the measurement of σ and/or the effects of bars in the increase of σ in case of barred galaxies. We analysed the effects of any systematic bias in the M_{BH} estimates, the effects of galaxy orientation in the measurement of σ and the increase of σ due to the presence of bars and found that these effects are insufficient to explain the observed offset. The lack of bulge–M_{BH} co-evolution is also an indicative of secular evolution and is one of the probable reasons for the observed offset (Kormendy *et al.* 2013). A detailed study of the nature of the bulges and the role of dark matter in the growth of black holes in these systems are important.

4. Discussion

Seed BHs that are formed in the early universe grow by mass accretion to become the massive BHs that we observe in our local universe; the accretion is driven by galaxy mergers and interactions. LSB galaxies are usually isolated and lie close to the edge of voids. Simulations suggest that lighter seed BHs grow through slow accretion leading relatively low mass BHs and lie below the M_{BH}–σ_e correlation. These low mass and relatively pristine black holes can reveal important clues to the initial black hole mass function and help us to constrain the early evolution of black holes in galaxies. The black hole masses in our sample lie in the range 10^5–10^7 M$_\odot$ and most of them lie below the M_{BH}–σ_e correlation. Also, their Eddington ratios are not high, hence their black holes are not accreting at a high rate and they fall in the low luminosity AGN (LLAGN) class. Thus, the BHs in GLSB galaxies represent one of the best candidates for pristine black holes and a good place to study seed black holes in our local universe.

Acknowledgements

The authors would like to thank the SDSS team for making the data publicly available. SS acknowledges the financial support from IAU for the participation in the symposium.

References

Baldwin J. A., Phillips M. M., & Terlevich R., 1981, *PASP*, 93, 5

Bothun, G., Impey, C., & McGaugh, S., 1997, *PASP*, 109, 745

Bothun, G. D., Schombert, J. M., Impey, C. D., Sprayberry, D., & McGaugh, S. S. 1993, *AJ*, 106, 530

Cappellari M., Emsellem E., 2004, *PASP*, 116, 138

Ferrarese L. & Merritt D., 2000, *ApJ*, 539, L9

Gültekin, K., Richstone, D. O., Gebhardt, K. *et al.* 2009a, *ApJ*, 698, 198

Heckman, Timothy M., Kauffmann, Guinevere, Brinchmann, Jarle, Charlot, Stéphane, Tremonti, Christy, White, & Simon D. M., 2004, *ApJ*, 613, 109

Impey, C. & Bothun, Greg, 1997, *ARA&A*, 35, 2671

Impey, C., Burkholder, V., & Sprayberry, D., 2001, *AJ*, 122, 2341

Impey, C. D., Sprayberry, D., Irwin, M. J., & Bothun, G. D., 1996, *ApJS*, 105, 209

Jorgensen I., Franx M., & Kjaergaard P., 1995, *MNRAS*, 276, 1341

Kormendy J. & Ho L. C., 2013, *ARA&A*, 51, 511

McConnell, Nicholas J. & Ma, Chung-Pei, 2013, *ApJ*, 764, 184

Mei L., Yuan W., & Dong X., 2009, *Res. Astron. Astrophys.*, 9, 269

Morelli, L., Corsini, E. M., Pizzella, A., Dalla Bontá, E., Coccato, L., Méndez-Abreu, J., & Cesetti, M., 2012, *MNRAS*, 423, 962

Pizzella A., Corsini E. M., Dalla Bontà E., Sarzi M., Coccato L., & Bertola F., 2005, *ApJ*, 631, 785

Ramya, S., Prabhu, T. P., & Das, M., 2011, *MNRAS*, 418, 789

Reines, Amy E., Greene, Jenny E., & Geha, Marla, 2013, *ApJ*, 775, 116

Rosenbaum, S. D., Krusch, E., Bomans, D. J., & Dettmar, R.-J., 2009, *A&A*, 504, 807

Schombert, James M. & Bothun, Gregory D., 1988, *AJ*, 95, 1389

Schombert, James M., Bothun, Gregory D., Schneider, Stephen E., & McGaugh, Stacy S., 1992, *AJ*, 103, 1107

Silk, Joseph & Rees, Martin J., 1998, *A&A*, 331, 1

Sprayberry, D., Impey, C. D., Bothun, G. D., & Irwin, M. J, 1995, *AJ*, 109, 558

Tremaine S. *et al.*, 2002, *ApJ*, 574, 740

Volonteri, Marta, 2010, *A&ARv*, 18, 279

Star clusters and black holes in galaxies across cosmic time
Proceedings IAU Symposium No. 312, 2014
Y. Meiron, S. Li, F.-K. Liu & R. Spurzem, eds.
© International Astronomical Union 2016
doi:10.1017/S1743921315007425

Disentangling AGN and starburst activities in NGC 6240 from an X-ray perspective

Junfeng Wang

Department of Astronomy and Institute of Theoretical Physics and Astrophysics,
Xiamen University, Xiamen, Fujian 361005, China
email: jfwang@xmu.edu.cn

Abstract. The circum-nuclear region in an active galaxy is often complex with presence of high excitation gas, collimated radio outflow, and star formation activities, besides the actively accreting supermassive black hole. The unique spatial resolving power of Chandra X-ray imaging spectroscopy enables more investigations to disentangle the active galactic nuclei and starburst activities. For galaxies in the throes of a violent merging event such as NGC6240, we were able to resolve the high temperature gas surrounding its binary active black holes and discovered a large scale soft X-ray halo.

Keywords. black hole physics, galaxies: active, galaxies: interactions, galaxies: halos, X-rays: ISM

1. Introduction

The current picture of galaxy evolution advocates co-evolution of galaxies and their nuclear massive black holes (MBHs), through accretion and merging, which successfully explains much of the observed active galactic nuclei (AGN) statistics and the $M_{\rm BH}$–σ relation. Quasar pairs (a few kpc–100 kpc separation) and the sub-light-year separation MBHs exemplify the early through late stages of the gravitational interaction. The double active nuclei of few nearby galaxies with disrupted morphology and intense star formation demonstrate the importance of major mergers of equal mass spirals in this evolution, leading to formation of a massive elliptical galaxy. In studies of the hot interstellar medium in galaxies experiencing these key stages, the full exploitation of *Chandra* imaging potential coupled with its CCD spectroscopy capabilities gives us the opportunity to deeply investigate the physical state of the diffuse X-ray emission at increasing distance from the central AGN, overcoming the contamination from bright point sources.

2. The infrared luminous galaxy merger NGC 6240

In the local universe, NGC 6240 ($z = 0.02448 \pm 0.00003$) is a unique galaxy, in the throes of a violent merging event and on its way to becoming an elliptical galaxy (Engel *et al.* 2010). It is experiencing intense star formation (e.g., $61 \pm 30 M_\odot$ yr^{-1} in Yun & Carilli 2002). With $L_{FIR} \sim 10^{11.8} M_\odot$ just below $10^{12} M_\odot$, the luminosity threshold for the ultraluminous infrared galaxies (ULIRGs), NGC 6240 is expected to become an ULIRG when the galaxies coalesce and a final starburst is triggered (Engel *et al.* 2010). The *Chandra* X-ray images Komossa *et al.* 2003 of the central region of NGC 6240 revealed two gravitationally interacting active supermassive black holes (SMBHs) separated by $\sim$0.7 kpc. These sources are characterized by the highly absorbed hard X-ray spectra typical of Compton thick active galactic nuclei (AGNs). They each have an observed luminosity of $L_{2-8{\rm keV}} \sim 10^{42}$ erg s^{-1}, and show prominent neutral Fe Kα

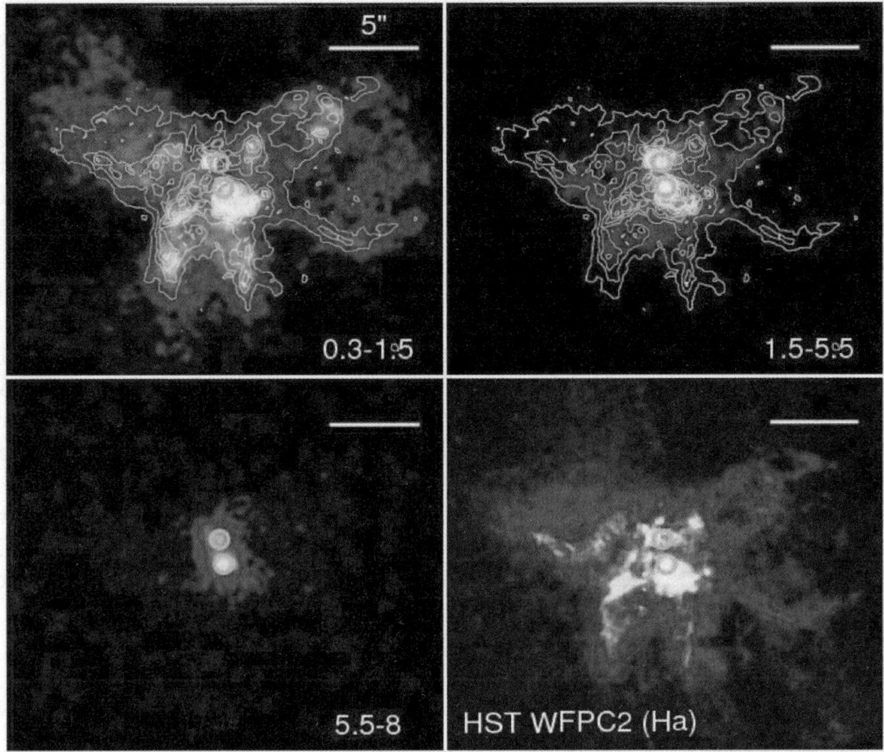

Figure 1. Comparison of multi-band X-ray emission and the HST/WFPC2 Hα image. Top left: 0.3–1.5 keV soft X-ray emission; top right: 1.5–5.5 keV X-ray emission; bottom left: 5.5–8 keV X-ray emission, dominated by the two nuclei; bottom right: HST/WFPC2 image of the Hα emitting gas, from which the white contours (shown in the top panels) were generated.

lines at 6.4 keV present in the spectra of both nuclei. Marginal evidence of an hydrogen-like iron line emission was found, which could be associated with a spatially resolved, multi-temperature hot gas outflow, powered by the starburst activity that dominates the 0.5-3 keV X-ray emission ($L_{0.1-2\mathrm{keV}} \sim 10^{43}$ erg s^{-1}; Huo *et al.* 2004). *XMM*-Newton observations (Netzer *et al.* 2005) resolved the Fe K line complex into three narrow lines, the neutral Fe Kα, Fe XXV, and a blend of Fe XXVI with Fe Kβ. These *XMM*-Newton data, however, do not have the resolution to resolve spatially the line-emitting regions.

3. Results

We report on a recent 150 ks long, sub-arcsecond resolution X-ray image of the nuclear region of the luminous galaxy merger NGC 6240 with *Chandra*, which resolves the X-ray emission from the pair of active nuclei and the diffuse hot gas in great detail (Figure 1; see Wang *et al.* 2014). We detect extended hard X-ray emission from $kT \sim 6$ keV ($\sim$70 million K) hot gas over a spatial scale of 5 kpc, indicating the presence of fast shocks with velocity of $\sim$2000 km s^{-1}. For the first time we obtain the spatial distribution of this highly ionized gas emitting Fe XXV, which shows a remarkable correspondence to the large scale morphology of H$_2$(1-0) S(1) line emission and Hα filaments. Propagation of fast shocks originated in the starburst driven wind into the ambient dense gas can account for this morphological correspondence.

With an observed $L_{0.5-8\mathrm{keV}} = 5.3 \times 10^{41}$ erg s^{-1}, the diffuse hard X-ray emission is ~ 100 times more luminous than that observed in the classic starburst galaxy M82. We estimate its total mass ($M_{\mathrm{hot}} = 1.8 \times 10^8 \mathrm{M_\odot}$) and thermal energy ($E_{\mathrm{th}} = 6.5 \times 10^{57}$ erg). The total iron mass in the highly ionized plasma is $M_{\mathrm{Fe}} = 4.6 \times 10^5 \mathrm{M_\odot}$. Both the energetics and the iron mass in the hot gas are consistent with the expected injection from the supernovae explosion during the starburst that is commensurate with its high star formation rate. However, a recent CO(1-0) interferometry observation of NGC 6240 (Feruglio $et\ al.$ 2013) implies a total kinetic power of the outflowing CO gas $E_{kin} = 3.8 \times 10^{57}$ erg, which leads to the plausible extra heating from the central double AGN.

We also detected an extended hot halo (Nardini $et\ al.$ 2013), with projected physical size of $\sim 110 \times 80$ kpc, and a single-component thermal model provides a reasonably good fit to the observed X-ray spectrum. The hot gas has a temperature of 7.5 MK and an estimated total mass of $10^{10} \mathrm{M_\odot}$, resulting in an intrinsic 0.4-2.5 keV luminosity of 4×10^{41} erg s^{-1}. The relative abundance of the main alpha-elements with respect to iron is several times the solar value, and nearly constant as well, implying a uniform enrichment by type II supernovae out to the largest scales. Taken as a whole, the observational evidence hints at widespread, enhanced star formation proceeding at steady rate over the entire dynamical timescale (about 200 Myr). Under favorable conditions, at least a fraction of it might be retained after the merger completion, and evolve into the hot halo of a young elliptical galaxy.

Acknowledgements

This work was supported by NASA grant GO1-12123X and the National Natural Science Foundation of China under grants 11443003 and 11473021.

References

Engel, H., Davies, R. I., Genzel, R., Tacconi, L. J., Hicks, E. K. S., Sturm, E., Naab, T., Johansson, P. H., Karl, S. J., Max, C. E., Medling, A., & van der Werf, P. P. 2010, $A\&A$, 524, A56

Feruglio, C., Fiore, F., Maiolino, R., Piconcelli, E., Aussel, H., Elbaz, D., Le Floc'h, E., Sturm, E., Davies, R., & Cicone, C. 2013, $A\&A$, 549, A51

Huo, Z. Y., Xia, X. Y., Xue, S. J., Mao, S., & Deng, Z. G. 2004, ApJ, 611, 208

Komossa, S., Burwitz, V., Hasinger, G., Predehl, P., Kaastra, J. S., & Ikebe, Y. 2003, ApJ (Letters), 582, L15

Nardini, E., Wang, J., Fabbiano, G., Elvis, M., Pellegrini, S., Risaliti, G., Karovska, M., & Zezas, A. 2013, ApJ, 765, 141

Netzer, H., Lemze, D., Kaspi, S., George, I. M., Turner, T. J., Lutz, D., Boller, T., & Chelouche, D. 2005, ApJ, 629, 739

Wang, J., Nardini, E., Fabbiano, G., Karovska, M., Elvis, M., Pellegrini, S., Max, C., Risaliti, G., U, V., & Zezas, A. 2014, ApJ, 781, 55

Yun, M. S. & Carilli, C. L. 2002, ApJ, 568, 88

Star clusters and black holes in galaxies across cosmic time
Proceedings IAU Symposium No. 312, 2014
Y. Meiron, S. Li, F.-K. Liu & R. Spurzem, eds.
© International Astronomical Union 2016
doi:10.1017/S1743921315007437

Relativistic Fe Kα line and ensemble spins of black holes in narrow-line Seyfert 1 galaxies

Weimin Yuan, Zhu Liu, Youjun Lu and Xin-Lin Zhou

National Astronomical Observatories, Chinese Academy of Sciences, Beijing 100012, China
email: `wmy@nao.cas.cn`

Abstract. While a broad line of the Fe Kα emission is commonly found in the X-ray spectra of typical Seyfert galaxies, the situation is unclear in the case of Narrow Line Seyfert 1 galaxies (NLS1s)—an extreme subset which are generally thought to harbor less massive black holes with higher accretion rates. We report results of our study of the assemble property of the Fe K line in NLS1s by stacking the X-ray spectra of a large sample of 51 NLS1s observed with XMM-Newton. We find in the stacked X-ray spectra a prominent, broad emission feature over 4–7 keV, which is characteristic of the broad Fe Kα line. Our results suggest that a relativistic broad Fe line may in fact be common in NLS1s. The line profile is used to study the average spin of the black holes in the sample. We find, for the first time, that their black holes are constrained to be likely spinning at averagely low or moderate rates as a population. The implications of the results are discussed in the context of the black hole growth in NLS1 galaxies.

Keywords. line: profile, galaxies: Seyfert, X-rays: galaxies

---

1. Introduction

A broad emission line at $E \sim 6$-7 keV has long been found in the X-ray spectra of about $\sim 40\%$ of Seyfert 1 galaxies (e.g., Nandra *et al.* 2007). It is generally interpreted to be the Fe Kα emission produced via the K-shell process in the proximity of supermassive black holes and is hence relativistically broadened (Fabian *et al.* 1989). The energy of the red wing of the broad line is determined by the inner radius of the accretion disc, which is commonly believed to be at the innermost stable circular orbit (ISCO). The ISCO is strongly dependent on the the black hole spin. As such, the profile of a relativistic broad Fe Kα line can be used to constrain the spin of the black hole (Brenneman & Reynolds 2006; Dauser *et al.* 2010). In observations, black hole spins have been measured in about two dozen individual bright AGNs (e.g., Patrick *et al.* 2012; Walton *et al.* 2013). However, for the vast majority of AGNs it is impossible to constrain the black hole spin with this method, which requires a very high signal-to-noise ratio of the X-ray spectra at the Fe Kα energies (de La Calle Pérez *et al.* 2010). Alternatively, spectral stacking is an effective way to obtain a composite spectrum with very high signal-to-noise for a certain class of objects, e.g., Corral *et al.* (2008); Iwasawa *et al.* (2012)

Narrow Line Seyfert 1 galaxies (NLS1s) are a subset of active galactic nuclei (AGNs) (Osterbrock & Pogge 1985; Goodrich 1989), which are generally thought to have relatively small black hole masses (Boroson 2002) and higher accretion rates close to the Eddington rate (Boroson & Green 1992). They show extreme properties among AGNs (Komossa 2008). Concerning the broad Fe Kα line in NLS1s, somewhat contradictory results have been presented in the literature. A few well-studied NLS1s show apparently a broad Fe Kα emission line, including *SWIFT* J2127.4+5654 (Miniutti *et al.* 2009; Marinucci *et al.* 2014) and 1H 0707-495 (Fabian *et al.* 2009). de La Calle Pérez *et al.* (2010) found that the broad Fe Kα lines were detected in 4 out of 30 NLS1s in their full sample and in

2 out of 4 in their flux limited sample. Ai *et al.* (2011) found evidence for significant Fe Kα lines in none of the objects in a sample of 13 extreme NLS1s that have very small broad-line widths (< 1200 km s^{-1}). It is not known whether a broad Fe Kα line is common in NLS1s. We investigate the ensemble property of the Fe Kα line of NLS1s by stacking the X-ray spectra of a large sample of NLS1s observed with *XMM-Newton*. We use the cosmological parameters $H_0 = 70$ km s^{-1} Mpc^{-1}, $\Omega_M = 0.27$ and $\Omega_\Lambda = 0.73$. All quoted errors correspond to the 90 per cent confidence level unless specified otherwise

2. Sample

Two NLS1s catalogues were used to compiled a sample of NLS1s that have *XMM-Newton* observations. The first is the largest NLS1s sample ($\sim$2000), which is homogeneously selected from the SDSS DR3 by Zhou *et al.* (2006) with well measured optical spectrometric parameters. The second includes about 400 NLS1s from the AGN catalogue compiled by Véron-Cetty & Véron (2006). Only sources with more than 100 net source counts in either the EPIC PN or the combined MOS detectors in the 2-10 keV band were selected. We excluded three objects that are previously known to show a significant broad Fe Kα line, namely, 1H 0707-495, Mrk 766, and NGC 4051. The sample includes 51 NLS1s that have a total of 68 observations.

The X-ray spectra are retrieved from the LEDAS website†. We combined the MOS1 and MOS2 spectra if both are available. The PN and the combined MOS spectra are treated independently. No significant broad Fe Kα line is found in individual spectra which is likely due to relatively low S/N in the Fe line region.

3. Stacked X-ray spectrum of NLS1s

A method introduced by Corral *et al.* (2008) was adopted to stack the X-ray spectra of objects with different redshifts from different observations. We outline the main procedures here and refer readers to the original paper for details and discussions about the method. We first fit 2-10keV spectra with an absorbed power-law model, ignoring 5-7 keV energy range. Then we reconstruct the source spectra by unfolding the observed spectra. The unfolded spectra are rescaled and de-redshifted to the source rest frame. Finally, the stacked spectrum is obtained by rebinning and averaging the rescaled and de-redshifted spectra.

Figure 1 (left-hand panel, blue points with error bars) shows the stacked 2–10 keV spectrum. A prominent broad emission feature in the 5–7 keV energy range superposing the power-law continuum is clearly shown. We test the reliability of the presence of this feature, by means of Monte Carlo simulations.

Using simulations similar to those in Corral *et al.* (2008), we check whether the broad emission feature is an artifact produced by the stacking procedure. For each source, we simulate 100 power-law plus absorption continuum spectra and produce 100 stacked continuum spectra to construct the confidence intervals for the underlying continuum of the composite source spectrum. Figure 1 (left-hand panel; open circles) shows the composite simulated spectrum as well as the 68 per cent (1σ) and 95 per cent (2σ) confidence intervals. It can be seen that the broad line feature of the composite source spectrum in the Fe Kα region fall at or out of the 2σ confidence levels. We also found that the method has almost no affect on a broad line profile, as in our case here. However, narrow lines remain unresolved in the unfolded spectra with a width comparable to the

† http://www.ledas.ac.uk/

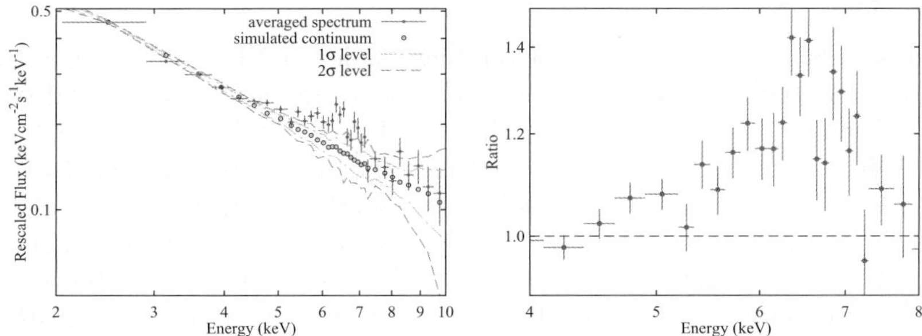

Figure 1. *Left-hand panel:* Stacked source spectrum (blue points with error bars), simulated power-law continuum (open circles), and its confidence ranges. The green dot-dashed line and the red dashed line are the 1σ and 2σ levels for each bin, respectively. *Right-hand panel:* Ratio of the stacked source spectrum to the stacked simulated power-law continuum in the 4–8 keV range. Figures from Liu *et al.* (2015)

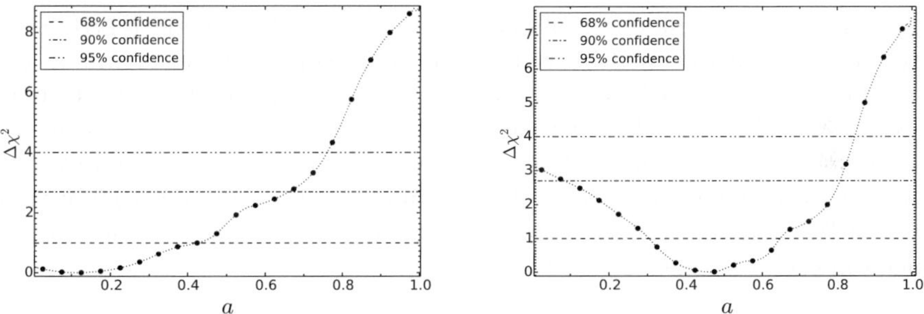

Figure 2. Error contour for the spin parameter, calculated using the RELLINE model. *Left-hand Panel:* Case A (with a 6.97 keV narrow emission line); *Right-hand Panel:* Case B (with a 6.67 keV narrow absorption line). Figures from Liu *et al.* (2015)

intrinsic instrumental energy resolution, which is energy dependent (Corral *et al.* 2008; Iwasawa *et al.* 2012). We thus conclude that the broad emission feature in 5-7 keV in the composite source spectrum should be real.

4. Broad Fe line and black hole spin

The ratio of the stacked source spectrum to the stacked simulated power-law continuum is shown in the right-hand panel in Figure 1. It's clearly shown a broad line profile. In addition, there is an indication of a high ionization emission (6.97 keV)/absorption (6.67 keV) Fe K line. Hereafter, we referred emission line scenario as Case A and absorption line scenario as Case B. We fit the composite spectrum with broad Fe Kα line models, i.e., `power-law+Gaussian(6.4 keV)+Gaussian(6.67/6.97 keV) +relline`. The line energy of the highly ionized Fe K line is fixed at 6.97 keV or 6.67 keV with line width of 85 eV for Case A and Case B, respectively. Only the broad line energy and the spin of BH are free parameters in the `relline` model with emissivity index fixed at 3.0 and the disk inclination fixed at 30 degree. This model leads an acceptable fit ($\chi^2/\mathrm{d.o.f} = 23/21$ for both Case A and B). In both cases, the spin can only be well constrained if we fixed the photon index to that obtained from the stacked simulated power-law spectra: in case A $a < 0.66$ (90 per cent significance level, see Figure 2, left panel) with a best-fit value of 0.10 and in case B $a = 0.47^{+0.34}_{-0.39}$ (90 per cent significance level, Figure 2, right panel). The inferred black hole spin is not very high in either case.

5. Discussion

Our results indicate that the bulk of the black holes of NLS1s in the nearby universe (a median redshift 0.085) either have averagely low or moderate spins, or distribute in a wide range of spins, from low spins to high ones. This is the first time that the ensemble black hole spin parameters are constrained for NLS1s as a population. In Case A, the spin $a < 0.66$ may suggest that the black holes in NLS1s may grow via mainly the process of chaotic accretion, which lead to low values of the final average spins, e.g., $a = 0.1 - 0.3$ as found by King *et al.* (2008). In Case B, the black hole spin parameter is constrained to be widely distributed ($a < 0.81$). If this is ture, it suggest that the growth of each of these black holes is due to chaotic accretion with many episodes, in each episode the accretion is prolonged with a significant increase of black hole mass and significant spin evolution.

6. Summary

In this work, we study the ensemble property of the Fe Kα line in NLS1s by stacking the X-ray spectra. A broad line feature is found in the composite spectrum of NLS1s and can be well fitted with a relativistic Fe line. Our results suggest that a relativistic Fe line may be common in NLS1s. There are tentative indications for low or intermediate spin of BHs in NLS1. This may imply that the BHs in NLS1s may grow mainly via chaotic accretion if the spin is low. Otherwise, if the spin is intermediate and widely distributed, the growth of BHs should be due to chaotic accretion with many episodes, the accretion is prolonged in each episode.

References

Ai Y. L., Yuan W., Zhou H. Y., Wang T. G., & Zhang S. H., 2011, *ApJ*, 727, 31
Boroson T. A., 2002, *ApJ*, 565, 78
Boroson T. A. & Green R. F., 1992, *ApJS*, 80, 109
Brenneman L. W. & Reynolds C. S., 2006, *ApJ*, 652, 1028
Corral A., *et al.* 2008, *A&A*, 492, 71
Dauser T., Wilms J., Reynolds C. S., & Brenneman L. W., 2008, *MNRAS*, 409, 1534
de La Calle Pérez I., *et al.* 2010, *A&A*, 524, A50
Fabian A. C., Rees M. J., Stella L., & White N. E., 1989, *MNRAS*, 238, 729
Fabian A. C., *et al.* 2009, *Nature*, 459, 540
Goodrich R. W., 1989, *ApJ*, 342, 224
Iwasawa K., *et al.* 2012, *A&A*, 537, A86
King A. R., Pringle J. E., & Hofmann J. A., 2008, *MNRAS*, 385, 1621
Komossa S., 2008, in *Rev. Mex. Astron. Astrofis. Conf. Ser.* Vol. 32, p. 86
Marinucci A., *et al.* 2014, *MNRAS*, 440, 2347
Liu Z., Yuan W., Lu Y., & Zhou X. L., 2015, *MNRAS*, 447, 517
Miller J. M., 2007, *ARA&A*, 45, 441
Miniutti G., *et al.*, 2009, *MNRAS*, 398, 255
Nandra K., O'Neill P. M., George I. M., & Reeves J. N., 2007, *MNRAS*, 382, 194
Osterbrock D. E. & Pogge R. W., 1985, *ApJ*, 297, 166
Patrick A. R., *et al.* 2012, *MNRAS*, 426, 2522
Tanaka Y., *et al.* 1995, *Nature*, 375, 659
Turner T. J., *et al.* 2002, *ApJL*, 574, L123
Véron-Cetty M.-P., & Véron P., 2006, *A&A*, 455, 773
Walton D. J., Nardini E., Fabian A. C., Gallo L. C., & Reis, R. C. 2013, *MNRAS*, 428, 2901
Zhou H., Wang T., Yuan W., Lu H., Dong X., Wang J., & Lu Y., 2006, *ApJS*, 166, 128

Star clusters and black holes in galaxies across cosmic time
Proceedings IAU Symposium No. 312, 2014
Y. Meiron, S. Li, F.-K. Liu & R. Spurzem, eds.

© International Astronomical Union 2016
doi:10.1017/S1743921315007449

Tidal disruption as a probe for supermassive black hole binaries

Shuo Li[1], Fukun Liu[2,5], Peter Berczik[1,3,4] and Rainer Spurzem[1,5,3]

[1]National Astronomical Observatories of China, Chinese Academy of Sciences, Beijing, China
email: lishuo@nao.cas.cn
[2]Astronomy Department, Peking University, 100871 Beijing, China
[3]Astronomisches Rechen-Institut, Zentrum für Astronomie, Universität Heidelberg,
Mönchhofstr. 12-14, D-69120 Heidelberg, Germany
[4]Main Astronomical Observatory, National Academy of Sciences of Ukraine, 27 Akademika
Zabolotnoho Street, 03680 Kyiv, Ukraine
[5]Kavli Institute for Astronomy and Astrophysics, Peking University, 100871 Beijing, China

Abstract. Supermassive black hole binaries (SMBHBs) are the products of frequent galaxy mergers. It is very hard to be detected in quiescent galaxy. By using one million particle direct N-body simulations on special many-core hardware (GPU cluster), we study the dynamical co-evolution of SMBHB and its surrounding stars, specially focusing on the evolution of stellar tidal disruption event (TDE) rates before and after the coalescence of the SMBHB. We find a boosted TDE rate during the merger of the galaxies. After the coalescence of two supermassive black holes (SMBHs), the post-merger SMBH can get a kick velocity due to the anisotropic GW radiations. Our results about the recoiling SMBH, which oscillates around galactic center, show that most of TDEs are contributed by unbound stars when the SMBH passing through galactic center. In addition, the TDE light curve in SMBHB system is significantly different from the curve for single SMBH, which can be used to identify the SMBHB.

Keywords. galaxies: evolution, galaxies: kinematics and dynamics, methods: numerical

1. Introduction

It is believed that massive galaxies are assembled by smaller galaxies through multiple mergers (e.g. Springel *et al.* 2005). And observations indicate the existence of a supermassive black hole (SMBH) in the galactic center for almost all of the massive galaxies. Besides, people find that the merger of two galaxies may drive plenty of gas infuse to the galactic center, prompt the star burst, feed the central SMBH and thus invoke the active galactic nuclei (AGNs). As a result, the feedback from accreting SMBH will limit the growth of the itself and affect the evolution of the host galaxy, which induce a close connection between the SMBH and its host galaxy (e.g. Croton *et al.* 2006). For this reason, the evolution of two SMBHs is very important.

In a merging system with two galaxies, their SMBHs will firstly approach each other to form a SMBHB through dynamic friction, and then continually dissipate their orbit energy mainly through gas dynamics or gravitational three body interactions with surrounding stars (e.g. Begelman *et al.* 1980). After they get close enough, the gravitational waves (GWs) will be very efficient to coalesce the SMBHB, accompanied with a strong GW burst and a recoiling velocity to the remnant SMBH (e.g. Campanelli *et al.* 2007). In the past decade, there are some indirectly observational evidences, though all of them are in gaseous environment, to confirm this scenario. However, in quiescent galaxies, the observational evidence for SMBHBs is missing. Whether the SMBHB can keep as a binary system or finally coalescence is still under debate.

Fortunately, there is a good probe can be used. A SMBH can tidally disrupt star accompanied with very strong emission flare lasting for several months or even years, which can temporally light up the quiescent SMBH (e.g. Rees 1988; Komossa & Bade 1999). Studies about the tidal disruption event (TDE) from SMBHB, both for statistic research on event rate and specific research on light curve, can help us to get a better understanding of SMBHB. Besides, the recoiling remnant SMBH after the coalescence can also produce TDE or other special signatures (e.g. Gualandris & Merritt 2008; Komossa & Merritt 2008). For this reason, we have finished series investigations focusing on the TDEs in SMBHB, which includes the TDE rate evolution of merging SMBHB and recoiling SMBH, and the variation of TDE flare light curve in closely bound SMBHB system.

2. Methods

We use direct N-body simulation with simplified tidal disruption scheme to investigate the dynamical co-evolution between SMBHB and its host galaxy. For simplicity, two merging galaxies with SMBHs are identical. We trace the evolution of TDE rate from two separated SMBHs to bound binary and finally the recoiling remnant. The integrations have been done by a GPU accelerated parallel direct N-body code (φ-GRAPE and φ-GPU) (e.g. Harfst *et al.* 2007), on the *laohu* GPU cluster in National Astronomical Observatories of China (NAOC) (more details can be found in Li *et al.* 2012).

To specifically investigate the flare light curve of TDE in closely bound SMBHB, we have done series of scattering experiments. According to analytical and SPH simulation results (e.g. Evans & Kochanek 1989), the fluid elements of disrupted star have approximate test particle orbits, which enables us to simulate the debris through scattering experiments. Finally, we can convert the evolution of mass falling rate of debris into the variation of light curve (more details can be found in Liu *et al.* 2009).

3. Results

TDE in merging galaxies. The evolution of TDE rate during galaxy merger can be divided into three phases. In phase I, r_{bh}, the separation of two SMBHs, is far more larger than their gravitational radii, which means that their TDE rates are similar to single SMBH in normal galaxy. As r_{bh} reduce to r_{inf}, the influence radius of the SMBH, the bound SMBHB is forming, which corresponding to phase II. Finally, the SMBHB will sink into galactic center and the binary evolved to phase III. The left panel of Fig. 1 shows the evolution of TDE rate and r_{bh}.

Our statistical research about r_{apo}, the last apocenter prior to the disruption, for all of disrupted stars indicates that the distribution of r_{apo} has a significant peak, corresponding to where the most of disrupted stars come from. Based on our simulation results, the most of disrupted stars come from the region around r_{inf} in phase I. That is consistent with the prediction of classical loss cone theory. As r_{bh} decreasing, the primary r_{apo} peak is moving inward. Then the TDEs are dominated by stars inside SMBHB orbit, in another word, those bound stars perturbed by one of the SMBH. As shown in the left panel of Fig. 1, that leads to a very efficient loss cone refilling rate and a significantly enlarged TDE rate compare to PI. In phase III, as r_{bh} continually shrinking, the most of stars inside SMBHB orbit have been consumed through three-boy scattering and tidal disruption. Thus, due to the triaxial stellar distribution in merged galactic core, the contribution from those unbound stars outside r_{inf} become dominating. The TDE rate in this stage is between PI and PII. Our extrapolation results show that, for a galaxy similar to M32, the TDE rate in phase I is $\sim 3 \times 10^{-5} \mathrm{M}_\odot \, \mathrm{yr}^{-1}$, and there is an order of

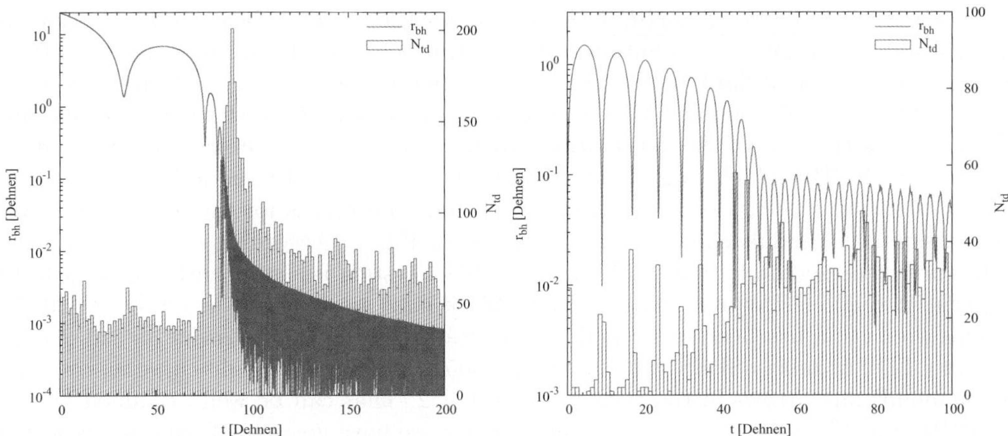

Figure 1. *Left panel*: evolution of TDE rate for merging galaxies. Red solid line represents $r_{\rm bh}$ and blue box denotes $N_{\rm td}$, number count of TDEs in every time bin. *Right panel*: similar to the left panel, here is the evolution of TDE rate for a recoiling SMBH. (See the electronic edition with color.)

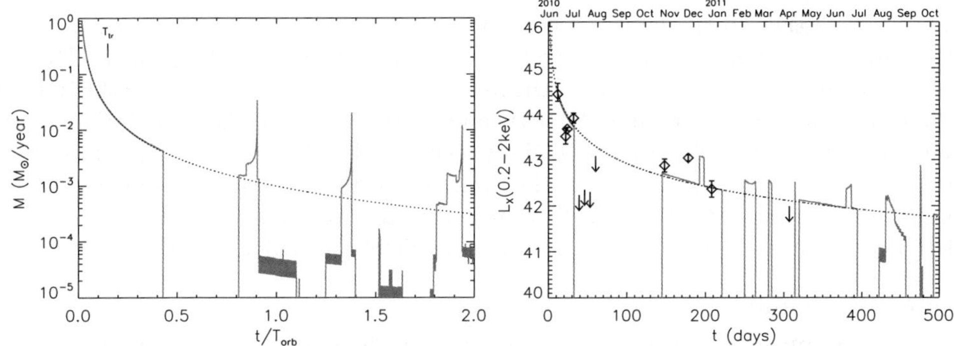

Figure 2. *Left panel*: Accretion rates of debris *vs.* time for a TDE in SMBHB, by Liu *et al.* 2009. Red Solid line is simulation result and doted line denotes the theoretical rate $\propto t^{-5/3}$ for single BH. $T_{\rm tr}$ marks the interruption time. *Right panel*: A rebuild X-ray light curve with red solid line for the SMBHB TDE in SDSS J120136.02+300305.5 (by Liu *et al.* 2014). Diamonds mark the observational data from XMM-Newton and Swift, and arrows represent the non-detection upper limit (see more details in Saxton *et al.* 2012). (See the electronic edition with color.)

magnitude boosted rate $\sim 5 \times 10^{-4} {\rm M}_\odot\ {\rm yr}^{-1}$ in phase II. The detailed treatment will be presented by Li *et al.* 2015, in preparation.

TDE for recoiling remnant SMBH. After SMBHB coalescence, the remnant SMBH may get a recoil velocity. For most of the cases, it is not strong enough to make the SMBH escape away from its host galaxy. Thus the recoiling SMBH will oscillate around galactic center and damp its energy through dynamic friction. Similar to Gualandris & Merritt 2008, we have divided the trajectory evolution of the recoiling SMBH into three phases, the phase I with fast damping oscillation outside core, the phase II for SMBH oscillating inside the core region with very slow damping, and then phase III the equilibrium status. Our simulation with tidal disruption has shown strongly boosted TDE rate every time when the recoiling SMBH passing through the galactic center (as shown in the right panel of Fig. 1). That indicates the captured stars dominate the TDE when SMBH return back to the center region (see more detail in Li *et al.* 2012).

Tidal flare light curve of SMBHB. According to the results above, it is possible to statistically investigate the evolution of SMBHB in gas poor environment. In addition, by analyzing the tidal flare light curve, we can even identify a specific SMBHB system. For a TDE in closely bound SMBHB, due to gravitational perturbation from another black hole (BH), some of the debris from disrupted star can not return and accreted by the primary SMBH which caused the disruption. As a result, the periodic perturbation from secondary BH will excite the truncations and recurrences for the flare light curve, as our simulation result shown in the left panel of Fig. 2. According to this figure, the light curve represented by red solid line in SMBHB system is significantly different from the case in single BH represented by dotted black line (more details in Liu *et al.* 2009).

This prediction has been confirmed by observation recently. Saxton *et al.* 2012 have caught a TDE in galaxy SDSS J120136.02+300305.5. It has special light curve represented by diamonds and arrows in the right panel of Fig. 2. That can be well explained by our SMBHB model in 2009. After a series simulations, we have got sets of parameters which can perfectly fit to the observation, as shown in the right panel of Fig. 2 for example (more details in Liu *et al.* 2014).

4. Conclusion

As we have shown in this letter, TDE can be a very powerful probe for investigating SMBHB in quiescent galaxy. As the two BHs evolving in merging galaxy, their TDE rates are also evolving. There is a boosted TDE rate during the formation of the SMBHB. And the TDE rate for recoiling remnant SMBH can be also enhanced when the SMBH passing through galactic center. With more TDEs which may be observed by LSST in the near future, we can statistically investigate the evolution of SMBHB through our results here. Besides, by using our flare light curve evolution model, we can specifically find out some SMBHB candidates, as we have already done for SDSS J120136.02+300305.5.

Acknowledgements

This work has been supported by NAOC through the Silk Road Project, the National Natural Science Foundation of China (NSFC11073002 and NSFC11303039) and the grant $ZDYZ2008 - 2$ from Ministry of Finance of People's Republic of China. It also partly supported by the National Science Foundation under Grant No. NSF PHY11-25915.

References

Begelman, M. C., Blandford, R. D., & Rees, M. J. 1980, *Nature*, 287, 307
Berczik, P., Merritt, D., Spurzem, R., & Bischof, H.-P. 2006, *ApJ* (Letters), 642, L21
Campanelli, M., Lousto, C., Zlochower, Y., & Merritt, D. 2007, *ApJ* (Letters), 659, L5
Chen, X., Liu, F. K., & Magorrian, J. 2008, *ApJ*, 676, 54
Croton, D. J., Springel, V., White, S. D. M., *et al.* 2006, *MNRAS*, 365, 11
Evans, C. R. & Kochanek, C. S. 1989, *ApJ*, 346, L13
Gould, A. & Rix, H.-W. 2000, *ApJ* (Letters), 532, L29
Gualandris, A. & Merritt, D. 2008, *ApJ*, 678, 780
Harfst, S., Gualandris, A., Merritt, D., *et al.* 2007, *New Astron.*, 12, 357
Komossa, S. & Bade, N. 1999, *A&A*, 343, 775
Komossa, S. & Merritt, D. 2008, *ApJ*, 683, L21
Li, S., Liu, F. K., Berczik, P., Chen, X., & Spurzem, R. 2012, *ApJ*, 748, 65
Liu, F. K., Li, S., & Chen, X. 2009, *ApJ* (Letters), 706, L133
Liu, F. K., Li, S., & Komossa, S. 2014, *ApJ*, 786, 103
Preto, M., Berentzen, I., Berczik, P., & Spurzem, R. 2011, *ApJ* (Letters), 732, L26

Rees, M. J. 1988, *Nature*, 333, 523
Saxton, R. D., Read, A. M., Esquej, P., Komossa, S., Dougherty, S., *et al.* 2012, *A&A*, 541, A106
Springel, V., White, S. D. M., Jenkins, A., *et al.* 2005, *Nature*, 435, 629

Star clusters and black holes in galaxies across cosmic time
Proceedings IAU Symposium No. 312, 2014
Y. Meiron, S. Li, F.-K. Liu & R. Spurzem, eds.

© International Astronomical Union 2016
doi:10.1017/S1743921315007450

Formation of discs around super-massive black hole binaries

Felipe G. Goicovic[1], Jorge Cuadra[1] and Alberto Sesana[2]

[1]Instituto de Astrofísica, Pontificia Universidad Católica de Chile, Santiago, Chile
email: fgarrido@astro.puc.cl

[2]Max-Planck-Institut für Gravitationsphysik, Potsdam-Golm, Germany

Abstract. We model numerically the evolution of $10^4 M_\odot$ turbulent molecular clouds in near-radial infall onto $10^6 M_\odot$, equal-mass supermassive black hole binaries, using a modified version of the SPH code GADGET-3. We investigate the different gas structures formed depending on the relative inclination between the binary and the cloud orbits. Our first results indicate that an aligned orbit produces mini-discs around each black hole, almost aligned with the binary; a perpendicular orbit produces misaligned mini-discs; and a counter-aligned orbit produces a circumbinary, counter-rotating ring.

Keywords. black hole physics, accretion discs, hydrodynamics

1. Introduction

Supermassive black holes (SMBHs) are ubiquitous in galactic nuclei (Richstone *et al.* 1998), and binaries of these massive objects are a likely product of the hierarchical evolution of structures in the universe. After a galaxy merger, where both progenitors host a SMBH, different mechanisms are responsible for the evolution of the binary orbit depending on its separation (see review by Colpi 2014). Dynamical interaction with stars appears to be efficient to bring the SMBHs down to parsec scales only, what is known as the "last parsec problem" (Begelman *et al.* 1980, Yu 2002). A possible way to overcome this barrier and merge the SMBHs within a Hubble time is interaction with gas. Many theoretical and numerical studies have focused on the orbital evolution of a sub-parsec binary surrounded by a circumbinary disc (e.g. Armitage & Natarajan 2005, Cuadra *et al.* 2009, Amaro-Seoane *et al.* 2013, Nixon *et al.* 2013, Roedig & Sesana 2014). However, the exact mechanism that would produce such discs is still unclear; it is necessary an efficient transport of gas from thousands or hundreds of parsecs to the central parsec.

Turbulence and gravitational instabilities in the interstellar medium, through the formation of clumps, allow portions of gas to travel almost unaffected by its surrounding, enhancing the probability of reaching the galactic nuclei (Hobbs *et al.* 2011). A possible manifestation of these events is the putative molecular cloud that resulted in the unusual distribution of young stars orbiting our Galaxy's SMBH. In particular, the simulation of Bonnell & Rice (2008) shows a spherical, turbulent cloud falling with a very low impact parameter ($\sim$0.1 pc) onto a one million solar masses SMBH.

Assuming that these accretion events are common in galactic nuclei, the goal of our work is to model such an event onto a binary instead of a single SMBH. In particular, we are interested on the properties of the discs that will form given different relative orientations between the orbital angular momenta of the cloud and the binary. Notice that this study is complementary to that shown in Dunhill *et al.* (2014), as we are modeling clouds with very low orbital angular momentum.

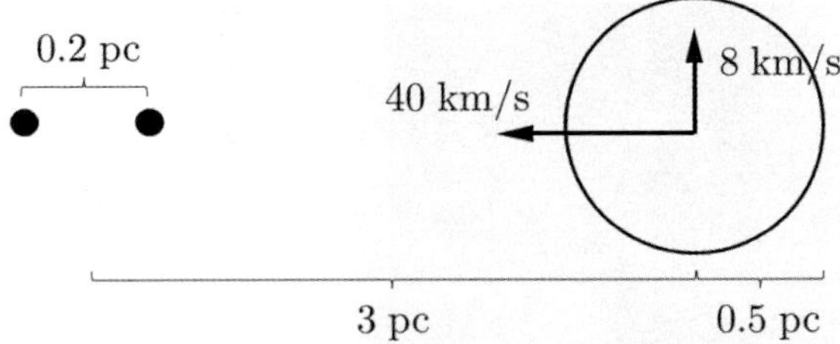

Figure 1. Initial setup of the simulations. The circle on the right represents the cloud, while the black, solid circles are the SMBHs.

2. The numerical model

We model the interaction between the binaries and clouds using a modified version of the SPH code GADGET-3 (Springel 2005). The cloud is represented using over 4×10^6 gas particles with a total mass of $10^4 M_\odot$, an initial turbulent velocity field and uniform density. The SMBHs are modeled as two equally-massive sink particles that interact only through gravity and can accrete SPH particles. The total mass of the binary is $10^6 M_\odot$, and its initial orbit is Keplerian and circular.

The physical setup of the simulation is shown in Figure 1. The initial velocity of the cloud yields a highly eccentric ($e \approx 0.98$), bound orbit where the pericenter distance is $r_{\mathrm{peri}} \approx 0.07$ pc, which is less than the binary radius, making the interaction between the gas and SMBHs very strong. As we expect clouds approaching the binary from different directions, we model systems with three different inclinations between the cloud and binary orbits: aligned, perpendicular and counter-aligned.

3. Results: formation and properties of the gaseous discs

In this section we present the main results of the simulations with the different inclinations, in particular the discs that form around the binary and each SMBHs.

Aligned orbits. On the left panel of Figure 2 we show the column density map of the simulation at different times, where we can see how the interaction develops. As the gas falls almost radially onto the binary, around 80% of the cloud is accreted by the SMBHs. Most of the remaining gas is pushed away due to an efficient slingshot. The bound material forms a tail that get stretched and diluted over time, feeding "mini-discs" that form around each SMBH.

To measure the alignment between the binary orbit and the mini-discs, we compute its angular momentum on the corresponding black hole reference frame. We show the time evolution of the direction of both discs on the Hammer projection of Figure 2. Here we observe that they tend to align with the orbit of the binary, as expected, although one disc is slightly tilted with respect to the aligned position and also precesses around that position. This behavior could have distinctive electromagnetic signatures. For example, the misalignment could affect the variability of spectral lines, or each disc have different polarization. The precession could be observed if jets are launched from the SMBHs and align with the mini-discs.

Perpendicular orbits. With this inclination, as in the previous case, around 80% of the cloud mass is added to the SMBHs. However, the interaction between the gas and the binary, that we can see in Figure 3, is completely different respect to the aligned case. Due to a less efficient slingshot, most of the remaining material stays bound to the system and it retains its original angular momentum, forming an unstable structure

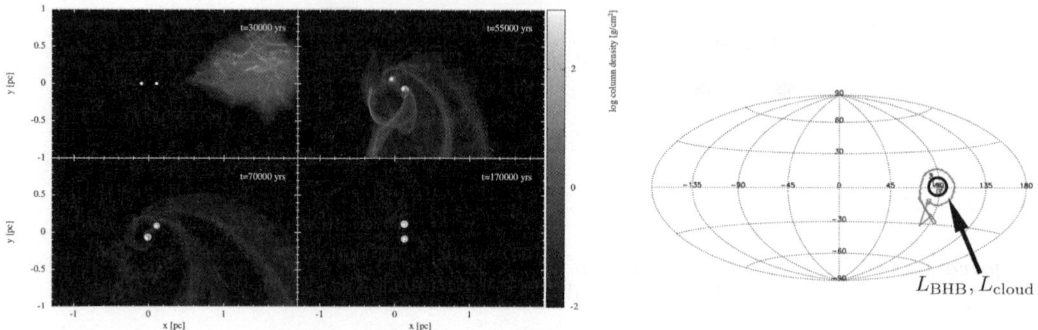

Figure 2. Left: Column density map of the aligned simulation at different times. For reference, the cloud and the binary move on the x-y plane. (For animations check `http://www.astro.puc.cl/ fgarrido/animations`) Right: Time evolution of the angular momentum direction of the mini-discs formed around each black hole (red and blue lines), in a Hammer projection. The triangle shows the formation of the disc and the square the end of the simulation.

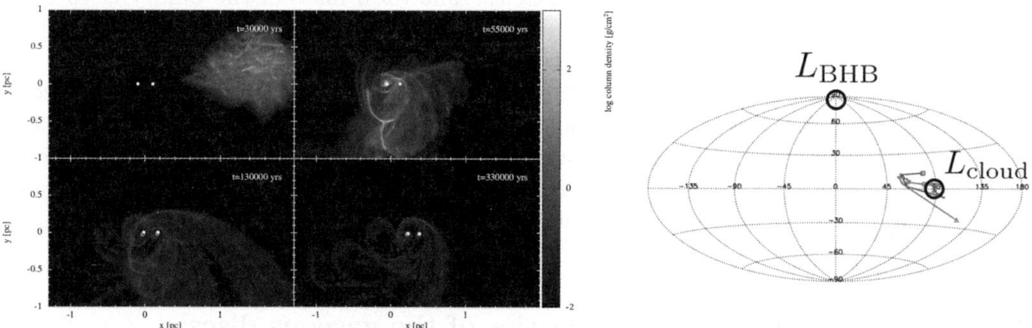

Figure 3. As Fig. 2, but for the model with perpendicular orbits. In this case the cloud moves on the x-y plane while the binary is on the x-z plane.

around the binary. The gas that reaches the SMBHs also produce mini-discs, but they are less massive and more intermittent than in the aligned case.

The direction of the mini-discs, shown on the right panel of Figure 3, shows that they tend to follow the original direction of the cloud, which makes them completely misaligned respect to the binary orbit. As well as the previous case, this could have distinctive signatures on the variability of lines or the direction of possible jets.

Counter-aligned orbits. In this case we have that the interaction of the binary with the gas produces shocks that cancel angular momentum, allowing the SMBHs to accrete even more material than in the previous cases; around 90% of the cloud is swallowed. The remaining material forms a tail that, due to the gravitational torques, produces a circumbinary ring. In this case we do not observe mini-discs during the entire length of the simulation.

Finally, we compute the eccentricity distribution of the gas on three different stages, as shown on the right panel of Figure 4. It is interesting that, if there is star formation in the ring, the stars would have very different orbits around the binary depending on the formation time-scale. For a very rapid star formation we could have highly eccentric orbits (solid line), while for a slow process the stars could be distributed on a narrow ring with nearly circular orbits (dashed line).

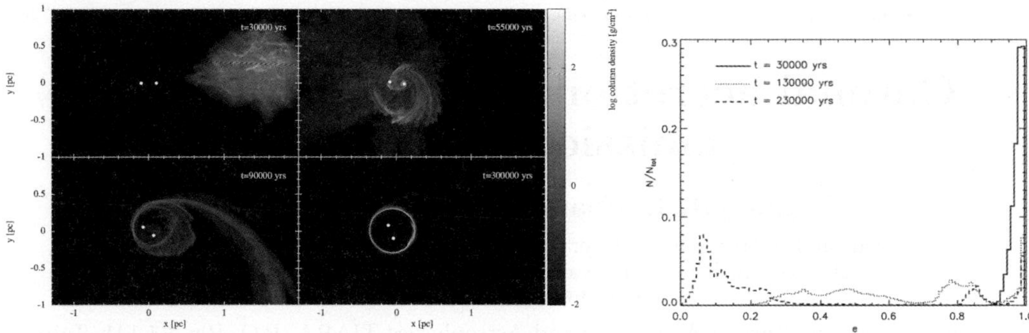

Figure 4. Left: As Fig. 2, but for the model with the counter-aligned orbits. In this case, the cloud and the binary move on the x-y plane, but the latter is rotating clockwise. Right: Eccentricity distribution of the gas for three different times in the simulation.

4. Conclusions and outlook

The preliminary results presented here show that accretion events onto binaries result in very different disc morphologies, depending on the relative inclination between the cloud and the binary orbits. Our work in progress is extending this study to larger impact parameters, showing lower accretion rates, and exploring the effect of the interaction on the black holes orbit and spin evolution. These results are likely to have important implications on the multi-messenger future studies of SMBH binaries and on the long-term evolution of these systems.

Acknowledgements

Column density maps were created with SPLASH by Price (2007). We acknowledge support from CONICYT-Chile through PCHA/Doctorado Nacional, FONDECYT (1141175), Basal (PFB0609), Anillo (ACT1101), Redes (120021) and Exchange (PCCI130064) grants.

References

Amaro-Seoane, P., Brem, P., & Cuadra, J. 2013, *ApJ*, 764, 14
Armitage, P. J. & Natarajan, P. 2005, *ApJ*, 634, 921
Begelman, M. C., Blandford, R. D., & Rees, M. J. 1980, *Nature*, 287, 307
Bonnell, I. A. & Rice, W. K. M. 2008, *Science*, 321, 1060
Colpi, M. 2014, *SSR*, 183, 189
Cuadra, J., Armitage, P. J., Alexander, R. D., & Begelman, M. C. 2009, *MNRAS*, 393, 1423
Dunhill, A. C., Alexander, R. D., Nixon, C. J., & King, A. R. 2014, *MNRAS*, 445, 2285
Hobbs, A., Nayakshin, S., Power, C., & King, A. 2011, *MNRAS*, 413, 2633
Nixon, C., King, A., & Price, D. 2013, *MNRAS*, 434, 1946
Price D. J., 2007, *PASA*, 24, 159
Richstone, D., *et al.* 1998, *Nature*, 395
Roedig, C. & Sesana, A. 2014, *MNRAS*, 439, 3476
Springel, V. 2005, *MNRAS*, 364, 1105
Yu, Q. 2002, *MNRAS*, 331, 935

Star clusters and black holes in galaxies across cosmic time
Proceedings IAU Symposium No. 312, 2014
Y. Meiron, S. Li, F.-K. Liu & R. Spurzem, eds.

© International Astronomical Union 2016
doi:10.1017/S1743921315007462

Coronal accretion: the power of X-ray emission in AGN

B.-F. Liu[1], R. E. Taam[2], E. Qiao[1] and W. Yuan[1]

[1]National Astronomical Observatories, Chinese Academy of Sciences,
20A Datun Road, Chaoyang District, Beijing 100012, China
email: bfliu@nao.cas.cn

[2]Academia Sinica Institute of Astronomy and Astrophysics-TIARA, P.O. Box 23-141, Taipei,
10617 Taiwan

Abstract. The optical/UV and X-ray emissions in luminous AGN are commonly believed to be produced in an accretion disk and an embedded hot corona respectively. We explore the possibility that a geometrically thick coronal gas flow, which is supplied by gravitational capture of interstellar medium or stellar wind, condenses partially to a geometrically thin cold disk and accretes via a thin disk and a corona onto the supermassive black hole. We found that for mass supply rates less than about 0.01 (expressed in Eddington units), condensation does not occur and the accretion flow takes the form of a corona/ADAF. For higher mass supply rates, corona gas condenses to the disk. As a consequence, the coronal mass flow rate decreases and the cool mass flow rate increases towards the black hole. Here the thin disk is characterized by the condensation rate of hot gas as it flows towards the black hole. With increase of mass supply rate, condensation becomes more efficient, while the mass flow rate of the coronal gas attains values of order 0.02 in the innermost regions of the disk, which can help to elucidate the production of strong X-ray with respect to the optical and ultraviolet radiation in high luminosity AGN.

Keywords. accretion, accretion disks — black hole physics — X–rays: galaxies—galaxies: active

1. Introduction

Accretion of gas onto a supermassive black hole is central to our understanding of active galactic nuclei (AGN). Observations of optical/UV and X-ray emissions in high luminosity AGNs point to the co-existence of hot and cool accretion flows, which have been described in terms of a hot corona lying above and below a cold standard geometrically thin disk. However, theoretical model involving the coexistence of a disk and corona as applied to luminous, radio quiet AGNs is inadequate. In particular, the inverse Compton scattering of photons from the geometrically thin disk by hot electrons in the corona quickly cools the coronal gas, if the mass supply to the accretion is dominated by a disk-like flow. In this case, the corona is too weak to emit sufficiently high X-ray radiation (Liu *et al.* 2012; Meyer *et al.* 2012). This is in contrast to the observational spectra showing that the coronal fraction in luminous AGNs can often be as high as a few tens of percent (e.g., Mushotzky *et al.* 1993; Elvis *et al.* 1994; Yuan *et al.* 1998). To alleviate this problem, we suggest that the physical properties of the gas fuel can play an important role in producing strong X-ray emission in AGNs. Unlike in black hole X-ray binaries where gas transferred from a companion star is cool and is mostly constrained to lie in the orbital plane, the accreting gas in AGN reflects the gravitational capture of gas from the interstellar medium or from the stellar winds emitted by evolved stars in the central regions of a galaxy (Ho 2008). In such an environment, large differences in the radiation spectrum are possible, if a thin disk exists initially.

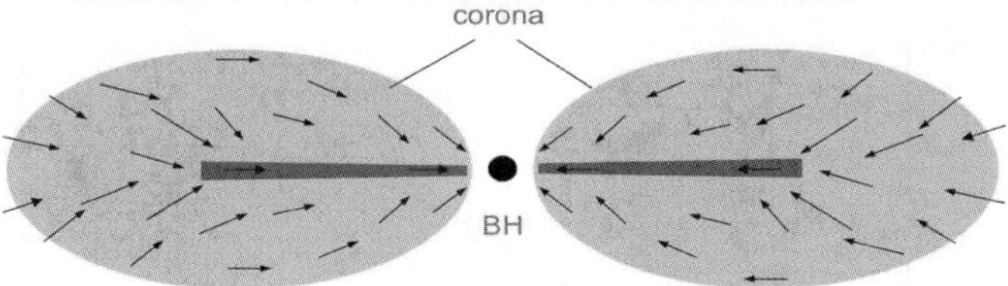

Figure 1. A schematic description of the mass flowing in the disk and corona with hot gas supply.

2. The physical model

We consider an accretion state in which a thin, cold disk initially surrounds a black hole. The gas supply to the accretion is from gravitational capture of the interstellar medium or the stellar winds. Such a Bondi accretion flow is hot since the accretion energy is primarily converted into internal energy due to compressional heating. As the gas flows toward the black hole, the hot gas partially condenses to the underlying disk as a consequence of disk corona interaction. Thus, the thin disk remains steady because the condensation of the hot gas continuously supplies gas to the accretion of the thin disk. On the other hand, a certain amount of hot gas remains in the corona, responding for the X-ray emission. A schematic description of the mass flowing in the disk and corona with hot gas supply are shown in Figure 1.

We study the dynamic and radiative interaction between the pre-exist disk and the hot accretion flow with mass supply from interstellar medium/stellar winds in detail. For a black hole, $M = 10^8 M_\odot$, viscosity parameter, $\alpha = 0.3$, and magnetic field parameter (the ratio of gas pressure to total pressure), $\beta = 0.8$, our calculations show that for a given mass supply rate to the corona there exists a critical distance, R_d, within which the coronal gas condenses to a cool disk. For distances greater than R_d, disk gas will evaporate to the corona. As there is only gas supply to the corona, any existing gas in a thin disk in the outer region ($R > R_d$) will be evacuated by the evaporation process and hence we take R_d as the outer boundary of the disk. The tendency for condensation resulting from efficient Compton cooling is mitigated by the thermal energy left in the corona by the condensed gas and the continuous hot gas supply preventing collapse of the corona. Thus, the accretion takes place as an ADAF with a nearly constant accretion rate in the outer region ($R > R_d$), while in the inner region, a fraction of hot gas condenses to the disk as it approaches the black hole. This leads to a decreasing mass flow rate in the corona and a corresponding increase of the flow rate in the cool disk toward the black hole.

In Figure 2. we show the spatial distribution of the mass flow rate in the accretion flows for mass supply rates of 0.03, 0.05, 0.08, and 0.1. It is found that the accretion rate in the inner corona can be roughly described by a linear function, $\dot{m} \approx 0.02 + 3 \times 10^{-4} \frac{R}{R_S}$, while the mass flow rate in the innermost region of the corona is nearly independent of the mass supply rate, which converges to $\dot{m} \approx 0.02$. (in Eddington units) . Such a characteristic can be understood as follows. For a higher mass supply rate, the density in the hot flow is higher, leading to the condensation of more gas to the disk by Compton cooling. Thus, the disk emission becomes stronger, which promotes more efficient Compton cooling. As a consequence, more hot gas condenses into the disk. This process stabilizes as the density of the hot flow decreases as a result of condensation and the electron temperature decreases (by efficient Compton cooling) to restrain the further increase of the Compton cooling rate. Hence, an equilibrium is established. We find that the density and mass flow rate remaining in the corona is nearly constant for the different supply rates.

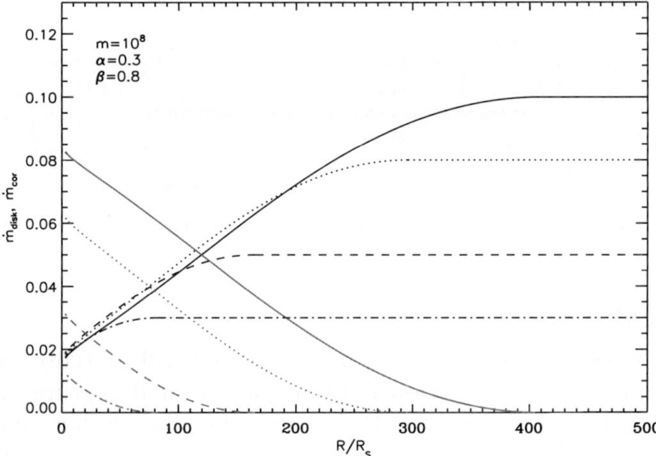

Figure 2. The spatial distribution of the mass flow rate in the disk and corona for a black hole of $10^8 M_\odot$. Curves in red (black) refer to the mass flow rate in the disk (corona). From bottom to top, the dash-dotted lines refer to a hot mass supply rate of $\dot{m} = 0.03$, dashed lines for $\dot{m} = 0.05$, dotted lines for $\dot{m} = 0.08$ and solid lines for $\dot{m} = 0.1$. The mass flow rate in the corona converges to $\dot{m} \sim 0.02$ in the innermost region, indicating that a nearly constant corona coexists with a variable disk.

We point out that the above picture applies for sufficiently high accretion rates, $\dot{m} > 0.02$ in the hot flow. For lower hot gas supply rates, R_d is interior to the innermost stable circular orbit, implying that the coronal gas cannot condense to a disk. Instead, the interaction between the disk and corona results in the evaporation of disk gas, which evacuates the disk. Thus, at low rates of accretion, only hot accretion describes the flow, that is, the advection-dominated accretion flow.

3. Emissions from the disk and the corona

With the structure of the corona and disk determined, we calculate the radiation emitted from the disk and corona for $\dot{m} = 0.03, 0.05, 0.08, 0.1$. It is found that the relative importance of the corona emission depends on the mass supply rate to the corona, $\dot{m}$. With an increase of $\dot{m}$, condensation takes places at a larger distance. The ratio of corona and disk luminosity decreases with increasing mass supply rate.

The spectra are shown in Figure 3, which is composed of soft photons from the disk, Compton scattering of these soft photons, and the Compton scattering of photons produced by the synchrotron and bremsstrahlung processes in the corona. It can be seen that the strength of the emission from the hot corona relative to the disk decreases with an increase of the hot mass supply rate or bolometric luminosity. This is a consequence of the increasing condensation rate to the inner disk. The predicted value of α_{ox}, which is calculated from the luminosities at $2500\mathring{A}$ and 2keV, varies from 0.9 to 1.3 with the mass supply rate increasing from 0.03 to 0.1 respectively. The photon index of the hard X-ray spectrum increases, varying from 1.8 to 2.3 in the 2-10 keV energy band.

4. Conclusion

We propose that an accretion flow is supplied to a supermassive black hole by the gravitational capture of interstellar medium or stellar wind material. It is found that the accretion flow is described by a hot accretion flow in outer region while is characterized

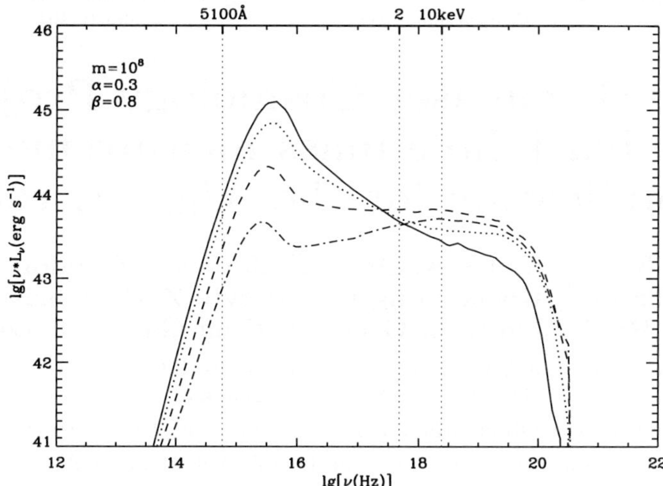

Figure 3. The spectra emitted from a disk and corona with a hot mass supply. Curves from bottom to top refer to mass supply rates of 0.03, 0.05, 0.08, and 0.1. The thin vertical lines denote the luminosity at 5100Åand 2-10keV, showing the relative strength of the optical to hard X-ray luminosity. With an increase of the mass supply rate, the hard X-ray spectrum becomes soft (steep) and the contribution of emission from the corona decreases.

by both hot, ionized and cool gas components in the central regions of galaxies. In this picture, the hot gas partially condenses to an underlying cool disk as it flows toward the black hole, releasing accretion energy as X-ray emission and meanwhile enhancing the disk emission. Such a model can help to clarify the relative importance of the X-ray to optical/UV radiation in high luminosity AGNs, avoiding assumption that a large fraction of disk accretion energy is transferred to the corona. On the other hand, the model provides a natural interpretation of ADAF for low luminosity AGN where condensation is absent at low accretion rates.

Acknowledgements

Financial support for this work is provided by the National Natural Science Foundation of China (grants 11173029, and U1231203) and the Strategic Priority Research Program "The Emergence of Cosmological Structures" of the Chinese Academy of Sciences (Grant No. XDB09000000).

References

Elvis M. *et al.* 1994, *ApJS*, 95, 1
Ho, L. C. 2008, *ARA&A*, 46, 475
Liu, J. Y., Liu, B. F., Qiao, E. L., & Mineshige, S. 2012, *ApJ*, 754:81
Meyer-Hofmeister, E., Liu, B. F., & Meyer, F. 2012, *A&A*, 544, A87
Mushotzky, R., Done, C., & Pounds K. A. 1993, *ARA&A*, 31, 717
Yuan, W., Brinkmann, W., Siebert, J. & Voges, W. 1998, *A&A*,330, 108

Star clusters and black holes in galaxies across cosmic time
Proceedings IAU Symposium No. 312, 2014
Y. Meiron, S. Li, F.-K. Liu & R. Spurzem, eds.
© International Astronomical Union 2016
doi:10.1017/S1743921315007474

The Megamaser Cosmology Project: precise black hole mass measurement and the implication for the M_{BH}–$\sigma_\star$ relation

Cheng-Yu Kuo[1] , James A. Braatz[2], James J. Condon[2], Caterina M. V. Impellizzeri[2], Kwok-Yung Lo[2], Ingyin Zaw[3], Christian Henkel[4], Mark J. Reid[5], Jenny E. Greene[6], Feng Gao[7] and Wei Zhao[7]

[1] Academia Sinica Institute of Astronomy and Astrophysics, P.O. Box 23-141, Taipei 10617, Taiwan
email: cykuo@asiaa.sinica.edu.tw

[2] National Radio Astronomy Observatory, 520 Edgemont Road, Charlottesville, VA 22903, USA

[3] New York University Abu Dhabi, Abu Dhabi, UAE

[4] Max-Planck-Institut für Radioastronomie, Auf dem Hügel 69, 53121 Bonn, Germany

[5] Harvard-Smithsonian Center for Astrophysics, 60 Garden Street, Cambridge, MA 02138, USA

[6] Department of Astrophysical Sciences, Princeton University, Princeton, NJ 08544, USA

[7] Shanghai Astronomical Observatory, 80 Nandan Road, Shanghai 200030, China

Abstract. We made dynamical black hole mass measurements from nineteen Seyfert 2 galaxies which host sub-parsec H_2O maser disks using the H_2O megamaser technique. The nearly perfect Keplerian rotation curves in many of these maser systems guarantee the high accuracy and precision of the black hole mass measurements. With the stellar velocity dispersion ($\sigma_\star$) of the galaxy bulges measured with the Dupont 2.5 m telescope at Las Campanas Observatory in the South and the Apache Point Observatory (APO) 3.5m telescope in the North, we found that H_2O maser galaxies, most of which host pseudo bulges rather than classical bulges, do not all follow the $M_{BH}-\sigma_\star$ relation shown in the literature. This result is well consistent with the latest findings by Kormendy & Ho (2013) that only early type galaxies and galaxies with classical bulges follow a tight $M_{BH}-\sigma_\star$ relation. Such a tight correlation may not exist in pseudo bulge galaxies.

Keywords. accretion disks, water maser, Seyfert galaxies, black hole mass

---

1. Introduction

The primary goal of the Megamaser Cosmology Project (MCP; Braatz *et al.* 2009; Reid *et al.* 2009a; Braatz *et al.* 2010) is to determine the Hubble constant H_0 to $\sim 3\%$ accuracy in order to constrain the equation of state parameter w of dark energy. In our endeavor to determine an accurate Hubble constant, we measure direct angular-diameter distances to galaxies in the Hubble flow with the H_2O megamaser technique. This technique involves sub-milliarcsecond resolution imaging and acceleration measurements of H_2O megamasers from nearly edge-on, sub-parsec gas disks at the centers of active galaxies. Since the maser disk is usually significantly smaller than the "gravitational sphere of influence" (Barth 2003) of the black hole (BH) at the center, the kinematics of the water masers provide a direct probe of the gravitational potential of the BH, and the BH mass (M_{BH}) can be measured with high precision from the rotation curve of the maser disk. Measuring BH masses is therefore a second important product of this project in addition to the distance determination. While we still need acceleration measurements for H_2O masers to determine distances, the VLBI imaging alone is sufficient to measure accurate central BH masses, assuming distances to the maser galaxies. These accurate BH masses provide an important basis for testing the M_{BH}-$\sigma_\star$ relation, a tight correlation between BH mass and the velocity dispersion of stars ($\sigma_\star$) in the galaxy bulge (e.g. Ferrarese & Merritt 2000; Gebhardt *et al.* 2000; Gütekin *et al.* 2009, Kormendy & Ho 2013) and thought to be a manifestation of a causal connection between the formation and evolution of the black hole and the bulge. Since H_2O megamasers are usually found in late type galaxies and BH masses measured from H_2O usually range from $\sim 10^6 - 10^{7.5}$ $M_\odot$, these BH masses play an important role to test whether the M_{BH}-$\sigma_\star$ relation defined by early type galaxies also holds for late type galaxies.

2. The sample and observations

We searched for circumnuclear H_2O maser disks from a sample of Seyfert 2 galaxies drawn from the SDSS survey (Data Release 7), the 6dF survey, and the 2MRS survey. Since the start of the MCP in 2006, we have observed more than 3000 Seyfert galaxies with recession velocities below 15000 km s^{-1}. Water masers have been detected in 162 galaxies (see https://safe.nrao.edu/wiki/bin/view/Main/MegamaserCosmologyProject). Most of the megamasers originate in active galactic nuclei (Lo 2005) and about 30 galaxies show spectra suggestive of emission from sub-parsec scale, edge-on, circumnuclear disks. In the MCP, we conducted VLBI observations of nineteen megamaser disks with the Very Long Baseline Array augmented by the 100-m Green Bank Telescope and the Effelsberg 100-m telescope. We have obtained their images with sub-milliarcsecond resolution and use their rotation curves to measure BH masses.

The VLBI plays a crucial role in measuring BH masses with high precision especially in galaxies with lighter supermassive black holes (i.e. $M_\bullet \sim 10^6 - 10^7 M_\odot$). The critical advantage provided by VLBI is angular resolution two orders of magnitude higher than the best optical resolution. For any given galaxy with a nearly constant central mass density of stars, M_{BH} is proportional to R_{inf}^3, where R_{inf} is the radius of the gravitational sphere of influence of BH (Barth *et al.* 2003). So, a factor of 100 increase in resolution permits measurements of masses up to 10^6 times smaller. Similarly, the central density limits that can be set are up to 10^6 times higher, high enough to rule out extremely dense star clusters as the massive objects at the center of many megamaser disks presented here based on dynamical argument (see Kuo *et al.* 2011).

3. BH masses and their implication for the M–σ relation

Among the 19 maser galaxies for which we took VLBI data and have finished data analysis for BH mass measurements, we have published the VLBI maps (see Figure 1) of the water maser spots in seven systems (i.e. UGC 3789, NGC 1194, NGC 2273, NGC 2960, NGC 4388, NGC 6264, and NGC 6323; see also Kuo *et al.* 2011). Publications of VLBI maps for the other 12 maser galaxies are in preparation (Gao *et al.* in prep.; Zhao *et al.* in prep.). The typical size of the observed maser disks is typically less than 1 pc as expected and all well-defined maser disks show warped structure. To obtain the BH mass, we first fit a Keplerian rotation curve to each observed maser disk:

$$|v_{\mathrm{K}}| = v_1 \left(\frac{\theta}{1\ \mathrm{mas}} \right)^{-1/2}, \tag{3.1}$$

where $|v_{\mathrm{K}}|$ is the orbital velocity (after relativistic corrections) of the high velocity masers and v_1 is the orbital velocity at a radius 1 mas from the dynamical center. With the

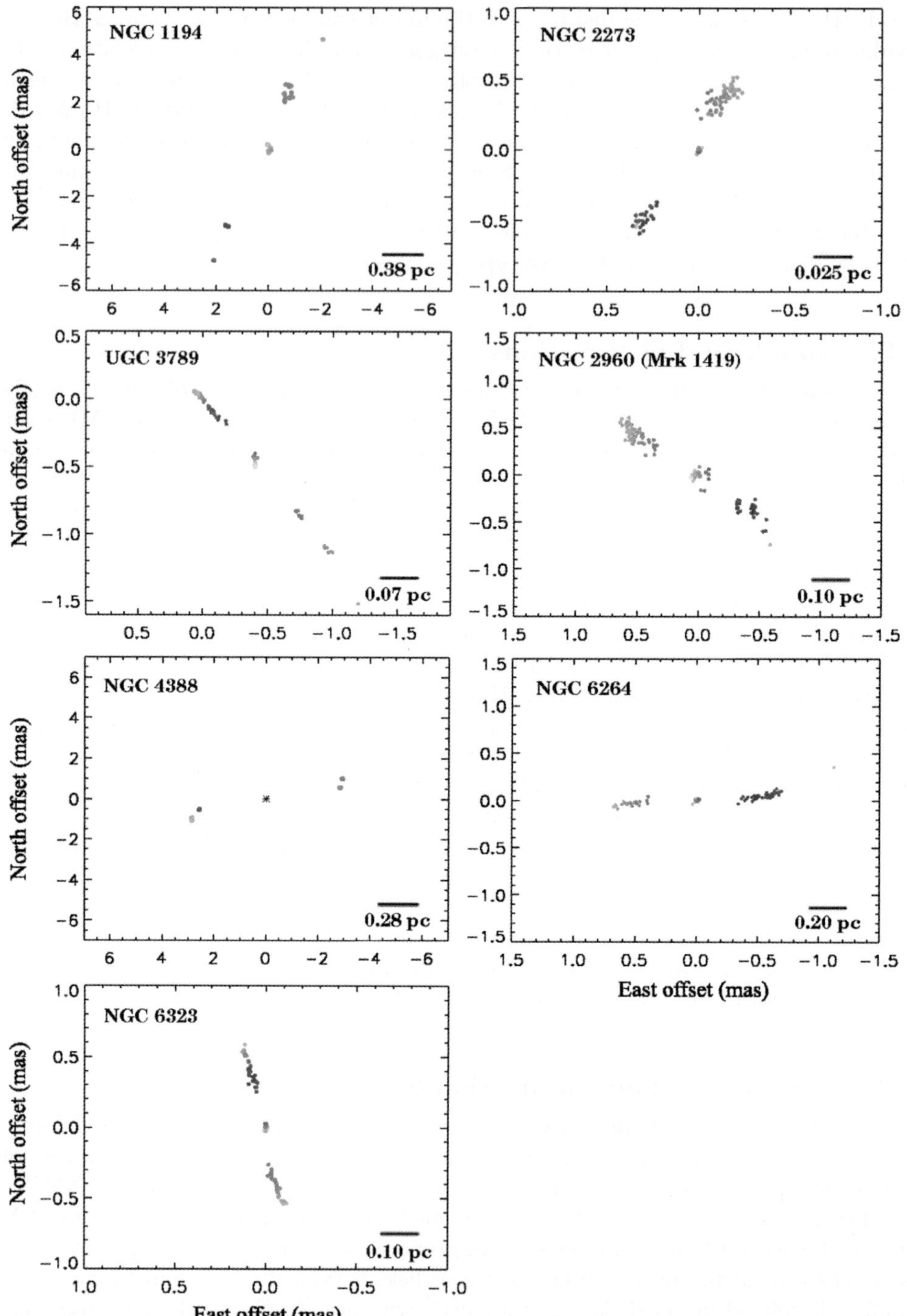

Figure 1. VLBI maps for the seven of 22 GHz H_2O masers megamasers analyzed. The maps are color coded to indicate redshifted, blueshifted, and systemic masers, where the "systemic" masers refer to the maser components having recessional velocities close to the systemic velocity of the galaxy. Except NGC 4388, maser distributions are plotted relative to the average position of the systemic masers. For NGC 4388, in which the systemic masers are not detected, we plot the maser distribution relative to the dynamical center determined by fitting the high velocity features with a Keplerian rotation curve.

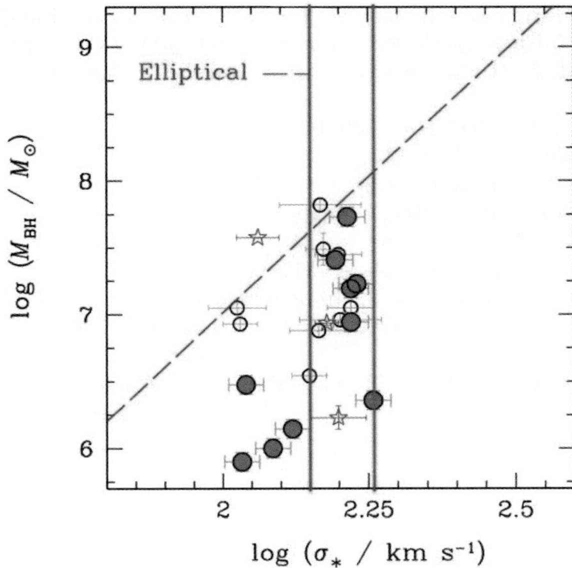

Figure 2. A plot showing the locations of H_2O megamaser galaxies in the M_{BH}-$\sigma_\star$ plane. The masers show no clear correlation. The dotted line shows the M_{BH}-$\sigma_\star$ relation determined from early type galaxies. The galaxies bound by the two green lines in this plot emphasizes the large scatter in the BH mass for galaxies with 140 km s^{-1} < $\sigma_\star$ < 190 km s^{-1}. The scatter in M_{BH} is nearly two orders of magnitude, too large to justify any correlation.

rotation curve, the BH mass is then calculated using the following equation:

$$M_{\rm BH} = \left(\frac{|v_{\rm K}|^2\theta}{G}\right)D_{\rm A} = \left(\frac{\pi v_1^2}{6.48 \times 10^8 G}\right)D_{\rm A} , \qquad (3.2)$$

where D_A is the angular diameter distance to the galaxy. In this work, we adopt the Hubble flow distance from the NASA/IPAC Extragalactic Database (NED) as $D_{\rm A}$ for the BH mass calculation.

Our new maser BH masses significantly increase the number of galaxies having dynamical BH masses $M_{\rm BH} \sim 10^7 M_\odot$. These measurements play a particularly important role in constraining the $M - \sigma_\star$ relation at the low-mass end of known nuclear BH masses. With the stellar velocity dispersion ($\sigma_\star$) of the galaxy bulges measured with the Dupont 2.5 m telescope at Las Campanas Observatory in the South and the Apache Point Observatory (APO) 3.5m telescope in the North (e.g. Greene *et al.* 2010), we found that H_2O maser galaxies in general do not follow the $M_{\rm BH}$–$\sigma_\star$ relation defined by early type galaxies (see Figure 2). From Figure 2, one can see that no correlation can be found for maser galaxies in the $M - \sigma_\star$ diagram, and the scatter in the BH mass for galaxies with 140 km s^{-1} < $\sigma_\star$ < 190 km s^{-1} is nearly two orders of magnitude, too large to justify any correlation.

This result is well consistent with the latest findings by Kormendy & Ho (2013) that only early type galaxies and galaxies with classical bulges follow a tight $M_\bullet$–σ relation. Such a tight correlation may not exist in pseudo bulge galaxies.

References

Barth, A. J. *Carnegie Observatories Astrophysics Series*, Vol. 1: Coevolution of Black Holes and Galaxies, 2003 ed. L. C. Ho (Pasadena: Carnegie Observatories)

Braatz, J., Condon, J., Reid, M., Henkel, C., Lo, K. Y., Kuo, C. Y., Impellizzeri, C., & Hao, L.
The Megamaser Cosmology Project Large Proposal, submitted to NRAO June 1, 2009
Braatz, J. A., Reid, M. J., Humphreys, E. M. L., Henkel, C., Condon, J. J., & Lo, K. Y. 2010,
ApJ, 718, 657
Ferrarese, L. & Merritt, D. 2000, *ApJ*, 539, 9
Gebhardt, K. *et al.* 2000, *ApJ*, 539, 13
Greene, J. E., Peng, C. Y., Kim M., Kuo, C.-Y., Braatz J. A.,Impellizzeri, C. M. V., Condon,
J.,Lo, F., & Henkel C. 2010, *ApJ*, 721, 26
Gültekin, K. *et al.* 2009, *ApJ*, 698, 198
ormendy, J. & Ho, L. C. 2013, *ARA&A*, 51, 511
Kuo, C. Y., Braatz, J. A., Condon, J. J., Impellizzeri, C. M. V., Lo, K. Y., Zaw, I., Schenker,
M., Henkel, C., Reid, M. J., & Greene, J. E. 2011, *ApJ*, 727, 20
Lo, K. Y. 2005 *ARA&A* 43 625
Reid, M. J., Braatz, J. A., Condon, J. J., Greenhill, L. J., Henkel, C., & Lo, K. Y. 2009, *ApJ*,
695, 287

Star clusters and black holes in galaxies across cosmic time
Proceedings IAU Symposium No. 312, 2014
Y. Meiron, S. Li, F.-K. Liu & R. Spurzem, eds.

© International Astronomical Union 2016
doi:10.1017/S1743921315007486

Multi-wavelength observations of the narrow-line Seyfert 1 galaxy RX J2314.9+2243

S. Komossa[1], I. Myserlis[1], L. Fuhrmann[1], D. Xu[2], D. Grupe[3], Z. Fan[2], S. Yao[2], E. Angelakis[1], V. Karamanavis[1], J. A. Zensus[1] and W. Yuan[2]

[1] Max-Planck-Institut für Radioastronomie, Auf dem Hügel 69, 53121 Bonn, Germany
email: `skomossa@mpifr.de`

[2] National Astronomical Observatories, Chinese Academy of Sciences, Beijing, 100012, China

[3] Space Science Center, Morehead State University, 235 Martindale Dr., Morehead, KY 40351, USA

Abstract. Narrow-line Seyfert 1 (NLS1) galaxies are a sub-class of active galactic nuclei (AGN) with relatively low-mass black holes, accreting near the Eddington rate. A small fraction of them is radio-loud and harbors relativistic jets. As a class, these provide us with new insights into the cause(s) of radio-loudness, the blazar phenomenon at low black hole masses, and the operation of radio-mode feedback. The NLS1 galaxy RXJ2314.9+2243 is remarkable for its multi-wavelength properties. We present new radio observations taken at Effelsberg, and a summary of the recent results from our multi-wavelength study. RXJ2314.9+2243 is radio-loud, luminous in the infrared, has a flat X-ray spectrum and peculiar UV spectrum, and hosts an exceptionally broad and blueshifted [OIII]λ5007 emission line, indicating the presence of a strong outflow. RXJ2314.9+2243 likely represents an extreme case of AGN induced feedback in the local universe.

Keywords. Black holes, quasars, jets

1. Radio-loud NLS1 galaxies

NLS1 galaxies exhibit many extreme properties among AGN, including, on average, super-strong FeII emission, rapid X-ray variability, near-Eddington accretion rates, and rapidly growing low-mass black holes (e.g., Grupe 2004, Xu *et al.* 2012, review by Komossa 2008). Their radio properties have only been studied systematically in recent years, revealing a number of surprises: as a class, they are less frequently radio-loud than broad-line Seyfert 1 galaxies (Komossa *et al.* 2006), but some of them are beamed (Yuan *et al.* 2008, Angelakis *et al.* 2015), γ-ray detected (Abdo *et al.* 2009, Foschini *et al.* 2011), and appear like classical blazars, and at least one of them is puzzling due to its exceptional host galaxy morphology (Zhou *et al.* 2007).

2. Results on RX J2314.9+2243

RX J2314.9+2243 (α_{2000}=23$^{\rm h}$ 14$^{\rm m}$ 55.7$^{\rm s}$; δ_{2000}=+22° 43′ 25″) is a radio-loud NLS1 galaxy (Komossa *et al.* 2006) at redshift z=0.169. In order to understand its nature, we have obtained multi-wavelength observations of RX J2314.9+2243 with *Swift* in the optical, UV and X-rays, with the Xinglong telescopes in the optical (monitoring and spectroscopy), and with the Effelsberg 100m telescope in the radio band at multiple frequencies and epochs (see Komossa *et al.* 2015 for details), including new Effelsberg radio data taken in October 2014.

Table 1. Radio measurements of RXJ2314.9+2243 performed with the Effelsberg 100m telescope. Not all frequencies ν were observed at all dates. RX J2314.9+2243 was also detected during the NVSS at 1.4 GHz with a flux density of 19±1 mJy.

ν [GHz]	flux density [mJy]					
	2013 Feb 03	2013 Feb 09	2013 July 7	2013 July 23	2014 Oct 18	2014 Oct 31
2.64	–	14 ± 3	12 ± 2	–	–	–
4.85	7 ± 1	7 ± 1	9 ± 2	7 ± 2	10 ± 2	8 ± 2
8.35	5 ± 1	5 ± 1	5 ± 1	5 ± 1	6 ± 1	5 ± 1
10.45	–	–	< 17	–	< 56	< 56
14.60	–	–	–	–	< 27	< 28
43.00	–	–	< 56	–	< 158	< 96

The spectral energy distribution (SED) of RX J2314.9+2243 shows a broad hump extending between the IR and FUV, a steep downturn in the UV towards shorter wavelengths, a steep radio spectrum and flat X-ray spectrum. Its steep UV spectrum ($\alpha_{\mathrm{UV}} = 1.6$) is likely intrinsic, since we do not find evidence for strong reddening in the optical spectrum. Its IR to FUV SED is consistent with a scenario, in which synchrotron emission from a jet dominates the broad-band emission, even though an absorption scenario cannot yet be fully excluded.

So far, no variability in any waveband was detected, with the exception of the X-ray band. In the radio regime, its spectrum is steep ($\alpha_{\mathrm{R}} = -0.76$), more reminiscent of CSS (compact steep-spectrum sources) than of blazars. Yet, its marginal γ-ray detection with *Fermi* (L. Foschini, priv. com.; Berton *et al.* 2015, in prep.), if confirmed, likely hints at flaring due to a relativistic jet. The observed radio emission likely represents emission from a quiescent jet component, and, indeed, the similar radio fluxes observed with NVSS and at a much later epoch with Effelsberg suggest, that RX J2314.9+2243 was not in a flaring state during the latter observations. New radio observations were carried out in October 2014 (Tab. 1), and the radio emission remains constant within the errors.

The high blueshift of its very broad [OIII] component, 1260 km/s, is consistent with a face-on view, with the jet (and outflow) pointing towards us. RXJ2314.9+2243 likely represents an extreme case of AGN induced feedback in the local universe.

Acknowledgements

This work was supported by grant NSFC 11273027. I.M. and V.K. are funded by the International Max Planck Research School (IMPRS) for Astronomy and Astrophysics at the Universities of Bonn and Cologne.

References

Abdo, A. A., *et al.* 2009, *ApJ*, 699, 976
Angelakis, E., Fuhrmann, L., Marchili, N., *et al.* 2015, *A&A*, 575, A55
Foschini, L., *et al.* 2011, *MNRAS*, 413, 1671
Grupe, D. 2004, *AJ*, 127, 1799
Komossa, S., *et al.* 2006, *AJ*, 132, 531
Komossa, S. 2008, *Revista Mexicana de Astronomía y Astrofísica Conference Series*, 32, 86
Komossa, S., *et al.* 2015, *A&A*, 574, 121
Xu, D., *et al.* 2012, *AJ*, 143, 83
Yuan, W., *et al.* 2008, *ApJ*, 685, 801
Zhou, H., *et al.* 2007, *ApJ*, 658, L13

Star clusters and black holes in galaxies across cosmic time
Proceedings IAU Symposium No. 312, 2014
Y. Meiron, S. Li, F.-K. Liu & R. Spurzem, eds.
© International Astronomical Union 2016
doi:10.1017/S1743921315007498

Radio-loud narrow-line Seyfert 1 galaxies with high-velocity outflows

S. Komossa[1], D. Xu[2] and J. A. Zensus[1]

[1] Max-Planck-Institut für Radioastronomie, Auf dem Hügel 69, 53121 Bonn, Germany
email: `skomossa@mpifr.de`

[2] National Astronomical Observatories, Chinese Academy of Sciences, Beijing, 100012, China

Abstract. We have studied four radio-loud Narrow-line Seyfert 1 (NLS1) galaxies with extreme optical emission-line shifts, indicating radial outflow velocities of up $2450\,\mathrm{km\,s^{-1}}$. The shifts are accompanied by strong line broadening, up to $2270\,\mathrm{km\,s^{-1}}$ in [NeV]. A significant ionization stratification (higher line shift at higher ionization potential) of most ions implies that we see a large-scale wind rather than single, localized jet-cloud interactions. The observations are consistent with a scenario, where the signatures of outflows are maximized because of a pole-on view into the central engine of these radio-loud NLS1 galaxies.

Keywords. Black holes, galaxies, emission lines, outflows, jets

1. Introduction

Powerful gaseous outflows in Active Galactic Nuclei (AGN) deposit mass, energy and metals in the interstellar medium of the host galaxy or on even larger scales (review by Fabian 2012). They therefore shape the structure and composition of the core environment, and may play an important role in unified models. The most powerful jets and outflows may significantly affect the co-evolution of galaxies and black holes by feedback processes, by regulating star formation, and possibly clearing the host galaxy off large fractions of its interstellar medium.

There is ample observational evidence for gaseous winds and outflows in AGN, which often manifest as kinematic shifts of optical emission lines. These are typically on the order of $100\,\mathrm{km\,s^{-1}}$ or less, but may be much higher especially in some high-redshift radio galaxies.

There are indications, that winds are especially strong in NLS1 galaxies, where accretion near the Eddington limit likely implies the presence of strong, radiation-pressure driven outflows (e.g. Komossa *et al.* 2008). Narrow-line Seyfert 1 galaxies are a subclass of AGN with extreme multi-wavelength properties. As a class, they are characterized by the narrow widths of their broad Balmer lines, super-strong FeII emission, ultra-steep X-ray spectra and enhanced variability. While NLS1 galaxies are on average more radio-quiet than broad-line AGN (Komossa *et al.* 2006), a small fraction of them is beamed and radio-loud (e.g. Komossa *et al.* 2006, Yuan *et al.* 2008).

2. Sample selection and results

In order to explore the presence and properties of powerful outflows in radio-loud NLS1 galaxies, we have analyzed four very radio-loud NLS1s from the sample of Yuan *et al.* (2008) which were initially noted for their shifted [OIII] emission lines, placing them in the class of "blue outliers". All four show extreme line shifts, among the highest measured so far. This is the first dedicated study of ionized gas outflows in a mini-sample

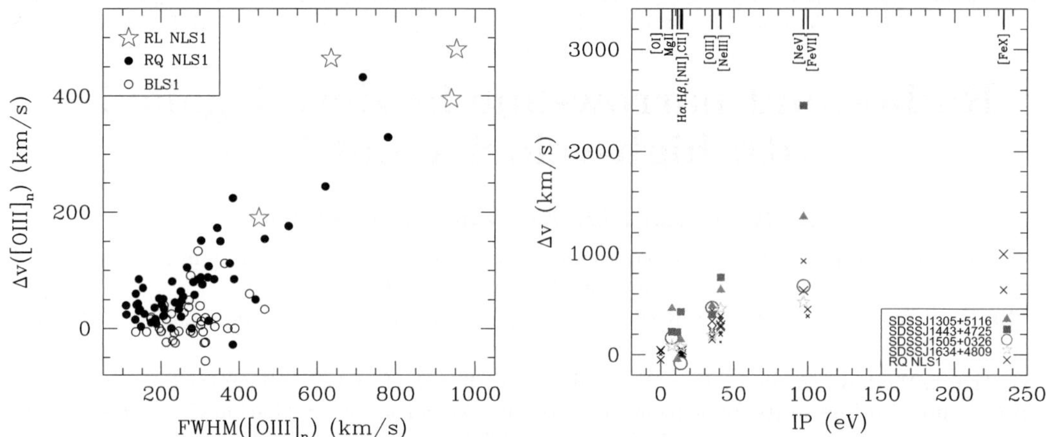

Figure 1. Left: Velocity shift – line width correlation of [OIII]λ5007. For comparison, the sample of Broad-line Seyfert 1 galaxies (open circles) and NLS1 galaxies (filled circles) of Komossa *et al.* (2008) is plotted. Right: Radial velocity as a function of ionization potential IP. The NLS1 galaxies of this study are marked by large grey symbols. For comparison, the nine "blue outliers" of Komossa *et al.* (2008) are shown (small black crosses).

of radio-loud NLS1 galaxies. In order to measure emission-line shifts and widths, we have analyzed the optical spectra of the four NLS1 galaxies, taken in the course of the Sloan Digital Sky Survey (SDSS DR7; Abazajian *et al.* 2009). Emission line shifts and widths were measured by fitting Gaussians to the lines (for details of the data preparation and analysis, see Komossa *et al.* 2015, in prep.).

Extreme emission-line blueshifts are present in all four galaxies (Fig. 1). In [OIII]λ5007, we measure a line width of FWHM([OIII]$_\mathrm{n}$) $=$ 950 km s^{-1} and a shift of Δv([OIII]$_\mathrm{n}$) $=$ 480 km s^{-1} (SDSSJ130522.75+511640.3), and Δv([OIII]$_\mathrm{n}$) $=$ 460 km s^{-1} (SDSSJ150506.48+032630.8). High-ionization lines of [NeV]λ3426 are present in all spectra, and show even higher shifts, with Δv([NeV]) $=$ 2450 km s^{-1} (SDSSJ144318.56+472556.7) and Δv([NeV]) $=$ 1360 km s^{-1} (SDSSJ130522.75+511640.3). The [OIII] emission lines follow a width-shift correlation, in the sense that more highly shifted lines are much broader (Fig. 1). Further, there is an overall trend among the observed emission lines of higher blueshift with higher ionization potential (Fig. 1). This finding implies that we see a large-scale flow rather than single, localized jet-cloud interactions.

Accreting close to the Eddington limit, NLS1 galaxies likely drive strong outflows. If high radio loudness in NLS1 galaxies is generally due to beaming, we expect a pole on view onto these galaxies, and so the effects of polar outflows are maximized, consistent with the findings presented in this work.

Acknowledgements

This work was supported by grant NSFC 11273027.

References

Abazajian, K. N., Adelman-McCarthy, J. K., Agüeros, M. A., *et al.* 2009, *ApJS*, 182, 543
Fabian, A. 2012, *ARAA*, 50, 455

Komossa, S., Voges, W., Xu, D., *et al.* 2006, *AJ*, 132, 531
Komossa, S., Xu, D., Zhou, H., Storchi-Bergmann, T., & Binette, L. 2008, *ApJ*, 680, 926
Yuan, W., Zhou, H. Y., Komossa, S., *et al.* 2008, *ApJ*, 685, 801

Star clusters and black holes in galaxies across cosmic time
Proceedings IAU Symposium No. 312, 2014
Y. Meiron, S. Li, F.-K. Liu & R. Spurzem, eds.

© International Astronomical Union 2016
doi:10.1017/S1743921315007504

Swift monitoring and *Suzaku* spectroscopy of the γ-ray detected narrow-line Seyfert 1 galaxy 1H 0323+342

S. Yao[1], W. Yuan[1], S. Komossa[2,1], D. Grupe[3,4], L. Fuhrmann[2] and B. Liu[1]

[1]National Astronomical Observatories, Chinese Academy of Sciences, 20A Datun Road, Chaoyang District, Beijing, China
email: yaosu@nao.cas.cn

[2]Max-Planck Institut für Radioastronomie, Auf dem Hügel 69, 53121 Bonn, Germany

[3]Space Science Center, Morehead State University, Morehead, KY, 40351

[4]Swift Mission Operation Center, 2582 Gateway Dr. State College, PA, 16801, USA

Abstract. 1H 0323+342 is one of the rare γ-ray detected narrow-line Seyfert 1 galaxies (NLS1s), a special subset of active galactic nuclei (AGN) owing to their hybrid behavior of both NLS1s and blazars. The rarity of such kind of sources makes their properties far from being understood. We analyze simultaneous X-ray and UV/optical monitoring observations of 1H 0323+342 performed by *Swift* over $\sim$ 7 years. The UV/X-ray correlation and the broad band SED reveal that the X-ray band is dominated by the disk/corona emission during the observations. The large normalized excess variance of the X-ray variability detected with *Suzaku* suggests a relatively small black hole mass of the order of $10^7 M_\odot$, consistent with the estimation based on the broad Hβ line in the optical band.

Keywords. galaxies: active — quasars: individual (1H 0323+342) — galaxies: particular — galaxies: jets — X-rays: galaxies.

---

1. Introduction and observations

An interesting discovery of *Fermi*/LAT in recent years is the detection of γ-ray emission from several narrow-line Seyfert 1 galaxies, which are usually considered to be radio-quiet and occupy a very different domain than blazars in taxonomy of AGN (Abdo *et al.* 2009a,b; D'Ammando *et al.* 2012). It raises the question of whether these sources form a previously unrecognized population or the downsizing counterpart of normal blazars (Komossa *et al.* 2006; Yuan *et al.* 2008). As one of these sources, 1H 0323+342 has been identified by Zhou *et al.* (2007) to be a prototype of radio-loud NLS1s and was detected by *Fermi*/LAT (Abdo *et al.* 2009b). 1H 0323+342 was monitored by *Swift* for 80 occasions from 2006 to 2013. It was also observed with *Suzaku* during an 84 ks exposure. These observations provide a good data set to study the multi-waveband emission and physics of this object. We analyze the X-ray and UV/optical data of 1H 0323+342 to study both the temporal and broad band spectral properties at these wavebands.

2. Results and discussion

The object is variable in both the UV and X-ray bands on timescales of days to years, and we find a statistically significant correlation between the UV flux and X-ray count rates at these timescales. A cross-correlation analysis suggests a time lag close to zero between the UV and the X-ray emission, however, with X-rays tentatively leading during

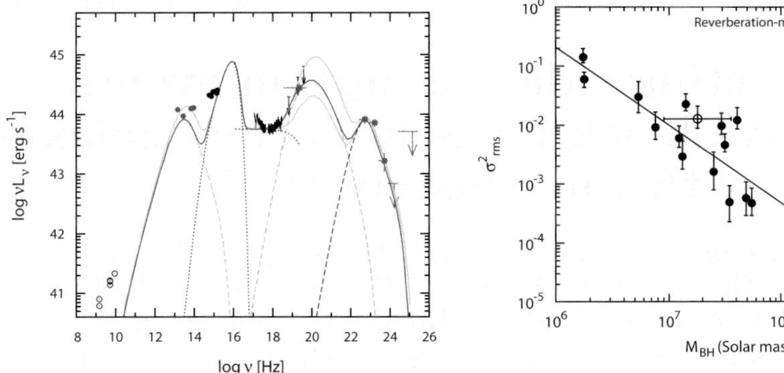

Figure 1. *Left panel*: The broad band SED of 1H 0323+342, consisting of *Swift* and *Suzaku* data used in this work. The model consist of accretion disk and corona (dotted lines) and jet (dashed line). *Right panel*: The log $M_{\rm BH}$-log $\sigma^2_{\rm rms}$ relation calibrated by the reverberation-mapped sample in Ponti *et al.* 2012 (solid line).

2010 October–November observations. This can be explained with the scenario that the UV emission is the result of reprocessing the primary X-ray emission from the accretion disk/corona. The X-ray spectrum obtained by *Suzaku* shows a soft excess below 1 keV and a power-law spectrum with a photon index of $\Gamma = 1.9$. The broad-band SED of 1H 0323+342 can be well modeled in the context of a one-zone leptonic jet model plus the accretion disk/corona, the latter is suggested to dominate the emission in the UV and X-ray bands (up to 10 keV) as observed with *Swift* and *Suzaku* (Figure 1, left panel).

Given their relatively small widths of their broad lines, the black hole masses $M_{\rm BH}$ of NLS1s are believed to be systematically lower than those of classical Seyfert 1s and quasars, resulting in high Eddington ratios (e.g., Grupe & Mathur 2004; Xu *et al.* 2012). However, there remains controversy as to whether their black hole masses are underestimated, if their broad line regions are planar and seen face on. As the X-ray below 10 keV is dominated by a disk/corona component during the *Suzaku* observation, an independent estimation of $M_{\rm BH}$ can be achieved by using the X-ray normalized excess variance $\sigma^2_{\rm rms}$ (Ponti *et al.* 2012). We find $\sigma^2_{\rm rms,2-4keV} = 12.3^{+8.1}_{-3.9} \times 10^{-3}$, which corresponds to a black hole mass of $M_{\rm BH} = 8.6^{+2.9}_{-2.7} \times 10^6 M_\odot$, using the relation in Ponti *et al.* (2012). This value is consistent with $M_{\rm BH} = 1.8 \times 10^7 M_\odot$ estimated from the broad Hβ line in Zhou *et al.* (2007) (see Figure 1, right panel). We conclude that the black hole mass is small in at least this particular γ-ray detected NLS1.

Acknowledgements

This work is supported by grants No. 11473035 & No. XDB09000000.

References

Abdo, A. A., Ackermann, M., Ajello, M., *et al.* 2009a, *ApJ*, 699, 976
—. 2009b, *ApJ*, 707, L142
D'Ammando, F., Orienti, M., Finke, J., *et al.* 2012, *MNRAS*, 426, 317
Grupe, D. & Mathur, S. 2004, *ApJL*, 606, L41
Komossa, S., Voges, W., Xu, D., *et al.* 2006, *AJ*, 132, 531
Ponti, G., Papadakis, I., Bianchi, S., *et al.* 2012, *A&A*, 542, A83
Xu, D., Komossa, S., Zhou, H., *et al.* 2012, *AJ*, 143, 83
Yuan, W., Zhou, H. Y., Komossa, S., *et al.* 2008, *ApJ*, 685, 801
Zhou, H., Wang, T., Yuan, W., *et al.* 2007, *ApJL*, 658, L13

Star clusters and black holes in galaxies across cosmic time
Proceedings IAU Symposium No. 312, 2014
Y. Meiron, S. Li, F.-K. Liu & R. Spurzem, eds.

© International Astronomical Union 2016
doi:10.1017/S1743921315007516

Detecting tidal disruption events of massive black holes in normal galaxies with the Einstein Probe

W. Yuan[1], S. Komossa[1,2], C. Zhang[1], H. Feng[3], Z.-X. Ling[1], D.H. Zhao[1], S.-N. Zhang[4], J.P. Osborne[5], P. O'Brien[5], R. Willingale[5], J. Lapington[5] and the Einstein Probe team

[1] Key Laboratory of Space Astronomy and Technology, National Astronomical Observatories, CAS, Beijing, 100012, China
email: wmy@nao.cas.cn

[2] Max-Planck-Institut für Radioastronomie, Auf dem Hügel 69, 53121 Bonn, Germany

[3] Department of Engineering Physics, Tsinghua University, Beijing 100084, China

[4] Institute of High Energy Physics, Chinese Academy of Sciences, 100049, Beijing, China

[5] Department of Physics and Astronomy, University of Leicester, Leicester, LE1 7RH, UK

Abstract. Stars are tidally disrupted and accreted when they approach massive black holes (MBHs) closely, producing a flare of electromagnetic radiation. The majority of the (approximately two dozen) tidal disruption events (TDEs) identified so far have been discovered by their luminous, transient X-ray emission. Once TDEs are detected in much larger numbers, in future dedicated transient surveys, a wealth of new applications will become possible. Here, we present the proposed Einstein Probe mission, which is a dedicated time-domain soft X-ray all-sky monitor aiming at detecting X-ray transients including TDEs in large numbers. The mission consists of a wide-field micro-pore Lobster-eye imager ($60° \times 60°$), and is designed to carry out an all-sky transient survey at energies of 0.5-4 keV. It will also carry a more sensitive telescope for X-ray follow-ups, and will be capable of issuing public transient alerts rapidly. Einstein Probe is expected to revolutionise the field of TDE research by detecting several tens to hundreds of events per year from the early phase of flares, many with long-term, well sampled lightcurves.

Keywords. black hole physics, X-rays: bursts, X-rays: general, space vehicles: instruments

1. Introduction

Tidal disruption events (TDEs) by massive black holes (MBH) are perhaps the most unique signature of the existence of MBHs in the cores of otherwise quiescent galactic nuclei (Rees 1988). So far only about two dozen TDE candidates were found, mostly in the declining phase with poorly sampled data, via searches from multi-wavelength surveys (see Komossa 2012 for a review). To detect TDEs in large numbers and to catch the 'smoking gun' at an early phase, it is desirable to monitor a large fraction of the sky with high sensitivity at high cadence, preferably in the soft X-ray band.

2. The Einstein Probe mission

Einstein Probe (EP) is a proposed small satellite to detect transients and monitor variable objects in the 0.5-4 keV band, aiming for launch around 2020. EP carries two scientific instruments (Fig. 1) and a fast alert downlink system to trigger follow-up observations. The primary instrument is a Wide-field X-ray Telescope (WXT) with a FoV of $60° \times 60°$, which is a lobster-eye type X-ray focusing telescope based on

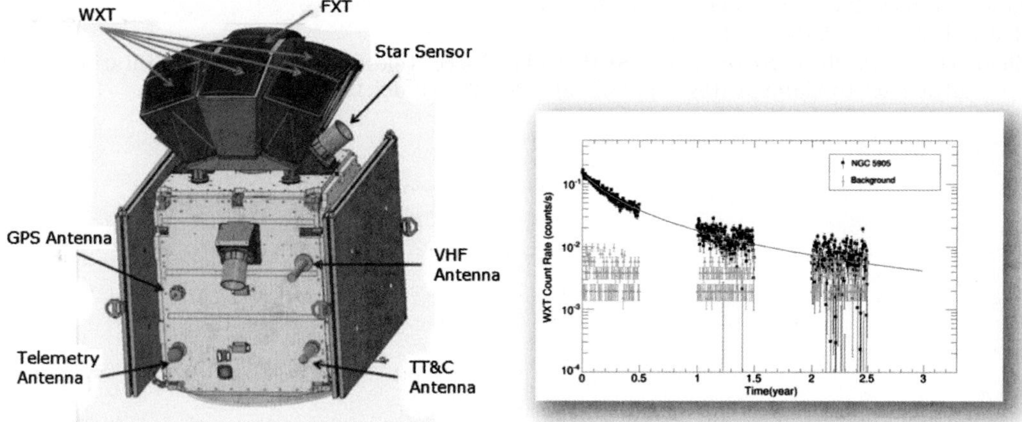

Figure 1. Left: layout of the Einstein Probe satellite. Right: simulated EP/WXT X-ray lightcurve of the TDE found in NGC 5905 (Komossa & Bade 1999).

Micro-Pore Optics (MPO). In addition, there is a Follow-up X-ray Telescope (FXT) of the same MPO technology, which has a much larger effective area ($\sim 60\,\mathrm{cm}^2$) but a smaller FoV ($1° \times 1°$) than WXT. EP will offer unprecedentedly high sensitivity and large grasp (see Zhao *et al.* 2014 for results of simulations of the instrument), which would supersede previous and existing X-ray all-sky monitors. The survey strategy of EP will consist of a series of pointings, and will be able to cover the entire night sky in about 3 orbits ($\sim$90 minutes each orbit). EP has been selected as one of the candidate missions in the Space Science programme of the Chinese Academy of Sciences, and is currently in the 'Advanced Study' phase for mission definition and technology development.

Benefiting from its large field-of-view, high sensitivity, high observing cadence, soft X-ray bandpass, fast alerting and follow-up capability, EP is an ideal mission to systematically search for and characterise TDEs. It is also expected to catch TDEs at an early phase, which is very important for studying the disruption processes, the formation and evolution of accretion discs, the launch of relativistic jets, as well as estimating the mass and spin of the black holes. EP is expected to detect TDEs at the peaks of the X-ray flares out to at least a few hundred Mpc, at an estimated rate from several tens to hundreds per year. Many of them will be caught at the rising phase of the flares, making it possible to observe the events from the very start in multiple wavebands. EP will be able to detect relativistic TEDs with jets similar to Sw J1644+57 (Burrows *et al.* 2011) out to redshifts $z > 1$. Einstein Probe is expected to revolutionise the field of TDE research by detecting and characterising TDEs in large numbers and catching them at the early phase of flares. This will greatly advance our understanding of the demography, formation and evolution of MBHs, as well as the physics of accretion and jet formation.

Acknowledgements

This work is supported by the Space Science Programme of CAS, Grant XDA04061100.

References

Burrows *et al.* 2011, *Nature*, 476, 421
Komossa, S. 2012, *EPJ Web of Conf.*, 39, id. 02001

Komossa, S. & Bade, N. 1999, *A&A*, 343, 775

Rees, M. J. 1988, *Nature*, 33, 523

Zhao, D., *et al.* 2014, *SPIE*, Proc. SPIE 9144, Space Telescopes and Instrumentation 2014: Ultraviolet to Gamma Ray, 91444E (July 24, 2014); doi:10.1117/12.2055434

Star clusters and black holes in galaxies across cosmic time
Proceedings IAU Symposium No. 312, 2014
Y. Meiron, S. Li, F.-K. Liu & R. Spurzem, eds.
© International Astronomical Union 2016
doi:10.1017/S1743921315007528

May PKS 1155+251 be the habitat of a binary black hole?

Xiaolong Yang[1,2] and Xiang Liu[1]

[1] Xinjiang Astronomical Observatory, Chinese Academy of Sciences, Urumqi, 830002, China;
email: yangxiaolong@xao.ac.cn

[2] Graduate University of Chinese Academy of Sciences, 100049, Beijing, China

Abstract. Close binary black holes (BBH) are important not only in astrophysics but they would be the strongest gravitational wave sources in the universe. Galaxy-galaxy merging systems are mostly found in optical and X-ray images. In radio, however, the VLBI can resolve the close binary system at pc scale, if their nuclei are radio loud. Recently we analyzed the archive VLBI data of PKS 1155+251, it shows twin core-jets like VLBI structure. In this poster, we present preliminary result from analyzing of the archive data. Further investigations with high frequency VLBI observations are required to confirm if it is a true BBH system.

Keywords. black hole physics: binary – galaxies: jets – quasars: general – radio continuum

1. Radio continuum images

PKS 1155+251 as a BBH candidate (Liu 2014), is a flat spectrum radio quasar, at $z = 0.2016$. The radio continuum images of PKS 1155+251 are obtained at 4 bands (C-band, X-band, U-band and S-band). Figure 1 shows naturally weighted 4.9 GHz and 15 GHz images. We can see that the 15 GHz image contains two bright and compact components, the two components are labeled as C1 and C2, other component labels can be seen from the images. Kellermann *et al.* (2004) has identified the bright component C1 to be the core and classified this source as a core-jet. Tremblay *et al.* (2008) observed this source in 2006, and identified it to be a CSO and C2 to be the core. However, we find the structure of this source show twin core-jets like structure, which can not be fully explained with the CSO scenario, i.e. it appears C1 to have a jet. We also find faint bridges among the bright components, this can rule out the possibility that one of the components to be a foreground or background source Rodriguez *et al.* (2006).

We set a logarithm contour level and set first level contour to the 3σ. The Fig.1 left shows large scale structure at 4.9 GHz, it exhibits east-west jet structure. The Fig.1 right shows small scale structure at 15 GHz with a higher resolution, interestingly, this image shows south-north jet morphology. The binary black holes interaction could be account for the jet orientation difference, which could be due to a dramatic change of the jet axis.

2. Component motions

Apparent motion of large scale structure. We obtained S-band data taken in three epochs 1996, 1997 and 2003. The apparent motion was estimated by settling component C1 as the reference center and all images were laid on Fig.2. We select the largest beam to restore all of three images to get the same resolution, then logarithm contour levels were used, first contour level was set to be 3σ to get acceptable signal to noise ratio, the σ was the largest one in the three images. The resulted images can be seen in Fig.2 left.

Small scale component motions. In order to obtain relative motions of components,

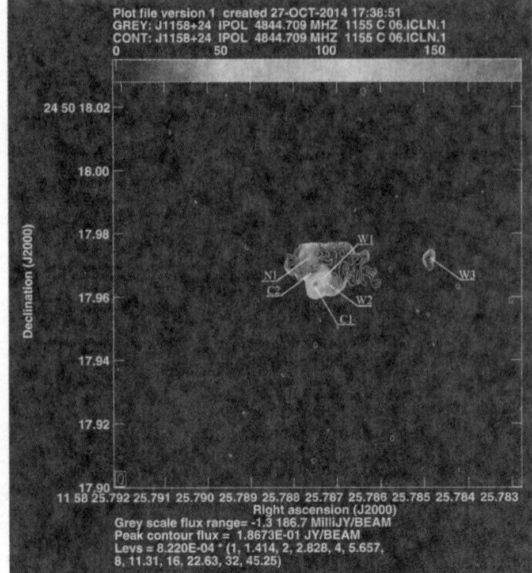 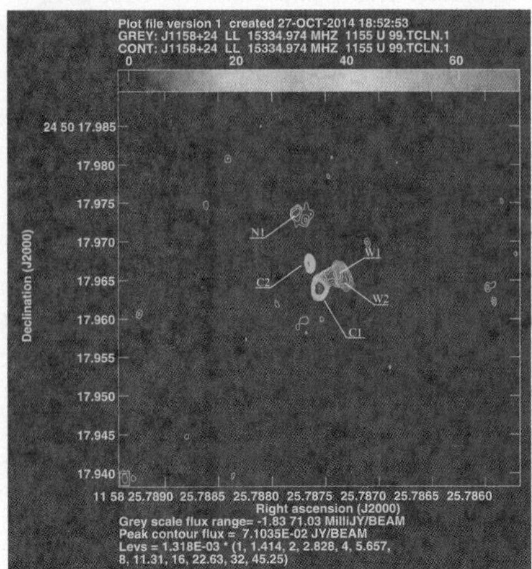

Figure 1. The VLBI morphologies of PKS 1155+251.

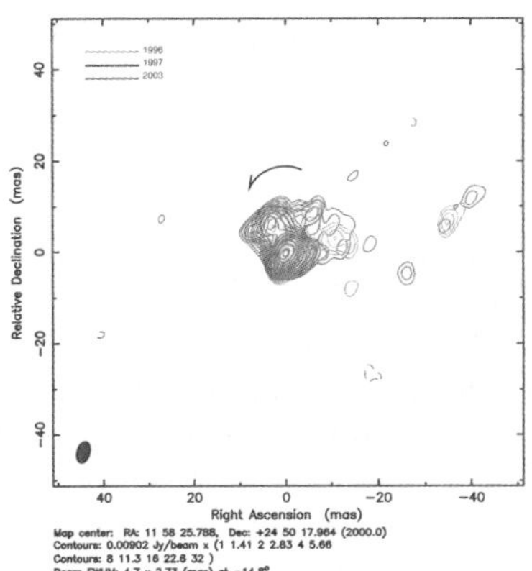 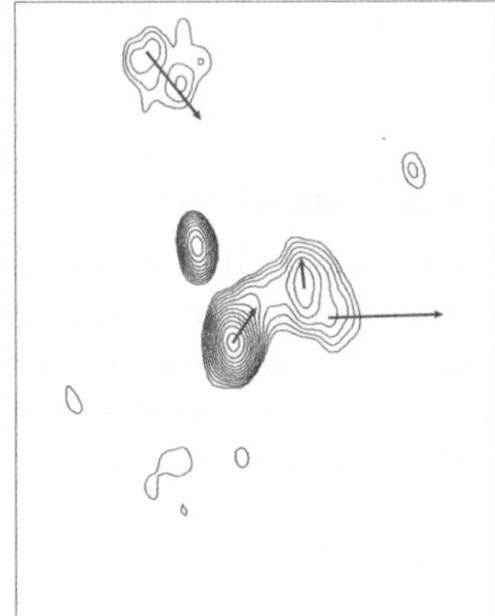

Figure 2. Component Motions.

we used fully calibrated U-band archive data, the time baseline is 2 years. Component motion studies were performed by fitting 5 Gaussian components in Difmap software to the 1999 visibility data. Then we used this model to fit the 2001 visibility data, all parameters were fixed at 1999 values except for position and flux density. We choose component C2 as the reference, the result has been showed in Fig.2 right. We add arrows at 1999 image to show the direction of motion of each component. This source appears to be shrinking, which may be due to the binary black holes interaction.

References

Kellermann, K. I., *et al.* 2004, *ApJ*, 609, 539

Liu, X. 2014, *J. Astrophys. Astr.*, Vol.35, No.3, 1

Rodriguez, C., *et al.* 2006, *ApJ*, 646, 49

Tremblay, S. E., *et al.* 2008, *ApJ*, 684, 153

Star clusters and black holes in galaxies across cosmic time
Proceedings IAU Symposium No. 312, 2014
Y. Meiron, S. Li, F.-K. Liu & R. Spurzem, eds.

© International Astronomical Union 2016
doi:10.1017/S174392131500753X

On the relationship between black hole mass and X-ray variability amplitude in the low-mass regime of active galactic nuclei

H. Pan[1], W. Yuan[1], X.-L. Zhou[1], X. Dong[2] and B. Liu[1]

[1]National Astronomical Observatories, Chinese Academy of Sciences, 20A Datun Road, Chaoyang District, Beijing, China
email: `panhaiwu@bao.ac.cn`

[2]Yunnan Observatories, Chinese Academy of Sciences, Kunming, Yunnan, China

Abstract. Recent studies of active galactic nuclei (AGN) found a statistical inverse scaling between the X-ray normalized excess variance $\sigma^2_{\rm rms}$ (variability amplitude) and the black hole mass spanning over $M_{\rm BH} = 10^6 - 10^9 \ M_\odot$. We present a study of this relation by including AGN with $M_{\rm BH} = 10^5 - 10^6 \ M_\odot$. It is found that the relation is no longer a simple extrapolation of the known inverse proportion, but starts to flatten around $10^6 \ M_\odot$. This behavior can be understood by the shape of the power spectrum density of AGN and its dependence on the black hole mass.

Keywords. galaxies: active, galaxies: nuclei, X-rays: galaxies

1. Introduction

Recent studies suggested an inverse linear correlation between black hole (BH) mass and excess variance $(\sigma^2_{\rm rms})$ in logarithmic space for active galactic nuclei (AGN) with $M_{\rm BH} = 10^6 - 10^9 \ M_\odot$ (Zhou *et al.* 2010; Ponti *et al.* 2012). This scaling relation is tight enough to provide a novel method to estimate the black hole mass of AGN. However, it is not clear if this relation can be extended to lower black hole masses. Our work is to study the relation in the low-mass regime using a sample of low-mass AGN with $M_{\rm BH} < 2 \times 10^6 \ M_\odot$ observed with *XMM-Newton* and *ROSAT*.

2. Result and discussion

We compiled a sample of 15 low-mass AGN from our sample of low-mass AGN with $M_{\rm BH} < 2 \times 10^6 \ M_\odot$ (Dong *et al.* 2012). There are 11 objects observed with *XMM-Newton* for a total of 15 observations, and 5 objects observed with *ROSAT* for 6 observations (one object observed with both satellites). The accretion rates in the Eddington units span the range $0.06 - 0.90$. We calculate the X-ray excess variances (Nandra *et al.* 1997) and normalize their values on a timescale of 80 ks, using the three power spectrum density (PSD) models in Table 1 (see Papadakis 2004 for the details about the PSD shape of AGN). The relation of the excess variance and black hole mass is plotted in Figure 1. It shows that for almost all the sources with $M_{\rm BH} < 2 \times 10^6 \ M_\odot$, the excess variances fall systematically below the extrapolation of the $M_{\rm BH} - \sigma^2_{\rm rms}$ relation derived from the high-mass sample (Ponti *et al.* 2012). In fact, the $M_{\rm BH} - \sigma^2_{\rm rms}$ relation can be derived from the dependence of the break frequency of the PSD, since excess variance is the integral of the PSD over frequency domain (van der Klis 1989, 1997). We fix the observed $\nu_{\rm br}(M_{\rm BH}, \dot{m})$ relations as their original form (McHardy *et al.* 2006; González-Martín & Vaughan 2012),

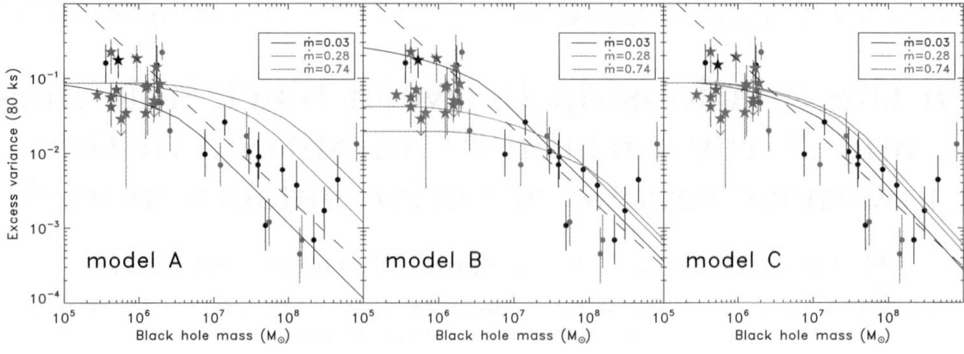

Figure 1. Relationship of the black hole mass and the excess variance for the objects in our sample (stars), together with the sample (dots) and the best-fit relations (dashed lines) from Ponti *et al.* (2012). The objects with $\dot{m}$ in the three $\dot{m}$ bins (0.01-0.1, 0.1-0.5, and 0.5-1.0) are plotted in black, red and blue, respectively. The solid lines represent the relations derived from the best-fit PSD models for three accretion rates.

Table 1. PSD models used

Model	Break Frequency Formalism	PSD Amplitude Suggested in Previous Work	PSD Amplitude Fitted in this Work	χ^2/dof
model A	$\nu_{\rm br} = 0.003\dot{m}(M_{\rm BH}/10^6\ M_\odot)^{-1}$ McHardy *et al.* (2006)	0.017 ± 0.006 Papadakis (2004)	$C_1 = 0.017 \pm 0.002$	$150/36$
model B	$\nu_{\rm br} = 0.003\dot{m}(M_{\rm BH}/10^6\ M_\odot)^{-1}$ McHardy *et al.* (2006)	$\alpha = 0.003^{+0.002}_{-0.001}$ $\beta = 0.8 \pm 0.15$ $C_1 = \alpha\dot{m}^{-\beta}$, Ponti *et al.* (2012)	$\alpha = 0.010 \pm 0.001$ $\beta = 0.40 \pm 0.04$	$106/35$
model C	$\nu_{\rm br} = 0.001\dot{m}^{0.24}(M_{\rm BH}/10^6\ M_\odot)^{-1}$ González-Martín & Vaughan (2012)	0.017 ± 0.006 Papadakis (2004)	$C_1 = 0.015 \pm 0.002$	$102/36$

and obtain the PSD amplitude by fitting the $M_{\rm BH} - \sigma^2_{\rm rms}$ relation. It can be seen that all the three models can reproduce the observed trend of the $M_{\rm BH} - \sigma^2_{\rm rms}$ relation well (Figure 1). In conclusion, the inversely proportional $M_{\rm BH} - \sigma^2_{\rm rms}$ relation established in the high-mass regime fails to extend to $M_{\rm BH}$ below $10^6\ M_\odot$. Our result is in good agreement with that obtained from a recent similar study by Ludlam *et al.* (2015). This is also consistent with the model prediction from our current understanding of the PSD of AGN and the dependence of the break frequency on $M_{\rm BH}$.

Acknowledgements

This work is supported by grants No. XDB09000000 & No. 11473035

References

Dong, X.-B., Ho, L. C., & Yuan, W., *et al.* 2012, *ApJ*, 755, 167

González-Martín, O. & Vaughan, S. 2012, *A&A*, 544, A80

Ludlam, R. M., Cackett, E. M., & Gultekin, K., *et al.* 2015, *MNRAS*, 447, 2112

McHardy, I. M., Koerding, E., Knigge, C., Uttley, P., & Fender, R. P. 2006, *Nature*, 444, 730

Nandra, K., George, I. M., Mushotzky, R. F., Turner, T. J., & Yaqoob, T. 1997, *ApJ*, 476, 70

Papadakis, I. E. 2004, *MNRAS*, 348, 207

Ponti, G., Papadakis, I., & Bianchi, S., *et al.* 2012, *A&A*, 542, A83

van der Klis, M. 1989, *Timing Neutron Stars*, 27

van der Klis, M. 1997, *Statistical Challenges in Modern Astronomy II*, 321

Zhou, X.-L., Zhang, S.-N., Wang, D.-X., & Zhu, L. 2010, *ApJ*, 710, 16

Star clusters and black holes in galaxies across cosmic time
Proceedings IAU Symposium No. 312, 2014
Y. Meiron, S. Li, F.-K. Liu & R. Spurzem, eds.

© International Astronomical Union 2016
doi:10.1017/S1743921315007541

A search of new samples of active galactic nuclei with low-mass black holes from SDSS

H. Liu[1], W. Yuan[1], H. Zhou[2,3] and X.-B. Dong[4]

[1] National Astronomical Observatory, Chinese Academy of Sciences, Beijing, 100012, China
email: `liuheyang@nao.cas.cn`

[2] Polar Research Institute of China, Jingqiao Road 451, Shanghai, 200136,China

[3] The University of Sciences and Technology of China, Hefei, Anhui, 230026, China

[4] Yunnan Observatory, Chinese Academy of Sciences, Kunming, Yunnan, 650011,China

Abstract. We report on the progress of our on-going work to search for low-mass black holes (LMBHs) in active galactic nuclei. The masses of black holes are estimated using the broad line width and luminosity obtained from one-epoch optical spectra. As the first step, we fitted the spectra of 1263 objects in the quasar catalog of the SDSS DR10 and obtained accurate measurement of the emission lines. Two AGNs are found to have $M_{\rm BH} \sim 10^6 \, M_\odot$. The next step is to analyze the spectra of the DR10 galaxy sample, from which a much larger sample of low-mass AGNs is expected to be obtained.

Keywords. galaxies:active, galaxies:nuclei, quasars:emission line

1. Introduction

As the link between stellar mass black holes and supermassive black holes, low-mass black holes (LMBHs) at the center of galaxies with masses ranging from $10^3 \, M_\odot$ to $10^6 \, M_\odot$ are important for research on black hole formation and co-evolution with galaxies. Current models indicate that low-mass black holes can give insight into the evolution of the first seed black holes (e.g. Volonteri *et al.* 2008). The common practice is to estimate the black hole masses of AGNs from the width and luminosity of the broad emission lines using the empirical scaling relation (Kaspi *et al.* 2000). A systematic search of LMBHs has been pioneered by Greene & Ho (2004, 2007) from the SDSS data finding more than 200 candidates. Recently we (Dong *et al.* 2012) carried out a systematic and homogeneous search from the SDSS DR4 data, resulting in 309 LMBH AGNs, many with low Eddington ratios. According to Yuan *et al.* (2014), a large population of low-mass black hole may exist in the local universe awaiting discovery. To enlarge the sample size, we perform an extended search for more LMBH AGNs from the SDSS DR10 data.

2. Preliminary results

It is difficult to search for AGNs with low-mass black holes since their spectra are dominated by starlight in general. Careful subtraction of the starlight and the continuum is essential for reliable measurement of the emission lines. It is also important to precisely separate the broad and narrow components of the Hα and Hβ lines as the broad lines in the spectra of LMBHs are relatively narrow and weak. To reach this goal, we have designed a set of elaborate codes and broad-line selection procedures as detailed in Dong *et al.* (2012).

As the first step, we start with objects classified as quasars with redshift $z \leqslant 0.5$ in the DR10 data, which have 1263 objects (Pâris *et al.* 2014). The spectra are analyzed

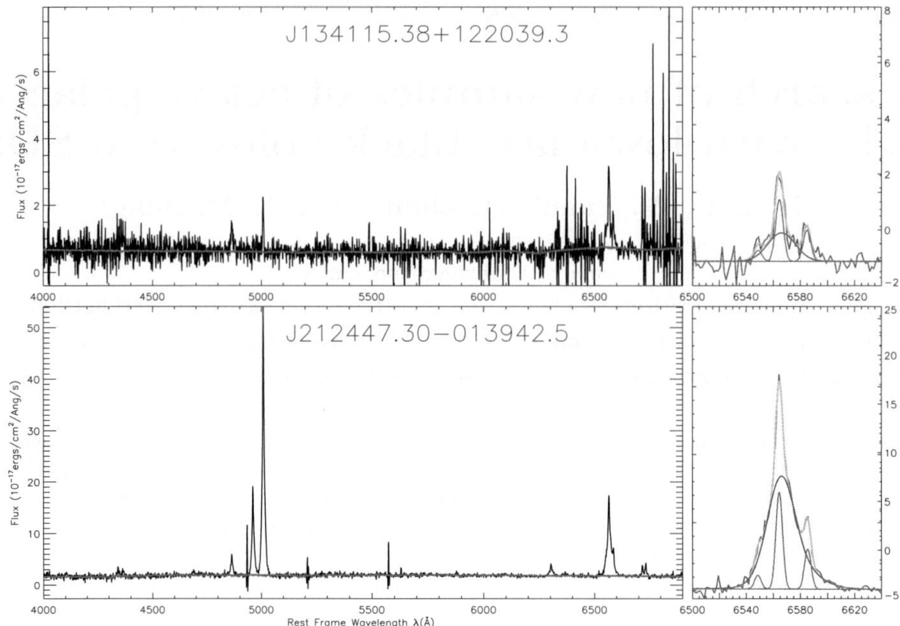

Figure 1. Emission line spectra and model fits for the spectra of two candidate AGNs with $M_{\rm BH} \sim 10^6$ M$_\odot$

Table 1. The basic parameters of LMBH candidates.

SDSS name	Redshift	FWHM(Hα) km s^{-1}	L(Hα) 10^{41} erg s^{-1}	$M_{\rm BH}$ 10^6 M$_\odot$	L_{bol}/L_{Edd}
J134115.38+122039.3	0.386	1094	1.535	1.53	0.22
J212447.30-013942.5	0.131	1180	1.245	1.63	0.17

following the procedures in Dong *et al.* (2012) and accurate measurements of the emission lines are obtained. Given the high luminosities of quasars, only two AGNs with $M_{\rm BH} \sim 10^6$ M$_\odot$ are found (Table 1), which is not surprising. Fig. 1 shows the best-fit for the emission line spectra of these two objects. Our next step is to analyze the spectra for the DR10 galaxy sample which have ~ 460000 objects with redshift $z \leqslant 0.5$. We expect that a large sample of LMBH AGNs can be obtained.

Acknowledgements

This work is supported by grants No. XDB09000000 & No. 11473035

References

Dong, X.-B., Ho, L. C., Yuan, W., *et al.* 2012, *ApJ*, 755, 167
Greene, J. E. & Ho, L. C. 2004, *ApJ*, 610, 722
Greene, J. E. & Ho, L. C. 2007, *ApJ*, 670, 692
Kaspi., S. Smith, P. S., Netzer, H., *et al.* 2000, *ApJ*, 533, 631
Pâris, I., Petitjean, P., *et al.* 2014, *A&A*, 563, A54
Volonteri, M., Lodato, G., & Natarajan, P. 2008, *MNRAS*, 383, 1079
Yuan, W., Zhou, H., Wang, L., Dou, *et al.* 2014, *ApJ*, 782, 55

Star clusters and black holes in galaxies across cosmic time
Proceedings IAU Symposium No. 312, 2014
Y. Meiron, S. Li, F.-K. Liu & R. Spurzem, eds.
© International Astronomical Union 2016
doi:10.1017/S1743921315007553

How to detect supermassive binary black holes at parsec scales

Xiang Liu

Xinjiang Astronomical Observatory of CAS, 150 Science 1-Street, Urumqi 830011, China
email: `liux@xao.ac.cn`

Abstract. It is difficult to find or identifying the binary black holes in parsec scales, since the dual AGN may be merged quickly. It is required to explore more possibilities to identifying binary black holes in parsec scales, we give some discussions, especially with the VLBI methods.

Keywords. black hole physics: binary – galaxies: jets – quasars: general – radio continuum

It is possible that supermassive binary black holes live in active galaxies through merging process. For the huge gravitational potential of massive black hole and the angular momentum losses via gas accreting and emission, the binary massive black holes will be gradually merged to form a larger black hole. It is not clear that the time scale of the evolution from far separation of two massive black holes to kpc scale separation, and the time scale from the kpc scale to pc scale separation. Statistical studies seem to suggest that the detection rate of pc-scale binary black holes (BBH) is much less than that of the kpc-scale dual AGN (Smith *et al.* 2010), this implies the inspiral of dual black holes may be faster in the pc scale than in the kpc scale (Blecha *et al.* 2013). It is difficult to find or identifying the binary black holes in the pc scale, since the dual AGN would be to merge quickly. It is required to explore more possibilities to identifying binary black holes in the pc-scales, we give some discussions in the following.

High spectral resolution to reveal the double-peaked broad lines. When supermassive binary black holes with similar masses have their own accretion disks, the two set of similar emission lines could be detectable. The double-peaked emission line AGN found by Wang *et al.* (2009), some of them were identified to show dual AGN in the kpc scale. When the dual AGN evolve into pc-scale, their narrow-line regions can be jointed, but the two broad-line regions may be not jointed yet, and their broad emission-lines could be double-peaked and identifiable. One needs higher spectral resolution and sensitivity to reveal the double-peaked broad lines. Furthermore, it may also be possible to associate the pc-scale BBH candidate with the line-of-sight radial velocity shifts of the broad lines if attributing the shift to the orbital motion of BBH, from multi-epoch spectral monitoring (Liu *et al.* 2014).

VLBI detection of double twin-jets from binary black holes at pc-scale. Very long baseline interferometry (VLBI) at radio can resolve the radio loud AGN at pc scales. Tens of thousands AGN have been imaged with the high resolution, but the binary black holes in AGN with the VLBI were not investigated systematically. Assuming radio loud AGN fraction is 10%, the fraction for both the dual AGN are radio loud will be 1% , this means for the 1% of dual AGN that we should be able to detect the double twin-jets from their binary black holes. Only one AGN has been identified with such double twin-jets/cores, PKS 0402+379 (Rodriguez *et al.* 2006), with two flat spectrum radio cores of $\sim$7 pc apart. Burke-Spolaor (2011) searched for flat spectrum radio cores from the geodetic VLBI database, found again only the PKS 0402+379. We have searched for binary black holes from the astrophysical VLBI databases, found 5 BBH candidates from

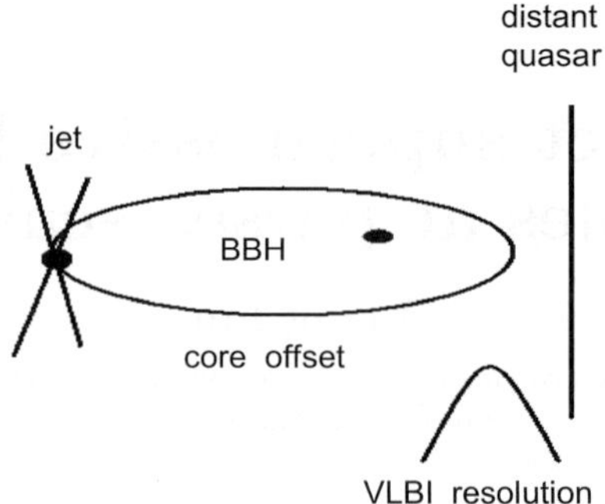

Figure 1. The schematic diagram of the phase-reference VLBI method to detect the core-offset of the radio loud core in a binary black hole system.

$\sim$2000 sources. Our strategy of searching for BBH candidates is looking for not only flat spectrum double cores, but also double twin-jets pair or their variant complex (Liu 2014). However, most of the VLBI-imaged jets we investigated are limited in spatial resolution at centimeter bands, further investigations to confirm the BBH candidates are ongoing through high and multi-frequency VLBI observations.

VLBI core-offset to be determined as due to the BBH orbital motion. For binary black hole AGN, with radio loud probability 10% of each hole, it is likely that only one BH is jetting and the other BH is radio quiet in most of dual AGN. In this case, one cannot detect twin jets from both black holes. The BL Lac object OJ287, for instance, is believed to host a BBH system for its nearly 12 years periodic optical/radio outbursts that attributing to the orbital motion of BBH. The VLBI image of OJ287 shows only one-sided jet, exhibiting large position-angle changes of inner-jets at high frequencies (Liu *et al.* 2012) and with the core-shifts between different frequencies in VLBI images as well. The core shifts are mainly caused by the opacity effect of inner jets. The position-angle changes of inner jets, however, may be caused by the precession or orbital motion of the BBH system. Therefore, with multi-epoch VLBI observations at *same* frequency, one would be able to detect the orbital motion of the radio loud BH in the BBH system if the other is radio quiet, the so-called 'core offset', with the phase-reference VLBI technique (see Fig. 1), as referenced to a distant high-redshift quasar (which has no detectable core-offset). This needs a long time to measure the core offset caused by the pc-scale orbital motion of BBH, but it is plausible to detect for the binary black holes at low redshift.

References

Blecha, L., Loeb, A., & Narayan, R. 2013, *MNRAS*, 429, 2594
Burke-Spolaor, S. 2011, *MNRAS*, 410, 2113
Liu, X., Shen, Y., Bian, F. Y., *et al.* 2014, *ApJ*, 789, 140
Liu, X. 2014, *J. Astrophys. Astr.*, Vol.35, No.3, 1
Liu, X., Mi, L.-G., Liu, B.-R., & Li, Q.-W. 2012, *Ap&SS*, 342, 465
Rodriguez, C., Taylor, G. B., Zavala, R. T., *et al.* 2006, *ApJ*, 646, 49
Smith, K. L., *et al.* 2010, *ApJ*, 716, 866
Wang, J.-M., *et al.* 2009, *ApJ (Letters)*, 705, L76

Star clusters and black holes in galaxies across cosmic time
Proceedings IAU Symposium No. 312, 2014
Y. Meiron, S. Li, F.-K. Liu & R. Spurzem, eds.
© International Astronomical Union 2016
doi:10.1017/S1743921315007565

Simulation of disc-bulge-halo galaxies using parallel GPU based codes

O. Veles[1,3], P. Berczik[2,1,3] and A. Just[3]

[1] Main Astronomical Observatory, National Academy of Sciences of Ukraine,
27 Akademika Zabolotnoho St., 03680, Kyiv, Ukraine
email: `veles@mao.kiev.ua`

[2] National Astronomical Observatories of China, Chinese Academy of Sciences,
20A Datun Rd., Chaoyang District, 100012, Beijing, China

[3] Astronomisches Rechen-Institut, Zentrum für Astronomie, University of Heidelberg,
Mönchhofstrasse 12-14, 69120, Heidelberg, Germany

Abstract. We compare the performance of the very popular Tree-GPU code BONSAI with the older Particle-(Multi)Mesh code SUPERBOX. Both code we run on a same hardware using the GPU acceleration for the force calculation. SUPERBOX is a particle-mesh code with high resolution sub-grid and a higher order NGP (nearest grid point) force-calculation scheme. In our research, we are aiming to demonstrate that the new parallel version of SUPERBOX is capable to do the high resolution simulations of the interaction of the system of disc-bulge-halo composed galaxy. We describe the improvement of performance and scalability of SUPERBOX particularly for the Kepler cluster (NVIDIA K20 GPU). A comparison was made with the very popular and publicly available Tree-GPU code BONSAI†.

Keywords. N-body galaxy simulation, parallel computing, GPGPU, supercomputer

1. Overview

- Direct gravitational N-body simulations: computational complexity $\sim N^2$
  * φGRAPE & φGPU codes (Harfst *et al.* 2007; Berczik *et al.* 2011, 2013)
- Tree methods: computational complexity $\sim N\log(N)$
  * bonsai GPU tree code (Bédorf *et al.* 2012)
- Particle mesh method: computational complexity $aN_{grid}^3\log(N_{grid}) + bN_{part}$
  * SUPERBOX code (Fellhauer *et al.* 2000)

2. Superbox code

The time-step cycle of SUPERBOX is divided into two main routines (98% of the total CPU time) (Fellhauer 2006):
1 - the FFT routine computes the potential on the grids (MPI or OpenMP parallel part)
2 - the PUSHER routine contains the force calculation, the position and velocity updating and collection of the output data (serial part) .

3. The SUPERBOX advantages

- The method is fast. The method is self-consistent.
- Due to the large number of particles statistical noise is extremely low.
- Effects of two body relaxation are negligible

† *http://castle.strw.leidenuniv.nl/software/bonsai-gpu-tree-code.html*

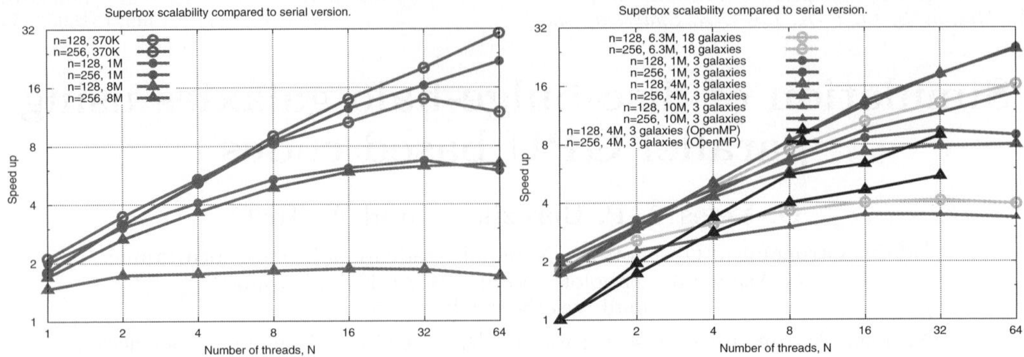

Figure 1. Scalability of parallel SUPERBOX version on Kepler cluster for single and multi galaxy models.

After optimization on Kepler cluster the parallel version of SUPERBOX show the increase of performance and good MPI scalability up to 64 CPU, for grid size = 256 and up to 20M particles.

- OpenMP version have good scalability and need less memory, but is limited by maximum CPU cores
- The particle positions update module have bad speedup with OpenMP

4. SUPERBOX vs BONSAI performance

Calculation time for one integration step show comparable performance (Bien *et al.* 2008):

- SUPERBOX (8 MPI CPU, 4M particles, grid-size=128) **3.4s**
- SUPERBOX (16-core OpenMP CPU, 4M particles, grid-size=128) **5.6s**
- BONSAI2 (2x GTX 570 GPU, 4M particles, θ =0.5) **4s**
- BONSAI2 (Tesla 20X GPU, 10M particles, θ =0.75) **1s**
- SUPERBOX (32 MPI CPU, 10.2M particles, grid-size=128) **8.3s**

Acknowledgements

OV & PB thank the International Astronomical Union for the grant support which gave the opportunity for him to participate on the IAU Symposium No. 312 in Beijing-2014.

References

Berczik, P., Nitadori, K., Zhong, S., *et al.* 2011, in *International conference on High Performance Computing*, Kyiv, Ukraine, October 8-10, 2011., p. 8-18, 8–18

Berczik, P., Spurzem, R., Wang, L., Zhong, S., & Huang, S. 2013, in *Third International Conference "High Performance Computing"*, HPC-UA 2013, p. 52-59, 52–59

J. Bédorf, E. Gaburov, & S. Portegies Zwart. *Bonsai: A GPU Tree-Code*. In R. Capuzzo-Dolcetta, M. Limongi, and A. Tornambè, editors, *Advances in Computational Astrophysics: Methods, Tools, and Outcome*, volume 453 of *Astronomical Society of the Pacific Conference Series*, page 325, July 2012.

R. Bien, A. Just, P. Berczik, & I. Berentzen. *High resolution in z-direction: The simulation of disc-bulge-halo galaxies using the particle-mesh code SUPERBOX*. *Astronomische Nachrichten*, 329:1029, December 2008.

M. Fellhauer, P. Kroupa, H. Baumgardt, R. Bien, C. M. Boily, R. Spurzem, & N. Wassmer. *SUPERBOX - an efficient code for collisionless galactic dynamics. New A*, 5:305–326, September 2000.

Fellhauer M., 2006, *Superbox manual, madf@ast.cam.ac.uk*

Harfst, S., Gualandris, A., Merritt, D., *et al.* 2007, *New. Astr.*, 12, 357

Star clusters and black holes in galaxies across cosmic time
Proceedings IAU Symposium No. 312, 2014
Y. Meiron, S. Li, F.-K. Liu & R. Spurzem, eds.
© International Astronomical Union 2016
doi:10.1017/S1743921315007577

Merging of unequal mass binary black holes in non-axisymmetric galactic nuclei

Peter Berczik[1,2,3]**, Long Wang**[4,1]**, Keigo Nitadori**[5] **and Rainer Spurzem**[1,4,3]

[1]National Astronomical Observatories of China, Chinese Academy of Sciences, 20A Datun Rd., Chaoyang District, Beijing 100012, China
email: berczik@nao.cas.cn

[2]Main Astronomical Observatory, National Academy of Sciences of Ukraine, 27 Akademika Zabolotnoho St., 03680, Kyiv, Ukraine

[3]Astronomisches Rechen-Institut, Zentrum für Astronomie, University of Heidelberg, Mönchhofstrasse 12-14, 69120, Heidelberg, Germany

[4]Kavli Institute for Astronomy and Astrophysics, Peking University, Beijing 100871, China

[5]RIKEN Advanced Sciennce Institute, Wako, Japan

Abstract. In this work we study the stellar-dynamical hardening of unequal mass massive black hole (MBH) binaries in the central regions of galactic nuclei. We present a comprehensive set of direct N-body simulations of the problem, varying both the total mass and the mass ratio of the MBH binary. Our initial model starts as an axisymmetric, rotating galactic nucleus, to describe the situation right after the galaxies have merged, but the black holes are still unbound to each other. We confirm that results presented in earlier works (Berczik *et al.* 2006; Khan *et al.* 2013; Wang *et al.* 2014) about the solution of the "last parsec problem" (sufficiently fast black hole coalescence for black hole growth in cosmological context) are robust for both for the case of unequal black hole masses and large particle numbers. The MBH binary hardening rate depends on the reduced mass ratio through a single parameter function, which quantitatively quite well agrees with standard 3 body scattering theory (see e.g., Hills 1983). Based on our results we conclude that MBH binaries at high redshifts are expected to merge with a factor of ~ 2 more efficiently, which is important to determine the possible overall gravitational wave signals. However, we have not yet fully covered all the possible parameter space, in particular with respect to the preceding of the galaxy mergers, which may lead to a wider variety of initial models, such as initially more oblate and / or even significantly triaxial galactic nuclei. Our N-body simulations were carried out on a new special supercomputers using the hardware acceleration with graphic processing units (GPUs).

Keywords. black holes, binary black holes, galactic nuclei, stellar dynamics

---

1. Introduction

Supermassive black holes (MBHs) are commonly observed at the centers of most galaxies, and their mass correlates to parameters of the host galaxy, such as its mass or velocity dispersion (see e.g., Gültekin *et al.* 2009). So, the presence of MBH binaries in galactic centers is an inevitable evolutionary phase during the merger history of galaxies, according to the popular hierarchical structure formation scenario in the Λ Cold Dark Matter (ΛCDM) paradigm. Binary black holes in galactic nuclei have been searched for by looking for special patterns in radio lobes or jets or quasi-periodic electromagnetic emission in light curves (Komossa 2006).

Originally it was expected that binary black holes would stall at some critical separation (called the "final parsec problem"), which is not small enough for gravitational

Table 1. Full set of our model runs.

$\frac{m_2}{m_1} \times 10^{-2}$	$\mu \times 10^{-2}$	025k	050k	100k	250k	400k	1M
0.01/1	0.0099	—	—	450	—	—	—
0.02/1	0.0196	—	—	450	—	—	—
0.03/1	0.0291	—	—	350	—	—	—
0.05/1	0.0476	—	—	350	—	—	—
0.10/1	0.0909	250	250	250*	250	250	150
0.20/1	0.1666	250	250	250*	250	250	150
0.50/1	0.3333	250	250	250*	250	250	150
1.00/1	0.5000	250	250	250*	250	250	150
0.02/2	0.0198	—	—	450	—	—	—
0.03/2	0.0295	—	—	450	—	—	—
0.05/2	0.0488	—	—	350	—	—	—
0.10/2	0.0952	—	—	350	—	—	—
0.20/2	0.1818	—	—	250	—	—	—
0.40/2	0.3333	—	—	250	—	—	—
1.00/2	0.6666	250	250	250	250	250	150
2.00/2	1.0000	250	250	250	250	250	150
0.04/4	0.0396	—	—	450	—	—	—
0.08/4	0.0784	—	—	450	—	—	—
0.10/4	0.0976	—	—	350	—	—	—
0.20/4	0.1905	—	—	350	—	—	—
0.40/4	0.3636	—	—	250	—	—	—
0.80/4	0.6666	—	—	250	—	—	—
2.00/4	1.3333	250	250	250	250	250	150
4.00/4	2.0000	250	250	250	250	250	150

Notes: First column: MBH masses in model units. Columns 3 – 8: Particle number in the stellar galactic nucleus. The first column gives the MBH masses and the second column the MBH reduced mass μ.

radiation induced orbit shrinking and coalescence to set in Begelman *et al.* (1980); Makino *et al.* (1993); Makino (1997); Hemsendorf *et al.* (2002); Berczik *et al.* (2005); Merritt (2006). However, there is a growing evidence that in the reality of galaxy mergers, where galaxies are axisymmetric or even triaxial, there is no problem to refill the loss-cone and the binary black hole hardening rate becomes independent of the particle number used in the simulation (Yu 2002; Merritt & Poon 2004; Berczik *et al.* 2006). This process will overcome the final parsec problem (Preto *et al.* 2009; Khan *et al.* 2011, 2013) and it seems that the previously studied spherically symmetric galactic nuclei actually were some worst case scenario (Preto *et al.* 2011). Berentzen *et al.* (2009) have demonstrated that the new more realistic models make it possible to bring an MBH binary to final relativistic coalescence in a short time scale.

2. Numerical model and initial conditions

The simulations presented in this work have been carried out using the direct N-body code φ-GPU (Berczik *et al.* 2011, 2013). The basic concepts of the code are based on the publicly available φ-GRAPE (Harfst *et al.* 2007) code, an Aarseth N-body 1-like code including an efficient MPI parallelization and support for the special-purpose hardware GRAPE (Fukushige *et al.* 1991) (currently Graphic Processing Units - GPU). The current version of the φ-GPU code has been fully rewritten in the C++ programming language for the most effective use of GPUs. It supports the use Hermite integration algorithms of 4th, 6th or 8th) order. Furthermore, it includes a hierarchical block

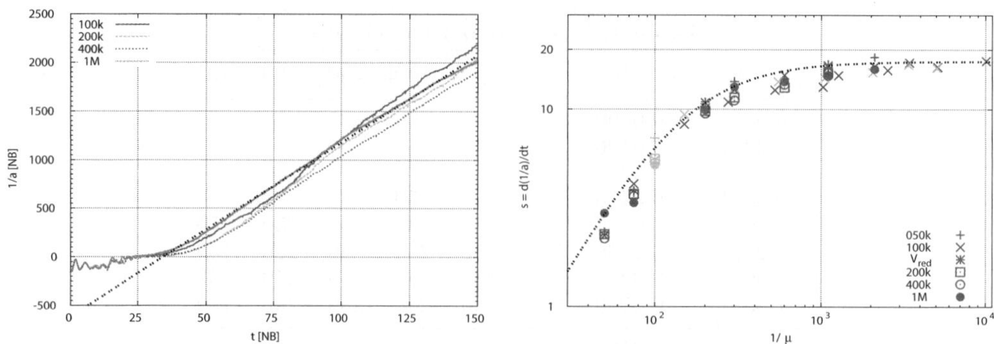

Figure 1. Evolution of the MBH binary hardening rate. The left panel shows the method of hardening determination based on the particular set of the runs. The right panel shows the hardening rate dependency from the reduced mass of the MBH binary system.

time-step scheme. Compared to the earlier φ-GRAPE (Harfst *et al.* 2007) implementation we obtain an additional speed-up a factor of $\times 2$ on the recent generation of NVIDIA GPUs (i.e., M2050 and K20 cards), depending on the specific number of particles and computing nodes used. Further details on this high-performance computing code will be described in papers (Berczik *et al.* 2011, 2013). The present code is well tested and already used to obtain important results in our earlier large scale few million body simulation (Khan *et al.* 2012).

We scale the numerical units of our initial models applying the standard N-body normalization (Aarseth *et al.* 1974) by setting both the gravitational constant G and the total mass of the stellar nucleus system to unity. The total energy of the system is set to $E = -\frac{1}{4}$.

The initial conditions of our N-body simulations presented here are based on the ones used in Berczik *et al.* (2006) and Berentzen *et al.* (2009). The initial stellar galactic nucleus follows a distribution function for a rotating King model (see, e.g., Einsel & Spurzem 1999, and references therein). The concentration and rotation parameter are set to $W_0 = 6$ and to $\omega_0 = 1.8$, respectively, in all the models. The total angular momentum vector of both the stellar nucleus and the MBHs are aligned with the z-axis of our coordinate frame. We place the two MBHs in the $z = 0$ mid-plane with initial coordinate components $x_{1,2} = 0$ and $y_{1,2} = \pm 0.3$, where the sub-scripts denotes the two black hole particles.

The MBH masses and particle numbers with which the stellar system has been realized are given in Table 1. The initial x-velocity of the MBHs in these simulations has been chosen to be $v_{x;1,2} = \pm V_{\mathrm{circ}}$, where V_{circ} is the circular velocity within the stellar background model. With our choice of initial parameter the circular velocity, in N-body units, is $V_{\mathrm{circ}} = 0.7$ at the initial distance. In Table 1., we also show with mark ($*$) the extra simulations which have been run, but this time with and initial MBH velocity of $v_{x;1,2} = \pm 0.1 \times V_{\mathrm{circ}} = 0.07$. We choose this value to have a faster pairing of the two MBH due to a more eccentric initial orbit.

3. Results and discussion

An example of the typical binary evolution in our simulations is shown in Fig. 1 (left panel) in terms of the inverse binary's semi-major a as a function of time. From the slope of a linear fit to $1/a(t)$ we then determine the hardening rate $d(1/a)/dt$ of the binary.

In Fig. 1 (right panel) we show the hardening rate as a function of reduced mass of the MBH binary μ. As we can see from the figure the hardening rate for our set of models and parameters are only a single function of the μ (or actually the $1/\mu$).

Our results have strong implications on the merger rate and history of MBHs and on the MBH growth in the cosmological context. Numerical ΛCDM simulations show that the galaxy merger rate in the early universe is higher than at low redshifts z. If at high redshift we have a higher galaxy merger rate and also the less massive MBH binaries (i.e., total binary mass and also a reduced mass are both smaller) we can get a significantly higher (few times) hardening rate at early universe. These rates (as we already see in Berczik *et al.* (2006); Berentzen *et al.* (2008)) are high enough to reach the relativistic regime and merge efficiently.

PB thanks the International Astronomical Union for the grant support which give him the opportunity to participate on the IAU Symposium No. 312 in Beijing-2014.

References

Aarseth, S. J., Henon, M., & Wielen, R. 1974, *Astron. and Astroph.*, 37, 183

Begelman, M. C., Blandford, R. D., & Rees, M. J. 1980, *Nature*, 287, 307

Berczik, P., Merritt, D., & Spurzem, R. 2005, *ApJ*, 633, 680

Berczik, P., Merritt, D., Spurzem, R., & Bischof, H.-P. 2006, *ApJL*, 642, L21

Berczik, P., Spurzem, R., Wang, L., Zhong, S., & Huang, S. 2013, in *Third International Conference "High Performance Computing"*, HPC-UA 2013, p. 52-59, 52–59

Berczik, P., Nitadori, K., Zhong, S., *et al.* 2011, in *International conference on High Performance Computing*, Kyiv, Ukraine, October 8-10, 2011., p. 8-18, 8–18

Berentzen, I., Preto, M., Berczik, P., Merritt, D., & Spurzem, R. 2008, *AN*, 329, 904

—. 2009, *ApJ*, 695, 455

Einsel, C. & Spurzem, R. 1999, *MNRAS*, 302, 81

Fukushige, T., Ito, T., Makino, J., *et al.* 1991, *Publ. Aston. Soc. Jpn.*, 43, 841

Gültekin, K., Richstone, D. O., Gebhardt, K., *et al.* 2009, *ApJ*, 698, 198

Harfst, S., Gualandris, A., Merritt, D., *et al.* 2007, *N. Astr.*, 12, 357

Hemsendorf, M., Sigurdsson, S., & Spurzem, R. 2002, *ApJ*, 581, 1256

Hills, J. G. 1983, *Astron. J.*, 88, 1269

Khan, F. M., Berentzen, I., Berczik, P., *et al.* 2012, *ApJ*, 756, 30

Khan, F. M., Holley-Bockelmann, K., Berczik, P., & Just, A. 2013, *ApJ*, 773, 100

Khan, F. M., Just, A., & Merritt, D. 2011, *ApJ*, 732, 89

Komossa, S. 2006, *Mem. Soc. Astron. Italiana*, 77, 733

Makino, J. 1997, *ApJ*, 478, 58

Makino, J., Fukushige, T., Okumura, S. K., & Ebisuzaki, T. 1993, *Publ. Aston. Soc. Jpn.*, 45, 303

Merritt, D. 2006, *Reports on Progress in Physics*, 69, 2513

Merritt, D. & Poon, M. Y. 2004, *ApJ*, 606, 788

Preto, M., Berentzen, I., Berczik, P., Merritt, D., & Spurzem, R. 2009, *Journal of Physics Conference Series*, 154, 012049

Preto, M., Berentzen, I., Berczik, P., & Spurzem, R. 2011, *ApJL*, 732, L26

Wang, L., Berczik, P., Spurzem, R., & Kouwenhoven, M. B. N. 2014, *ApJ*, 780, 164

Yu, Q. 2002, *MNRAS*, 331, 935

Star clusters and black holes in galaxies across cosmic time
Proceedings IAU Symposium No. 312, 2014
Y. Meiron, S. Li, F.-K. Liu & R. Spurzem, eds.

© International Astronomical Union 2016
doi:10.1017/S1743921315007589

Photometric and spectroscopic study of cD galaxies

S. N. Kemp[1], Ernesto Pérez-Hernández[1] and Víctor Hugo Ramírez-Siordia[1,2]

[1]Instituto de Astronomía y Meteorología, Universidad de Guadalajara, Av. Vallarta 2602, Col. Arcos Vallarta, 44130, Guadalajara, Jalisco, México
email: snk@astro.iam.udg.mx

[2]Centro de Radioastronomía y Astrofísica, Universidad Nacional Autónoma de México, Campus Morelia, Apartado Postal 3-72, 58090, Morelia, Michoacán, México

Abstract. We have carried out photometry and spectroscopy on a sample of 10 cD galaxies. The photometry shows, in general, fairly flat and red profile colours, implying an envelope with the same stellar population as the central galaxy. This may indicate a possible primordial origin for both structures, consistent with ideas of downsizing. Preliminary spectroscopic results are generally in agreement with the photometry, with for example younger populations at large radii for A2199, but A2589 has only younger populations.

Keywords. galaxies: evolution, galaxies: formation, galaxies: stellar content

---

cD galaxies are supergiant elliptical galaxies found usually in the centres of rich clusters. They have an extended halo-like component (envelope) and an underlying de Vaucouleurs-Sérsic elliptical galaxy-like component. The envelope can reach radial distances of < 500 kpc (Oemler 1976, Schombert 1988). There have been many theories to explain the formation of these envelopes. These include tidal stripping (Gallagher & Ostriker 1972), where material is stripped from neighbouring galaxies in the cluster, falling into the gravitational potential of the cD; mergers where the envelope is built up hierarchically by successive major and minor mergers with cluster galaxies; primordial origin (Merritt 1984): where the envelope is formed simultaneously with the rest of the elliptical galaxy (which appears to be related to recent theories of "downsizing" and early formation of the largest galaxies); and cooling flows (Fabian *et al.* 1982): in clusters with X-ray emission from hot gas there is frequently a minimum in T in the centre which can be interpreted as a flow of cooling gas towards the central regions, and if the gas cools sufficiently it could form stars, and hence the envelope populations.

The colours of the stars in the envelopes will be affected by their process of formation and subsequent evolution. We carried out a programme of deep surface photometry on a sample of 10 cD envelopes using the 2.1m and 1.5m telescopes at San Pedro Mártir, Baja California, México, obtaining 30-90 min total exposure per filter. For the majority of the galaxies we obtained very flat colour profiles within the errors, varying by no more than +/- 0.1 mag in $B - V$ (Kemp *et al.* 2009, Guzmán Jiménez 2006, Ramírez-Siordia 2009). The profiles of A193, A496, A2634, A2589, A2670 do not show appreciable gradients. About 30% of the galaxies do have gradients, generally with bluer centres, implying recent star formation activity, possibly as a result of mergers (A2199) or cooling flows (A426). The flat profiles imply similar stellar populations in the underlying galaxy and the envelope, maybe formed at the same time. This could favour the primordial origin or downsizing hypothesis, and implies that any subsequent evolution that the galaxy experiences does not usually dominate its global colours.

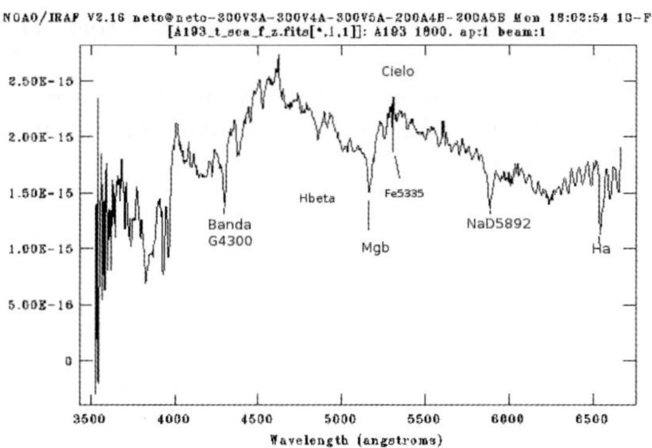

Figure 1. Spectrum of the centre of the cD galaxy in Abell 193

Variations in age and metallicity of the stellar populations with position could produce similar colours. We are carrying out a programme of medium-resolution spectroscopy of cDs at the 2.1m in San Pedro Mártir and with OSIRIS/GTC to obtain age and metallicity of stellar populations. We obtained spectra of the central regions of each galaxy and various radial positions along the slit up to r_e (Pérez-Hernández 2014). We compare the central spectra of A193, A2199 (both with multiple nuclei possibly indicating mergers) and A496, A2589 (both no evidence of multiple nuclei or mergers). A193 (Fig 1) and A496 both have populations of 10–12 Gyr and metallicities near solar, compatible with their flat, red photometric colour profiles, while A2589 has young populations 2–6 Gyr in all positions with metallicity of 0.3–0.4, in spite of also having a flat, red, colour profile. A2199 has younger ages at larger radii corresponding to the bluer colours seen in the photometry, and slightly sub-solar metallicities. There seem to be no major differences between galaxies with or without multiple nuclei, other factors have greater effects.

We are also carrying out analysis in 2 dimensions of galaxy profiles using GALFIT to fit components to the underlying galaxy and envelopes. The main result so far is that the extended halo component frequently has a Sérsic index of close to 1, i.e. an exponential fall off with radius.

Acknowledgements

We wish to thank CONACyT for funding the Basic Science project 'Stellar populations in early-type galaxies' which included grants for VHRS and EPH. SNK wishes to thank the PROSNI programme of the University of Guadalajara for assistance in attending this conference.

References

Fabian, A. C., Nulsen, P. E. J.., & Canizares, C. R. 1982, *MNRAS*, 201, 933
Gallagher, J. S. & Ostriker, J. P. 1972, *AJ*, 77, 288
Guzmán Jiménez, V. 2006, *Estudio fotométrico de los colores de las envolventes de una muestra de galaxias cD*, Masters thesis, University of Guadalajara
Kemp, S. N., Guzmán Jiménez, V., & Ramírez Beraud, P., *et al.* 2009, *New Quests in Stellar Astrophysics II: Ultraviolet Properties of Evolved Stellar Populations*, Springer, 91
Merritt, D. 1984, *ApJ*, 276, 26
Oemler, A. Jr. 1976, *ApJ*, 209, 693

Pérez-Hernández, E. 2014, *Espectroscopía de Halos de Galaxias tipo cD's*, Undergraduate thesis, University of Guadalajara

Ramírez-Siordia, V. H. 2009, *Poblaciones Estelares según los colores ópticos en Galaxias Gigantes Elípticas tipo cD*, Undergraduate thesis, University of Guadalajara

Schombert, J. M. 1988, *ApJ*, 328, 475

Star clusters and black holes in galaxies across cosmic time
Proceedings IAU Symposium No. 312, 2014
Y. Meiron, S. Li, F.-K. Liu & R. Spurzem, eds.

© International Astronomical Union 2016
doi:10.1017/S1743921315007590

The effect of binary stars on photometric redshift for galaxies at $z \sim 2.0$

Yu Zhang and Jinzhong Liu

Xinjiang Astronomical Observatory, Chinese Academy of Sciences, 150 Science 1-Street,
Urumqi, Xinjiang 830011, China
email: zhy@xao.ac.cn

Abstract. Evolutionary population synthesis (EPS) models play an important role in many studies on the formation and evolution of galaxy. The poorly calibration for some stellar evolution stages in EPS models can lead to uncertainty of the parameter determinations for galaxies. By means of the HyperZ code and a set of theoretical galaxy template, which are built on the EPS models with and without binary interactions, we present photometric redshift (photo-z) estimates for passive galaxy sample at redshift $z \sim 2.0$. The passive galaxy sample is selected from Fang *et al.*, and they also provide the redshift for these passive galaxies. By comparing the photo-z determined from EPS models with and without binary interactions, we find that the binary interactions have little effect in the photo-z determinations.

Keywords. galaxies: stellar content, galaxies: high-redshift, binaries: general.

1. Introduction

The binary stars are very common in nearby star clusters and galaxies. And binary interactions play an important role in evolutionary populations synthesis (EPS) models, they can make the overall shape of the spectral energy distribution (SED) of population to vary, especially, in ultraviolet (UV) passbands they can make the SED of population bluer by $2 - 3$ magnitude at $t \sim 1$ Gyr. At $z \sim 2.0$, the rest-frame UV emission is shifted to the optical wavelengths, which leads to that the effect of binary interactions is obvious for the observed-frame optical passbands. If the SEDs of galaxies at $z \sim 2.0$ in the observed-frame optical passbands are affected by the binary interactions (Zhang *et al.* 2012, 2013), their photo-z can also be affected. In this study we present the photo-z estimates for passive galaxies at $z \sim 2.0$ based on the HyperZ code and the EPS models with and without binary interactions, and investigate the effect of binary interactions on photo-z determinations.

2. The galaxy sample and method

We adopt the passive galaxies catalogue from Fang *et al.* (2012), which uses the color criteria $g - z$ and $z - K$ to select the passive galaxies from the All-wavelength Extended Groth Strip International Survey (AEGIS). The AEGIS is covers all major wavebands from X-ray to radio and optical spectroscopy over a large area of sky. The photometric data used in this work covers u, g, r, i, z, J, and K bands. Because the spectroscopic redshift data for passive galaxies at high redshift are very rare, in order to check the accuracy of the photometric redshift we derived, we compare our photo-z with that obtained by Fang *et al.* (2012).

In our work, the photo-z is obtained by the HyperZ code of Bolzonella *et al.* (2000), which is the first publicly available photo$-z$ code and has been widely used for photo-z

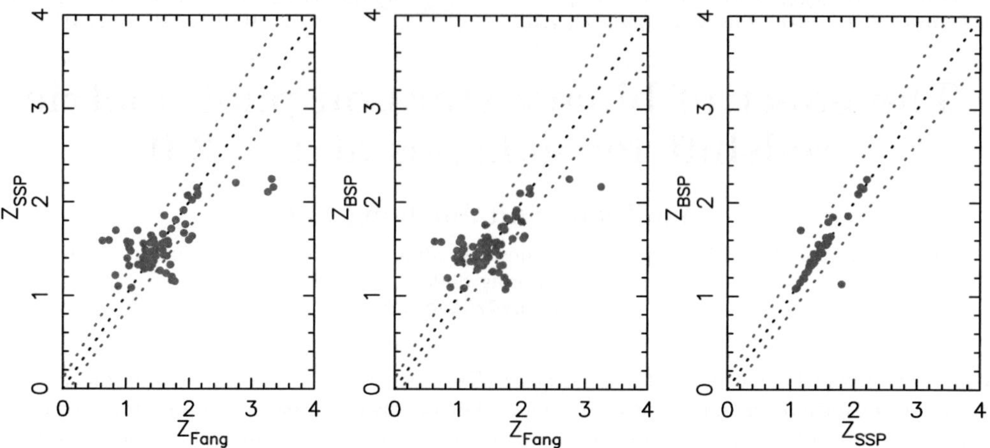

Figure 1. The left and middle panels show the comparisons between $Z_{\rm Fang}$ and $Z_{\rm phot}$ for SSP and BSP models. The right panel give the comparison for $Z_{\rm phot}$ estimated from SSP and BSP models. The blue dotted lines are for $(Z_{\rm SSP/BSP}-Z_{\rm Fang})/(1+Z_{\rm Fang})< 0.1$

estimations of galaxies. HyperZ is based on the standard SED fitting technique and on the detection of strong spectral features, such as 4000Å break, Balmer break, Lyman decrement or strong emission lines.

To check the binary interactions on the determination of photo-z for galaxies, we built theoretical template SEDs using the EPS models with (BSP model, Zhang $et\ al.$ 2005) and without (SSP model, Zhang $et\ al.$ 2004) binary interactions. Both of these models present the SEDs of populations with and without binary interactions. The galaxy template should not only comprise the SEDs of different ages but also include the star formation history. In this work, we built eight types of galaxies: the Burst galaxies with a delta burst SFR, the Irrs with a constant SFR, six types with the following exponentially declining SFR with character time decays, $\psi(t) \propto \exp(-t/\tau)$, where τ and t are the $e-$folding time scale and the age of population, respectively.

3. Result and disscussion

In Fig.1, we display the comparisons between photo-z obtained by Fang $et\ al.$ (2012, $Z_{\rm Fang}$) and photo-z for SSP ($Z_{\rm SSP}$) and BSP ($Z_{\rm BSP}$) models. From these comparisons, we find that the photo-z obtained from SSP and BSP models are in good agreement with $Z_{\rm Fang}$. In order to investigate the effect of binary interactions on the photo-z estimates, we also compare the $Z_{\rm SSP}$ and $Z_{\rm BSP}$ in the right panel of Fig.1. From the comparison, we see that the photo-z estimates from two models are almost the same, this verify that the binary interactions have no effect on the photo-z determination. Theoretically, the binary interactions can affect the photo-z determination, and we will do further study on this problem.

Acknowledgements

This work is supported by the program of the light in China's Western Region (LCWR) (No. XBBS201221) and Natural Science Foundation (No. 11303080).

References

Bolzonella, M., Miralles, J., & Pelló, R. 2000, $A\&A$, 363, 476

Fang, G., Kong, X., Chen, Y., & Lin X. 2012, *ApJ*, 751, 109
Zhang, F., Han, Z., Li, L., & Hurley, J. 2004, *A&A*, 415, 117
Zhang, F., Han, Z., Li, L., & Hurley, J. 2005, *MNRAS*, 357, 1088
Zhang, Y., Han, Z., Liu, J., Zhang, F., & Kang, X. 2012, *MNRAS*, 421, 1678
Zhang, Y., Liu, J., Zhang, F., & Han, Z. 2013, *A&A*, 554, 136

Star clusters and black holes in galaxies across cosmic time
Proceedings IAU Symposium No. 312, 2014
Y. Meiron, S. Li, F.-K. Liu & R. Spurzem, eds.
© International Astronomical Union 2016
doi:10.1017/S1743921315007607

Evolution of binary supermassive black holes and the final-parsec problem

Eugene Vasiliev

Lebedev Physical Institute, Moscow, Russia
email: eugvas@lpi.ru

Abstract. I review the evolution of binary supermassive black holes and focus on the stellar-dynamical mechanisms that may help to overcome the final-parsec problem – the possible stalling of the binary at a separation much larger than is required for an efficient gravitational wave emission. Recent N-body simulations have suggested that a departure from spherical symmetry in the nucleus of the galaxy may keep the rate of interaction of stars with the binary at a high enough level so that the binary continues to shrink rather rapidly. However, a major problem of all these simulations is that they do not probe the regime where collisionless effects are dominant – in other words, the number of particles in the simulation is still not sufficient to reach the asymptotic behavior of the system. I present a novel Monte Carlo method for simulating both collisional and collisionless evolution of non-spherical stellar systems, and apply it for the problem of binary supermassive black hole evolution. I show that in triaxial galaxies the final-parsec problem is largely non-existent, while in the axisymmetric case it seems to still exist in the limit of purely collisionless regime relevant for real galaxies, but disappears in the N-body simulations where the feasible values of N are still too low to get rid of collisional effects.

Keywords. galaxies: nuclei, galaxies: kinematics and dynamics, methods: numerical

1. Introduction

Most galaxies are believed to host central massive black holes (MBH) (Ferrarese & Ford 2005; Kormendy & Ho 2013). In the hierarchical merger paradigm, galaxies in the Universe typically experience several major and multiple minor mergers in their lifetime, and every such merger results in the eventual formation of MBH binary. The evolutionary stages of MBH binary were outlined in Begelman *et al.* (1980):

• The merger of two galaxies creates a common nucleus; both MBH sink to the center of the merger remnant due to dynamical friction and form a binary (distance $a \sim 10$ pc, all estimates are given for a $10^8 \, M_\odot$ black hole). The two black holes are considered a binary when they start to "feel" each other (their mutual gravitational force is larger than the force from the rest of the galaxy).

• Three-body interactions between the MBH binary and the stars of galactic nucleus eject stars from the vicinity of the binary, and form a depleted zone, or "mass deficit" (Milosavljević *et al.* 2002), in the nucleus (which indeed has been identified in several galaxies, see e.g. Dullo & Graham 2014). The initial stage of this process is very rapid and leads to a decrease of the binary's semi-major axis a by a factor of few to ten ($a \sim 1$ pc).

• Once the binary becomes hard and expels almost all stars that ever come close to the binary, its subsequent shrinking could slow down considerably, and could even last longer than the Hubble time – a situation that has been called the "final-parsec problem" (Milosavljevic & Merritt 2003a).

• If, despite all troubles, the binary shrinks to a separation $\lesssim 10^{-2}$ pc, the emission of gravitational waves (GW) becomes the dominant mechanism that extracts the energy from the binary, leading to its rapid coalescence ($a \sim 10^{-5}$ pc).

A star passing at a distance $\lesssim 2a$ from the binary will experience a complex 3-body interaction which results in ejection of the star with a larger velocity than it had before the interaction (the slingshot effect). These stars carry away energy and angular momentum from the binary, so that its semi-major axis a decreases: $\frac{d}{dt}(1/a) \approx 16G\,\rho/\sigma \equiv H_{\rm full}$ (Quinlan 1996). Thus, if density of field stars ρ remains constant (their velocity dispersion σ usually does), the binary hardens with a constant rate. However, the reservoir of low angular momentum stars (so-called loss cone – stars that have small enough pericenter distances and can be ejected) is finite and is depleted rather quickly. The subsequent evolution depends on the rate of the loss-cone repopulation.

2. Loss-cone theory in spherical and non-spherical systems

The classical loss-cone theory was developed in application to the problem of star capture by a single MBH (e.g. Frank & Rees 1976; Lightman & Shapiro 1977, see also a recent review in Merritt 2013). In the case of a binary MBH, the loss-cone boundary is defined by the radius of binary's orbit rather than the radius of capture of tidal disruption (and therefore is much larger). In terms of angular momentum L of a star, the boundary corresponds to $L_{\rm LC} \equiv \sqrt{2\kappa\,G\,M_{\rm bin}\,a}$, with $\kappa \simeq 1$; most stars entering the loss cone are on highly eccentric orbits ($L_{\rm LC} \ll L_{\rm circ}$, where the latter quantity is the angular momentum of a circular orbit with the same energy). The stars inside the loss cone are eliminated within one orbital period $T_{\rm orb}$. The crucial parameter for the evolution is the timescale for repopulation of the loss cone. In the absence of other processes, the repopulation time is a small fraction of the two-body relaxation time:

$$T_{\rm rep} \sim T_{\rm rel}\,\frac{L_{\rm LC}^2}{L_{\rm circ}^2}, \quad \text{where} \quad T_{\rm rel} = \frac{0.34\,\sigma^3}{G^2\,m_\star\,\rho_\star\,\ln\Lambda}. \tag{2.1}$$

If $T_{\rm rep} \lesssim T_{\rm orb}$, the loss cone is full (refilled faster than orbital period). In real galaxies, however, the opposite regime applies – the empty loss cone. In this case the hardening rate is much lower: $H \equiv \frac{d}{dt}(a^{-1}) \simeq H_{\rm full}\,T_{\rm orb}/T_{\rm rep}$. Relaxation is too slow for an efficient repopulation of the loss cone: in the absence of other processes the binary would not merge in a Hubble time.

There are several possible ways to enhance the loss-cone repopulation and hardening rate of the binary:

- *Brownian motion* of the binary enables its interaction with a larger number of stars (Milosavljević & Merritt 2001; Chatterjee *et al.* 2003; Milosavljević & Merritt 2003b).
- *Time-dependent solution* of the Fokker-Planck equation describing the diffusion of stars in angular momentum generally yields a higher loss-cone repopulation rate at the initial stage of evolution due to steeper gradients in the phase space distribution of stars (Milosavljević & Merritt 2003b).
- *Secondary slingshot* allows stars scattered by the binary with velocities lower than escape velocity to return to the galaxy center and interact with the binary more than once (Milosavljević & Merritt 2003b), although this effect seems to be suppressed in non-spherical systems due to non-radial trajectories of ejected stars (Vasiliev *et al.* 2014).
- *Gas* in the circumbinary accretion disk may increase the dissipation of binary's binding energy, however, its effect may be rather difficult to model and it is not guaranteed to solve the hardening problem (Lodato *et al.* 2009; Roškar *et al.* 2014); moreover, this factor does not apply for dry mergers (gas-poor galaxies) which are the subject of this study.
- *Perturbations* from a possible third black hole (e.g. Iwasawa *et al.* 2006; Amaro-Seoane *et al.* 2010) may increase the binary eccentricity due to the Kozai-Lidov

mechanism, thereby driving it into GW-dominated regime sooner; other massive objects such as giant molecular clouds (Perets & Alexander 2008) may substantially increase the two-body relaxation rate and efficiently scatter stars into the loss cone.

• *Non-spherical* shape of the merger remnant gives rise to torques that change stars' angular momenta and have been proposed as a possible mechanism that keeps the loss cone full (Yu 2002; Merritt & Poon 2004; Holley-Bockelmann & Sigurdsson 2006).

Let us explain the last point in more detail. If the distribution of stars and the gravitational potential that they generate is not exactly spherical, then the angular momentum L of any star is not conserved, but experiences oscillations due to torques from the non-spherical potential. The period of these oscillations is a few times T_{orb}, and their amplitude depends on the degree of flattening, as well as on the average value of L (stars with $L \ll L_{\mathrm{circ}}$ tend to have larger variations in L). Therefore, due to non-spherical torques, a much larger number of stars can attain low values of L and enter the loss cone at some point in their (collisionless) evolution, regardless of two-body relaxation. In the axisymmetric potential, one component of L parallel to the symmetry axis (L_z) is conserved, therefore the total angular momentum cannot become smaller than L_z; still, there are many more stars with $L_z < L_{\mathrm{LC}}$ than stars in the loss cone (with the instantaneous value of $L < L_{\mathrm{LC}}$), and a large fraction of them can eventually attain $L < L_{\mathrm{LC}}$. In the triaxial case, there is an entire population of centrophilic orbits such as pyramids and chaotic orbits (e.g. Poon & Merritt 2001, Figure 1g,a), that can achieve arbitrary low values of L – all of them could eventually enter the loss cone, although not all manage to do so in a Hubble time. These considerations have led to a conclusion that the loss cone should remain full in axisymmetric and especially triaxial systems.

3. *N*-body simulations and their limitations

Simulations of binary MBH evolution initially focused on the spherical case, starting the evolution from a near-equilibrium stellar system with two embedded black holes before or just after the formation of the binary. At present, it seems impossible to evolve an N-body system with a number of particles comparable to that of a real galactic nucleus ($N_\star \gtrsim 10^8$), therefore all studies to date used a much smaller number of particles $N \lesssim 10^6$ and extrapolated the results to a real galaxy. From the analytical estimates outlined above, it is clear that the evolution of the binary MBH in a real galaxy of spherical shape should occur in an empty-loss-cone regime. In this case, we expect the hardening rate of the binary (proportional to the repopulation rate of the loss cone) to scale with the number of particles in the simulation as $T_{\mathrm{rel}}^{-1} \propto N^{-1}$. However, it is worth noting that the empty-loss-cone regime is not so easy to achieve – N should be large enough so that even a very small fraction of the relaxation time is still much longer than the dynamical time (see Harfst *et al.* 2007, Figure 15, for quantitative estimates of the minimum required N). In the early studies (Quinlan & Hernquist 1997; Milosavljević & Merritt 2001), this was not possible and the hardening rate was almost independent of N – a signature of the full-loss-cone regime. Later simulations with a larger N have confirmed that the hardening rate drops with N if it is large enough (Makino & Funato 2004; Berczik *et al.* 2005; Merritt *et al.* 2007). At the same time, studies that considered a flattened model for the nucleus (Berczik *et al.* 2006) or a merger simulation that resulted in a non-spherical remnant (Preto *et al.* 2011; Khan *et al.* 2011) indicated no dependence of the hardening rate on N, which was found to be rather close to the full-loss-cone hardening rate. It was inferred that the departure from spherical symmetry was strong enough that the enhanced rate of angular momentum mixing sustains a full loss cone.

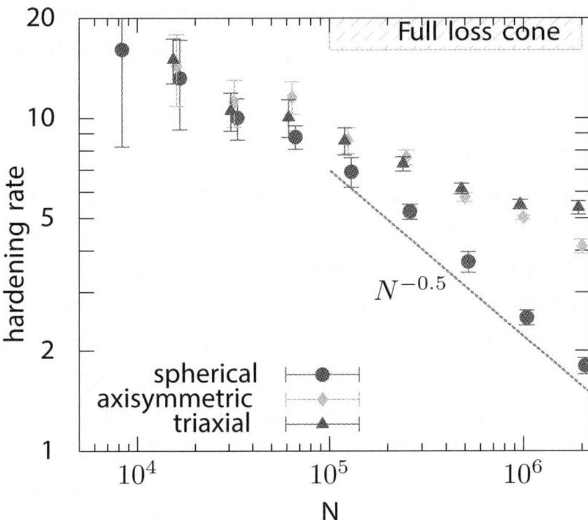

Figure 1. Binary hardening rates $d(a^{-1})/dt$ as functions of the number of particles N in simulations done in three geometries: spherical (red circles), axisymmetric (green squares) and triaxial (blue triangles). This figure is an updated version of the one presented in Vasiliev *et al.* (2014), which in additions includes the results of simulations with $N = 2 \times 10^6$. It demonstrates that in all three cases the hardening rate does depend on N, although less so for non-spherical systems; there is a hint for leveling off at the highest N in the triaxial case. In all three cases it remains substantially lower than the full-loss-cone rate for high enough N.

However, it remains to be proved that the apparent N-independence of the hardening rate in merger simulations is the result of non-spherical shape and not of some other physical process, such as an initially enhanced fraction of stars on radial orbits, or incomplete mixing of the large-scale structure of the merger remnant that could create time-dependent variations in the potential, again enhancing the angular momentum mixing. Recently, two papers attempted a "controlled experiment" by exploring the binary evolution in a system constructed to be initially in equilibrium. Rather disappointingly, they have reached conflicting conclusions: while Khan *et al.* (2013) have found the hardening rate in their axisymmetric models to be N-independent, Vasiliev *et al.* (2014) have shown the rate to depend on N quite significantly in the axisymmetric and even triaxial geometry, although not as strongly as in the spherical case (Figure 1). The reasons for this discrepancy remain obscure, but may have something to do with the different methods used for preparing the initial conditions for the models: "adiabatic squeezing" technique (Holley-Bockelmann & Sigurdsson 2006) in the former paper and Schwarzschild modeling in the latter one.

The latter paper also have demonstrated that it is indeed very difficult to properly isolate the effect of loss-cone repopulation caused by non-spherical torques, because in an N-body simulation with a feasible number of particles ($N \lesssim 10^6$) it is strongly contaminated by the artificially enhanced two-body relaxation, thought to be unimportant in real galaxies. Thus a robust extrapolation of the result of simulation to much higher N remains a very challenging task. For instance, can we trust the hints of leveling off of N-dependence of hardening rate in the triaxial model at $N \geqslant 10^6$ in Figure 1? These simulations have also been conducted for a much shorter interval of time than is necessary to lead the binary into the GW-dominated regime, and it is not certain that the hardening rate stays constant with time. Therefore, unless we can afford much higher values of N and longer simulation intervals, the answer to the question whether the final-parsec problem exists in non-spherical galaxies would remain elusive. It is clear that at present time conventional N-body simulations cannot reach the required regime in which collisional (relaxation) effects are minor compared to collisionless (non-spherical) ones.

4. The novel Monte Carlo method

Fortunately, there exist a technique that allows to reliably model the collisional evolution of stellar systems with much cheaper computational demand than a direct N-body simulation. This is the Monte Carlo method of stellar dynamics, pioneered by Spitzer & Hart (1971) and Hénon (1971). It is based on the standard two-body relaxation theory, which predicts the statistical effect of many individual encounters on the trajectory of a given star, without the need to simulate these encounters explicitly. These methods have been continuously developed up to the present time and are successfully applied to various problems of dynamical evolution of star clusters (Giersz *et al.* 2013; Pattabiraman *et al.* 2013) and galactic nuclei (Freitag & Benz 2002). The mentioned independent implementations are based on the Hénon's formulation of the Monte Carlo method, which relies substantially on the assumption of spherical geometry; the alternative Spitzer's formulation is free from this assumption, although historically has been used only in a spherical case.

We have developed a novel Monte Carlo method based on Spitzer's work, which can be applied in any geometry and can explicitly distinguish the collisionless and collisional effects. The detailed description of the method is given in Vasiliev (2015), here we outline its main features. The evolution of an N-body system is followed by a hybrid approach combining a temporally smoothed extension of the self-consistent field method (Hernquist & Ostriker 1992) with a prescription for relaxation in terms of position-dependent velocity perturbations, based on Spitzer's original formulation of the Monte Carlo method. The orbits of particles are followed in a smooth, non-spherical potential represented by a suitable basis-set expansion (cf. Brockamp *et al.* 2014; Meiron *et al.* 2014), with a high-accuracy adaptive-timestep orbit integrator. After each integrator timestep, the particle velocity is perturbed using velocity diffusion coefficients computed from a spherical isotropic distribution function that approximates the true distribution of particles. The amplitude of this perturbation can be scaled to mimic the relaxation rate of a target stellar system that is composed of $N_\star$ stars; this number is not related to the actual number of particles in the simulation and can be set at will, in particular, turning off the relaxation entirely corresponds to $N_\star = \infty$. Orbits of all particles are integrated in parallel, independently from each other, for an interval of time ("episode") that can be comparable or even much longer than T_{orb}, although should be $\ll T_{\mathrm{rel}}$. At the end of the episode, the system state (coefficients of expansion of the non-spherical potential, and the spherical approximation to the distribution function) is updated using trajectories of particles sampled during the entire episode. The main advantages of the algorithm are the much lower computational demand than a full N-body simulation (it scales linearly with N, since the motion of each particle depends only on the mean-field potential and the global distribution function, but not on explicit interactions with other particles), the possibility of scaling the relaxation rate to the desired level independently from N, and the ability to deal with potentials of arbitrary shape.

The application of this method to the problem of binary MBH evolution is based on the conservation approach (Sesana *et al.* 2007; Meiron & Laor 2012). During each episode, particles are moving in the combined potential of the stellar cluster and the time-dependent potential of the binary MBH. At each close encounter of the particle with the binary, we record the change of energy and angular momentum of the particle; at the end of the episode, these changes are summed up, and the orbit of the binary is adjusted using

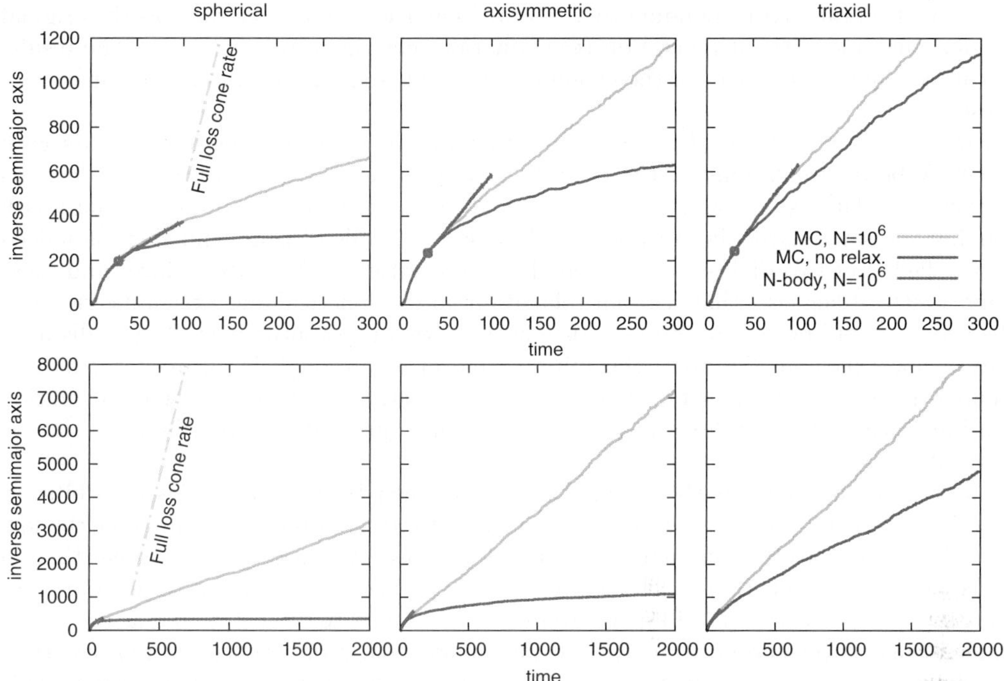

Figure 2. The evolution of inverse binary semi-major axis a^{-1} in N-body simulations of Vasiliev *et al.* (2014) (red curve, extending up to $t = 100$) and in Monte Carlo calculations which included relaxation at the amplitude corresponding to the $N = 10^6$ particle system from the N-body simulation (green curve, upper) or without relaxation (blue curve, lower). The thick red dot marks $t = 30$ as the initial moment for the Monte Carlo calculations. Top row shows short-term evolution, which demonstrates the qualitative agreement between Monte Carlo and N-body simulations; bottom row is for long-term evolution. Left, central and right panels show three geometries – spherical, axisymmetric, and triaxial. Only in the latter case the binary continues to shrink at an appreciable rate even without relaxation, although this rate is several times lower than the full-loss-cone rate shown as dot-dashed line. In the other two cases, the binary evolution without relaxation stalls at a separation just a few times smaller than the initial value of a, due to depletion of loss-cone orbits. On the other hand, with the relaxation rate corresponding to $N = 10^6$ (state-of-the-art N-body simulations), there is considerable evolution in all three geometries, demonstrating that the N-body simulations cannot yet reach the dynamical regime relevant for real galaxies.

the conservation law for energy and angular momentum. This procedure captures most relevant dynamical processes in the nucleus, such as the depletion and repopulation of the loss cone, secondary slingshot, and change of shape of the stellar potential, although it does not account for the influence of Brownian motion of the binary. In this preliminary study, we focus on equal-mass binaries on circular orbits and do not follow the evolution of the eccentricity, which is expected to remain low throughout the evolution for these conditions (e.g. Sesana 2010).

5. Results

We start our Monte Carlo simulations from snapshots taken at $t = 30$ in the N-body simulations of Vasiliev *et al.* (2014), when the hard binary has already formed. We follow the evolution of the system in three cases (spherical, axisymmetric, and triaxial, with the last two systems having initial axis ratios of 1:1:0.8 and 1:0.9:0.8), with two choices of the relaxation rate: corresponding to a $N = 10^6$ particle system (the same as the original simulations) and without relaxation (the situation relevant to real galaxies). The results of this calculation, in terms of the evolution of inverse semi-major axis of the binary, are presented in Figure 2.

First of all, it confirms that the Monte Carlo method yields consistent results with direct N-body simulations, if the relaxation amplitude is scaled to the same level as in these simulations. Second, in the presence of relaxation the evolution continues with roughly the same hardening rate for much longer intervals of time, and there is indeed only a factor of two difference between three geometries at $N = 10^6$. Third, and most importantly, it suggests that without relaxation, in the spherical and axisymmetric cases the evolution of binary semi-major axis a slows down and eventually stalls at a distance only a few times smaller than the initial value of a, although it occurs later and at smaller distance in the axisymmetric case. By contrast, in the triaxial case the evolution continues without any relaxation at a rate which is just twice lower than the rate at $N = 10^6$.

The difference between the geometries is explained by the properties of orbits: as mentioned above, in the axisymmetric case there are more orbits that can enter the loss cone due to the variation in angular momentum caused by non-spherical torques than in the spherical case, where no such torques exist. However, the minimum value of angular momentum for any orbit is still non-zero in the axisymmetric geometry, and as the size of the loss cone shrinks proportionally with the binary separation, eventually all such orbits are depleted. By contrast, in the triaxial case the population of genuinely centrophilic orbits, all of them being able to achieve arbitrary low values of angular momentum, is large enough to sustain a steady influx of stars into the loss cone. It is remarkable that the hardening rate in this case is about an order of magnitude lower than the full-loss-cone rate; nevertheless, scaled to real galaxies, it appears to be sufficient to drive the binary to the merger in a time much shorter than the Hubble time.

6. Conclusions

We have reviewed the evolution of binary massive black holes in gas-free galaxies, focusing on the final-parsec problem – the difficulty to keep the shrinking rate of the binary at a high enough level for it to merge in a Hubble time. This problem is connected to the rate of repopulation of the loss cone – region of phase space occupied by stars that are able to interact with the binary. In non-spherical galaxies this repopulation occurs both due to collisionless effects – variations of angular momentum of stars due to torques from the global potential, and to collisional effects – two-body relaxation. In real galaxies, collisional effects are supposed to play a very small role, however, in existing N-body simulations their contribution is greatly enhanced due to a limited number of particles.

We have introduced a novel Monte Carlo method that is suitable for the dynamical evolution of non-spherical galactic nuclei and can explicitly adjust the relaxation rate. Preliminary calculations for equal-mass binary on a circular orbit suggest that in the absence of relaxation, the shrinking of the orbit stalls (the final-parsec problem exists) in spherical and axisymmetric systems, but not in the triaxial ones. In the future, we are going to apply the method for unequal-mass binaries, which are expected to develop substantial eccentricity that accelerates the transition to the GW-dominated regime (Sesana 2010; Meiron & Laor 2012; Khan *et al.* 2012). An open question is why merger simulations seem to imply the evolution rate comparable with the full-loss-cone rate, while our isolated systems never approach this rate for sufficiently high N; we are going to address it by running Monte Carlo simulations using merger remnants as initial conditions.

Another interesting question is the rate of tidal disruptions of stars by either component of the binary at different evolutionary stages (e.g. Chen *et al.* 2009; Wegg & Bode 2011); this process can also be included in the Monte Carlo code.

This work was partly supported by the National Aeronautics and Space Administration under grant no. NNX13AG92G. I thank David Merritt for comments on the manuscript, and Alberto Sesana for fruitful discussions.

References

Amaro-Seoane, P., Sesana, A., Hoffman, L., *et al.* 2010, *MNRAS*, 402, 2308

Begelman, M. C., Blandford, R. D., & Rees, M. J. 1980, *Nature*, 287, 307

Berczik, P., Merritt, D., & Spurzem, R. 2005, *ApJ*, 633, 680

Berczik, P., Merritt, D., Spurzem, R., & Bischof, H. 2006, *ApJL*, 642, L21

Brockamp, M., Küpper, A., Thies, I., Baumgardt, H., & Kroupa, P. 2014, *MNRAS*, 441, 150

Chatterjee, P., Hernquist, L., & Loeb, A. 2003, *ApJ*, 592, 32

Chen, X., Madau, P., Sesana, A., & Liu, F.-K. 2009, *ApJL*, 697, L149

Dullo, B. & Graham, A. 2014, *MNRAS*, 444, 2700

Ferrarese, L. & Ford, E. 2005, *Space Sci. Revs*, 116, 523

Frank, J. & Rees, M. 1976, *MNRAS*, 176, 633

Freitag, M. & Benz, W. 2002, *A&A*, 394, 345

Giersz, M., Heggie, D., Hurley, J., & Hypki, A. 2013, *MNRAS*, 431, 2184

Harfst, S., Gualandris, A., Merritt, D., Spurzem, R., Portegies Zwart, S., & Berczik, P. 2007, *New Astron.*, 12, 357

Hénon, M., 1971, *Ap&SS*, 13, 284

Hernquist, L. & Ostriker, J. 1992, *ApJ*, 386, 375

Holley-Bockelmann, K. & Sigurdsson, S. 2006, *arXiv:astro-ph/0601520*

Iwasawa, M., Funato, Y., & Makino, J. 2006, *ApJ*, 651, 1059

Khan, F. M., Just, A., & Merritt, D. 2011, *ApJ*, 732, 89

Khan, F. M., Preto, M., Berczik, P., Berentzen, I., Just, A., & Spurzem, R. 2012, *ApJ*, 749, 147

Khan, F. M., Holley-Bockelmann, K., Berczik, P., & Just, A. 2013, *ApJ*, 773, 100

Kormendy, J. & Ho, L. 2013, *ARA&A*, 51, 511

Lightman, A. & Shapiro, S. 1977, *ApJ*, 211, 244

Lodato, G., Nayakshin, S., King, A., & Pringle, J. 2009, *MNRAS*, 298, 1392

Makino, J. & Funato, Y. 2004, *ApJ*, 602, 93

Meiron, Y. & Laor, A. 2012, *MNRAS*, 422, 117

Meiron, Y., Li, B., Holley-Bockelmann, K., & Spurzem, R. 2014, *ApJ*, 792, 98

Merritt, D., 2013, *Classical & Quantum Gravity*, 244005

Merritt, D., Mikkola, S., & Szell, A. 2007, *ApJ*, 671, 53

Merritt, D. & Poon, M.-Y. 2004, *ApJ*, 606, 788

Milosavljević, M. & Merritt, D. 2001, *ApJ*, 563, 34

Milosavljević, M., Merritt, D., Rest, A., & van den Bosch, F. 2002, *MNRAS*, 331, L51

Milosavljević, M. & Merritt, D. 2003a, in: J. Centrella & S. Barnes (eds.), *The Astrophysics of Gravitational Wave Sources*, AIP Conf. Proc.(Melville, NY: AIP), 686, p. 201

Milosavljević, M. & Merritt, D. 2003b, *ApJ*, 596, 860

Pattabiraman, B., Umbreit, S., Liao, W.-K., Choudhary, A., Kalogera, V., Memik, G., & Rasio, F. 2013, *ApJS*, 204, 15

Perets, H. & Alexander, T. 2008, *ApJ*, 677, 146

Poon, M. Y. & Merritt, D. 2001, 549, 192

Preto, M., Berentzen, I., Berczik, P., & Spurzem, R. 2011, *ApJL*, 732, L26

Roškar, R., Mayer, L., Fiacconi, D., Kazantzidis, S., Quinn, T., & Wadsley, J. 2014, *arXiv:1406.4505*

Quinlan, G. 1996, *New Astron.*, 1, 35

Quinlan, G. & Hernquist, L. 1997, *New Astron.*, 2, 533

Sesana, A., Haardt, F., & Madau, P. 2007, *ApJ*, 660, 546

Sesana, A. 2010, *ApJ*, 719, 851
Spitzer, L. & Hart, M. 1971, *ApJ*, 164, 399
Vasiliev, E. 2015, *MNRAS*, 446, 3150
Vasiliev, E., Antonini, F., & Merritt, D. 2014, *ApJ*, 785, 163
Wegg, C. & Bode, J. N. 2011, *ApJL*, 738, L8
Yu, Q. 2002, *MNRAS*, 331, 935

Star clusters and black holes in galaxies across cosmic time
Proceedings IAU Symposium No. 312, 2014
Y. Meiron, S. Li, F.-K. Liu & R. Spurzem, eds.

© International Astronomical Union 2016
doi:10.1017/S1743921315007619

Resonant motions of supermassive black hole triples

Wei Hao[1]**, Rainer Spurzem**[2,3,4,1]**, Thorsten Naab**[1]**, Long Wang**[4,5]**,
M. B. N. Kouwenhoven**[4]**, Pau Amaro-Seoane**[6]
and Rosemary A. Mardling[7,8]

[1]Max-Planck-Institut für Astrophysik, Karl-Schwarzschild-Str. 1 85741 Garching, Germany
email: `elvis@mpa-garching.mpg.de`

[2]National Astronomical Observatories of China, Chinese Academy of Sciences, 20A Datun
Rd., Chaoyang District, 100012, Beijing, P.R. China

[3]Astronomisches Rechen-Institut, Zentrum für Astronomie, Univ. of Heidelberg,
Mönchhof-Strasse 12-14, 69120 Heidelberg, Germany

[4]Kavli Institute for Astronomy and Astrophysics, Peking University, Yi He Yuan Lu 5,
Haidian District, Beijing 100871, P.R. China

[5]Department of Astronomy, School of Physics, Peking University, Yi He Yuan Lu 5, Haidian
District, Beijing 100871, P.R. China

[6]Max Planck Institut für Gravitationsphysik (Albert-Einstein-Institut), D-14476 Potsdam,
Germany

[7]School of Mathematical Sciences, Monash University, Victoria 3800, Australia

[8]Astronomy Department of the University of Geneva, Geneva Observatory, 51 Ch. des
Maillettes, CH1290 Versoix, Switzerland

Abstract. Triple supermassive black holes (SMBH) can form during the hierarchical mergers of
massive galaxies with an existing binary. Perturbations by a third black hole may accelerate the
merging process of an inner binary, for example through the Kozai mechanism. We analyze the
evolution of simulated hierarchical triple SMBHs in galactic centers, and find resonances in the
evolution of the semi-major axis, the eccentricity and the inclination, for both the inner and the
outer orbits of the triple system, which are not only Kozai like. Through resonant oscillations,
SMBH can trigger a significant increase of the inner SMBH binary eccentricity shortening the
merger timescale expected from gravitational wave (GW) emission. As hierarchical triple SMBHs
may be frequent in massive galaxies, the influence of orbital resonances is of great importance to
our understanding of black hole coalescence and gravitational wave detection. Although Kozai
mechanism is believed to play an important role in this process, detailed studies on the pattern
of these resonances is necessary.

Keywords. black hole physics, kinematics and dynamics, oscillations, etc.

1. Introduction

Supermassive black holes (SMBHs) are common in the centers of massive galaxies (e.g.
Kormendy & Ho 2013). Within the framework of a cold dark matter cosmology, these
SMBHs would form SMBH binaries as a consequence of hierarchical mergers of their
host galaxies White & Rees 1978, Begelman, Blandford & Rees 1980. Massive galaxies
are typically the product of multiple mergers of galaxies (e.g. De Lucia *et al.* 2006). The
long time span from 0.1 to 1 Gyr for the coalesce of SMBH binaries would largely increase
the possibility of a secondary merger before the final coalesce of these SMBH binaries
(Khan *et al.* 2013), if the merger timescale is much larger than the timescale between two
mergers. Since most massive galaxies should have experienced multiple mergers during

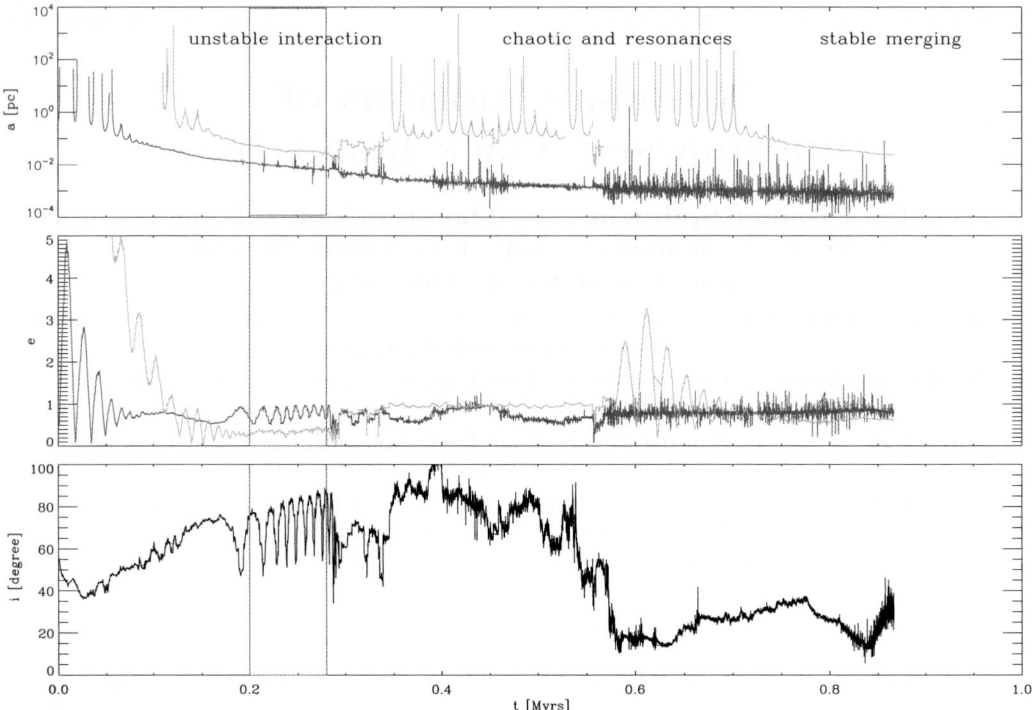

Figure 1. Full evolution of the semi-major axis, eccentricity, and relative inclination with time in this example simulation. This plot shows how semi-major axis A, eccentricity ECC and inclination angle difference between inner and outer binary $i_{\rm in} - i_{\rm out}$ as a function of time. Red curve represents inner binary and green curve represents outer binary, and the same bellow. The discrete part of the semi-major evolution, especially those of the outer binary, suggesting the three body system is unbound sometimes, but they could be pulled back by the surround stars afterwards.

their lifetime, a triple or more SMBH system would frequently form. If we take the central inner SMBH binary as a whole respect to the third SMBH on wider orbit, we can define them as outer binary, and this inner binary together with outer binary is called a hierarchical triple system.

The impact of the third SMBH on the merging SMBH binary can be crucial, either by ejection of one component of the three or by perturbing the orbit of the inner binary (Valtonen, Mikkola & Heinamaki 1994). Besides, repeated encounters between the third SMBH and the progenitor SMBH binary would possibly excite eccentricity of the binary (Makino & Ebisuzaki 1994). For a hierarchical triple system, if the inclination angle between the orbital planes of inner and outer binary is higher than $39.2°$, these binaries would oscillate with a Kozai resonance (Kozai 1962). This process could greatly excite the inner binary's eccentricity, which is important for the merging SMBH binary since the GW emission efficiency is a strongly depend on their semi-major axis and eccentricity (Peters 1964).

Although Kozai (-like) mechanism is believed to play a pivotal role during the evolution of hierarchical triple SMBHs in galaxy centers, it is not clear how exactly it could impact the merger of the inner binary in the environment of galactic nuclei.

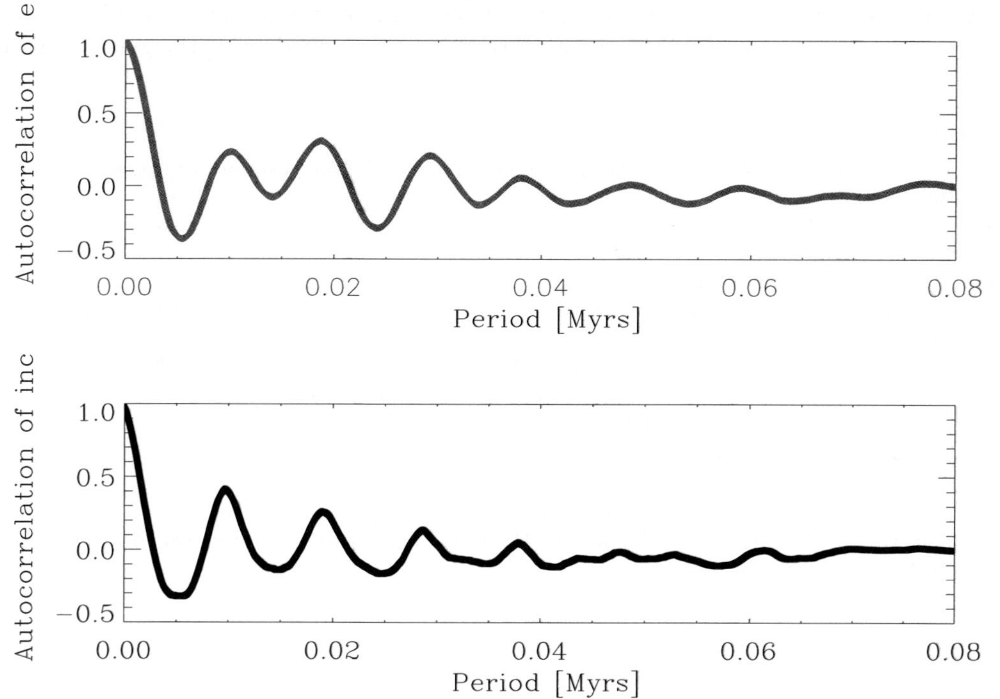

Figure 2. Auto-correlation result of inner eccentricity and inclination angle

2. Simulation setup

Our simulation has a total particle number of 64 000, including three SMBHs with 0.01 unit mass each and the 63 997 surrounding equal mass star clusters with a total mass of 0.97 unit mass. All our particles are non-evolving point mass. Plummer model with $n = 5$ polytrope is applied for the stellar system. We start the simulation from parsec scale, and terminate our simulation once the inner binary reach to a separation of milliparsec scale, which is the typical distance for triggering immediate merger by GW emission. We used a special version of NBODY6 to model this process (Amaro-Seoane *et al.* 2010 Aarseth, S., 2003).

3. Results

Figure 1 shows an example of Kozai-like resonance found in one of our simulation data. When looking at the range in the magenta box, we can see clear resonance between eccentricity of inner binary and the inclination angle. Figure 2 shows the result of auto-correlation for eccentricity of the inner binary and inclination angle, indicating it is oscillating with a fine period of 0.01 Myr, which is about a hundred times higher than standard Kozai timescale. This Kozai timescale is given by Kiseleva & Chernin (1988)

$$\tau = \frac{2T_{\mathrm{out}}^2}{3\pi T_{\mathrm{orb}}} \frac{m_1 + m_2 + m_3}{m_3} \left(1 - e_{\mathrm{out}}^2\right)^{3/2} . \tag{3.1}$$

Although this oscillation is not a pure Kozai resonance, but it played the similar role on exciting the eccentricity of the inner binary. And this would accelerate the merger process. Besides, several other pattern of oscillations are found as well, (e.g., oscillations between outer eccentricity and inclination,) indicating more investigation should be apply to reveal the complexity of this hierarchical triple system embedded in the galaxy centers.

References

Aarseth, S., 2003, *Gravitational N-body Simulations (Cambridge, Cambridge University Press,* 2003, 430

Amaro-Seoane P., Sesana A., Hoffman L., Benacquista M., Eichhorn C., Makino J., & Spurzem R., 2010, *MNRAS*, 402, 2308

Begelman M. C., Blandford R. D., & Rees M. J., 1980, *Nature*, 287, 307

De Lucia G., Springel V., White S. D. M., Croton D., & Kauffmann G., 2006, *MNRAS*, 366, 499

Khan F. M., Holley-Bockelmann K., Berczik P., & Just A., 2013, *ApJ*, 773, 100

Kiseleva L. G. & Chernin A. D. ., 1988, *Soviet Astronomy Letters*, 14, 412

Kormendy J. & Ho L. C., 2013, *ARAA*, 51, 511

Kozai Y., 1962, *AJ*, 67, 591

Makino J. & Ebisuzaki T., 1994, *ApJ*, 436, 607

Peters P. C., 1964, *Physical Review*, 136, 1224

Valtonen M. J., Mikkola S., & Heinamaki P., 1994, *American Astronomical Society Meeting Abstracts No.184*, 26, 913

White S. D. M. & Rees M. J., 1978, *MNRAS*, 183, 341

Star clusters and black holes in galaxies across cosmic time
Proceedings IAU Symposium No. 312, 2014
Y. Meiron, S. Li, F.-K. Liu & R. Spurzem, eds.

© International Astronomical Union 2016
doi:10.1017/S1743921315007620

Large scale direct galaxy collision simulations with central supermassive binary black holes

Margaryta Sobolenko[1], Peter Berczik[2,1,4] and Rainer Spurzem[2,3,4]

[1]Main Astronomical Observatory, National Academy of Sciences of Ukraine, 27 Akademika Zabolotnoho St., 03680, Kyiv, Ukraine
email: `sobolenko@mao.kiev.ua`

[2]National Astronomical Observatories of China, Chinese Academy of Sciences, 20A Datun Rd., Chaoyang District, 100012, Beijing, China

[3]Kavli Institute for Astronomy and Astrophysics, Peking University, Beijing 100871, China

[4]Astronomisches Rechen-Institut, Zentrum für Astronomie, University of Heidelberg, Mönchhofstrasse 12-14, 69120, Heidelberg, Germany

Abstract. We present a set of, large scale direct N-body simulations of the galaxy collision with the central Supermassive Black Hole Binary (SMBHB) system. Based on our simulations which include the accurate Post Newtonian (PN) relativistic dynamical corrections we can estimated the merging time for the real astrophysical object. Each galaxy initially was represented as a set of particles (up to N=500k) with Plummer distribution. The SMBHBs system is described using the two special high mass, i.e. "relativistic", particles. The interaction between these two particles have an extra PN correction terms (up to 3.5PN). Merging time upper limit was obtained for the closely interacting galaxy system NGC 6240.

Keywords. black hole physics, methods: N-body simulations, galaxies: nuclei, individual (NGC6240)

1. Introduction

After first discovering the AGN it was proposed that central power engine of this objects can be the massive black holes (Lynden-Bell 1969; Lynden-Bell & Rees 1971; Wolfe & Burbidge 1970; Longair 1996). Nowadays we know that at the cores of many active galaxies indeed we have a massive black holes (Ferrarese & Ford 2005; Collin 2006). As an close example we can see our Galaxy, which contains the massive central black hole with the mass $M_\bullet \approx 4 \times 10^6$ M$_\odot$ (Gillessen *et al.* 2009). Usually the central supermassive black holes (SMBH) have a mass range $10^6 - 10^9$ M$_\odot$.

Mass of the SMBH correlate with various physical parameters of the host galaxies stellar spheroid. The most clear correlations we can see are between the SMBH mass and the bulge luminosity $L_{\rm bul}$, the bulge mass $M_{\rm bul}$, the bulge velocity dispersion σ, the mass of the galaxy $M_{\rm gal}$, and even the Sérsic index of the surface brightness profile (Hartmann *et al.* 2014, reference therein).

During their evolution galaxies can merge with central BHs and can create also galaxies with SMBHB. At the first stage (Begelman *et al.* 1980) the separation between components can shrink due to dynamical friction in the stellar background and later can eventually form the "hard" binary. Further the binary can become more hard due to the three-body scattering of single stars, mainly inside the influence radius of the SMBHB system. At the last stage of merging the binary relativistic (with the emission of gravitational waves) in-spiral can lead to the final fast coalescence of the SMBHB.

We can observe the SMBHB at the different stages of merging as different objects: as

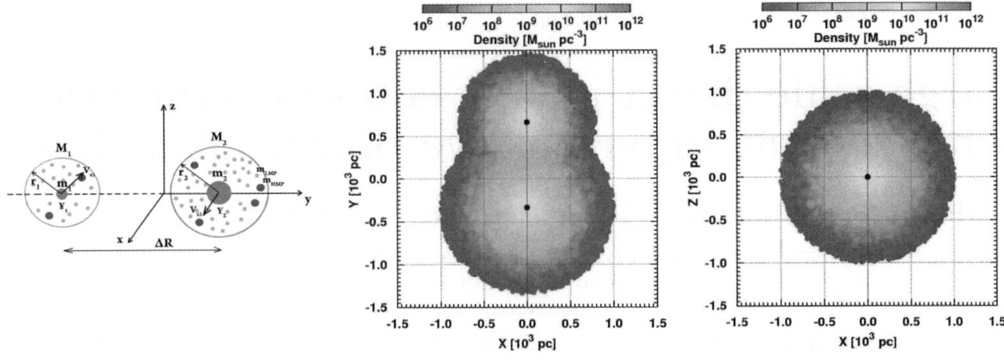

Figure 1. *left panel*: Initial configuration for system with two BHs at the centres. *middle panel*: Density cut at the XY plane at view. *right panel*: Density cut at the XZ plane at view.

SMBH at the galaxy pairs ($\Delta r \sim 10 - 100$ kpc), as dual SMBHs ($\Delta r \sim 1$ kpc), as binary SMBHs ($\Delta r < $ pc) or as a recoiling SMBHs after merging. This type of classification, of course, is quite arbitrary. For all the objects that we observe now and generally call as "binary SMBHs" with the separation more than 1 pc, we decide to use the term "SMBH binary (SMBHB)".

One of the first discovered SMBHB was a interacting galaxy NGC6240 (Komossa *et al.* 2003) situated at z=0.0243 (Solomon *et al.* 1997). The two sources at the LINER (Nakai *et al.* 2002; Véron-Cetty & Véron 2010, see references) galaxy are resolved in hard X-rays by the Chandra X-Ray Observatory with separation $\Delta\theta = 1.56''$ (corresponding $\Delta r = 0.756$ kpc, assumed $H_0 = 68$ kms^{-1}Mpc^{-1}, $1'' = 504$ pc). The lower limit for south-western BH mass $M_{\bullet S} = 8.7 \pm 0.3 \times 10^8$ M$_\odot$ was measured using stellar kinematics (Medling *et al.* 2011; Engel *et al.* 2010). The dynamical mass of the nuclei should be $M_{\rm dyn} = 2 - 8 \times 10^9$ from the stellar velocity field in central 500 pc (Tecza *et al.* 2000).

2. Numerical model and initial conditions

For simulations we took the next physical value for system: total mass of the galaxies is $M_{\rm tot} = 1.3 \times 10^{11}$ M$_\odot$, total masses of the SMBHB are $M_{\bullet\rm tot} = 1.3 \times 10^9$ M$_\odot = 1\% M_{\rm tot}$ and $M_{\bullet\rm tot} = 2.6 \times 10^9$ M$_\odot = 2\% M_{\rm tot}$, BHs and galaxies mass ratio is $q = M_{\bullet\rm SW}/M_{\bullet\rm NE} = 0.5$, initial separation is $\Delta R = 1$ kpc, initial eccentricity is $e_0 = 0.5$. We represent two galaxies as two Plummer (non-rotating) spheres with two special particles at the centres called BHs (Fig. 1). Also spheres contain another two type of particles called "light mass particles" (LMP) with $90\% N_{\rm tot}$ and "high mass particles" (HMP) with $10\% N_{\rm tot}$. The mass ratio between HMP and LMP was 1:9. We set the total number of the particles as $N_{\rm tot} = 500$k.

We use the publicly available φGPU code † (Berczik *et al.* 2011, 2013) with N-body 4$^{\rm th}$ order Hermite integration of the motions for all the particles using the blocked hierarchical individual time step scheme. This Hermite scheme requires us to know the acceleration and its first time-derivative, called *jerk*, including the $\mathcal{PN}$ corrections. In our φGPU code we use the generalized "Aarseth" type criteria for the timestep definition (Nitadori & Makino 2008):

$$\Delta t = \eta_p \left(\frac{A^{(1)}}{A^{(p-2)}} \right)^{1/(p-3)},$$

(2.1)

† `ftp://ftp.mao.kiev.ua/pub/berczik/phi-GPU/`

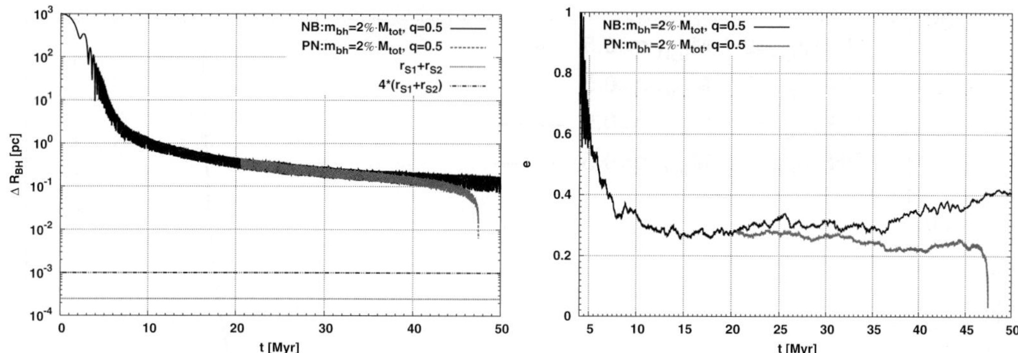

Figure 2. Evolution of separation between the SMBHs (left panel) and eccentricity (right panel) for model with $M_{\bullet\mathrm{tot}} = 2\% M_{\mathrm{tot}}$

where

$$A^{(k)} = \sqrt{|\boldsymbol{a}^{(k-1)}||\boldsymbol{a}^{(k+1)}| + |\boldsymbol{a}^{(k)}|^2}. \tag{2.2}$$

Here, $\boldsymbol{a}^{(k)}$ is the kth derivative of acceleration, p is the order of the integrator, η_p is the accuracy parameter. For 4th-order Hermite scheme the timestep looks like:

$$\Delta t = \eta_4 \frac{A^{(1)}}{A^{(2)}}, \tag{2.3}$$

where

$$A^{(1)} = \sqrt{|\boldsymbol{a}||\boldsymbol{a}^{(2)}| + |\boldsymbol{a}^{(1)}|^2}, \qquad A^{(2)} = \sqrt{|\boldsymbol{a}^{(1)}||\boldsymbol{a}^{(3)}| + |\boldsymbol{a}^{(2)}|^2}. \tag{2.4}$$

For all our runs we use the $\eta_4 = 0.1$ parameter for our two black holes. The gravitational softening ε used in form:

$$\varepsilon_{ij}^2 = \alpha(\varepsilon_i^2 + \varepsilon_j^2)/2, \tag{2.5}$$

where $\varepsilon_\bullet = 0$, $\varepsilon_{\mathrm{HMP}} = 10^{-4}$, $\varepsilon_{\mathrm{LMP}} = 10^{-5}$. If one of the interacting particle was a BH we use the coefficient $\alpha = 10^{-4}$, for other cases $\alpha = 1$.

3. Results and discussion

We use two distinct hardening regimes for obtaining the merging time for object NGC6240. First classical regimes caused by hardening from stellar-dynamical effects (S_{NB}) (Khan *et al.* 2012). The next regime is a relativistic regime with hardening by gravitational waves emission (S_{GW}). At the classical regime we neglect the S_{GW}) and estimate S_{NB} =const. GW emission start dominate at time when $S_{\mathrm{NB}} = S_{\mathrm{GW}}$. We start to calculate the S_{GW} at time when $S_{\mathrm{GW}} = 0.03\% S_{\mathrm{NB}}$. At Figure 2 we show the relative separation ΔR between the two black holes during a galaxy merger and eccentricity e evolution for model with $M_{\bullet\mathrm{tot}} = 2\% M_{\mathrm{tot}}$. Eccentricity stays almost constant during a whole period of time.

The estimated merging time for model with $M_{\bullet\mathrm{tot}} = 1\% M_{\mathrm{tot}}$ is $T_{\mathrm{merge}1\%} = 40$ Myr and for the model with $M_{\bullet\mathrm{tot}} = 2\% M_{\mathrm{tot}}$ is $T_{\mathrm{merge}2\%} = 48$ Myr (Fig. 3).

Acknowledgements

MS & PB thanks the International Astronomical Union for the grant support which gave the opportunity to him participate on the IAU Symposium No. 312 in Beijing-2014.

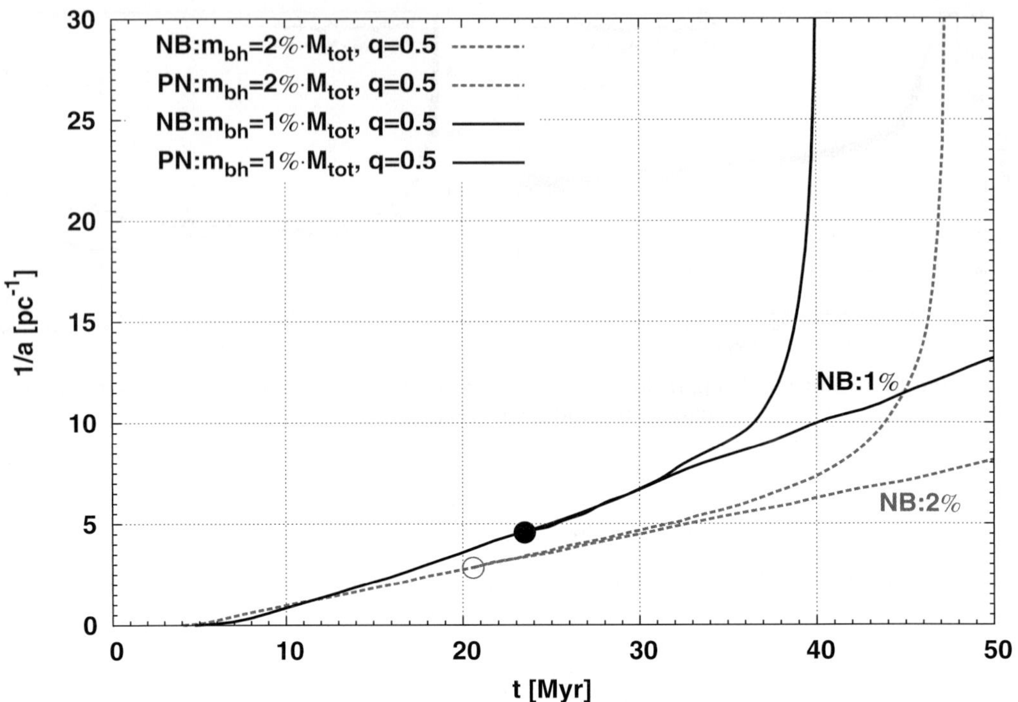

Figure 3. Evolution of SMBH binary inverse semi-major axis for two models with
$M_{\bullet tot} = 1\% M_{tot}$ and $M_{\bullet tot} = 2\% M_{tot}$

References

Begelman, M. C., Blandford, R. D., & Rees, M. J. 1980, *Nature*, 287, 307

Berczik, P., Spurzem, R., Wang, L., Zhong, S., & Huang, S. 2013, in *Third International Conference "High Performance Computing"*, HPC-UA 2013, p. 52-59, 52–59

Berczik, P., Nitadori, K., Zhong, S., *et al.* 2011, in International conference on High Performance Computing, Kyiv, Ukraine, October 8-10, 2011., p. 8-18, 8–18

Collin, S. 2006, in *American Institute of Physics Conference Series*, Vol. 861, Albert Einstein Century International Conference, ed. J.-M. Alimi & A. Füzfa, 587–595

Engel, H., Davies, R. I., Genzel, R., *et al.* 2010, *A&A*, 524, A56

Ferrarese, L., & Ford, H. 2005, *Space Sci. Rev.*, 116, 523

Gillessen, S., Eisenhauer, F., Trippe, S., *et al.* 2009, *Astrophys. J.*, 692, 1075

Hartmann, M., Debattista, V. P., Cole, D. R., *et al.* 2014, *Mon. Not. R. Astron. Soc.*, 441, 1243

Khan, F. M., Preto, M., Berczik, P., *et al.* 2012, *Astrophys. J.*, 749, 147

Komossa, S., Burwitz, V., Hasinger, G., *et al.* 2003, *Astrophys. J., Lett.*, 582, L15

Longair, M. S. 1996, *Our evolving universe* (Cambridge, New York: Cambridge University Press)

Lynden-Bell, D. 1969, *Nature*, 223, 690

Lynden-Bell, D., & Rees, M. J. 1971, *Mon. Not. R. Astron. Soc.*, 152, 461

Medling, A. M., Ammons, S. M., Max, C. E., *et al.* 2011, *Astrophys. J.*, 743, 32

Nakai, N., Sato, N., & Yamauchi, A. 2002, *Publ. Aston. Soc. Jpn.*, 54, L27

Nitadori, K., & Makino, J. 2008, *N. Astr.*, 13, 498

Solomon, P. M., Downes, D., Radford, S. J. E., & Barrett, J. W. 1997, *Astrophys. J.*, 478, 144

Tecza, M., Genzel, R., Tacconi, L. J., *et al.* 2000, *Astrophys. J.*, 537, 178

Véron-Cetty, M.-P., & Véron, P. 2010, *A&A*, 518, A10

Wolfe, A. M., & Burbidge, G. R. 1970, *Astrophys. J.*, 161, 419

Star clusters and black holes in galaxies across cosmic time
Proceedings IAU Symposium No. 312, 2014
Y. Meiron, S. Li, F.-K. Liu & R. Spurzem, eds.

© International Astronomical Union 2016
doi:10.1017/S1743921315007632

Star accretion onto supermassive black holes in axisymmetric galactic nuclei

Shiyan Zhong[1], Peter Berczik[1,2,3] and Rainer Spurzem[1,2,4]

[1] National Astronomical Observatories of China, Chinese Academy of Sciences, 20A Datun Rd., Chaoyang District, 100012, Beijing, China

[2] Astronomisches Rechen-Institut, Zentrum für Astronomie, University of Heidelberg, Mönchhofstrasse 12-14, 69120, Heidelberg, Germany

[3] Main Astronomical Observatory, National Academy of Sciences of Ukraine, 27 Akademika Zabolotnoho St., 03680, Kyiv, Ukraine

[4] Kavli Institute for Astronomy and Astrophysics, Peking University, Beijing, China

Abstract. Tidal Disruption (TD) of stars by supermassive central black holes from dense rotating star clusters is modeled by high-accuracy direct N-body simulation. We study the time evolution of the stellar tidal disruption rate and the origin of tidally disrupted stars. Compared with that in spherical systems, we found a higher TD rate in axisymmetric systems. The enhancement can be explained by an enlarged loss-cone in phase space which is raised from the fact that total angular momentum $\mathbf{J}$ is not conserved. As in the case of spherical systems, the distribution of the last apocenter distance of tidally accreted stars peaks at the classical critical radius. However, the angular distribution of the origin of the accreted stars reveals bimodal features. We show that the bimodal structure can be explained by the presence of two families of regular orbits, namely short axis tube and saucer orbits.

Keywords. galaxies: kinematics and dynamics, galaxies: nuclei, methods: numerical, quasars: supermassive black holes, stars: kinematics and dynamics

1. Introduction

A large fraction of galaxies show evidence of supermassive black holes (henceforth SMBH) residing in their center. They are typically embedded in nuclear star clusters (NSC). NSCs have size similar to galactic globular clusters, but they are much heavier and brighter (Böker *et al.* 2002; Böker *et al.* 2004). In massive galaxies NSCs may not be significant or even do not exist, however, the SMBHs are still surrounded by enormous number of stars. SMBH residing in these NSCs can tidally disrupt stars that come close to its tidal radius and eventually accrete the gaseous debris, which can light up the central SMBH for a period of time (Rees 1988; Evans & Kochanek 1989). This kind of event is a useful tool to examine the relativistic physics near SMBH since the disruption occurs at a place very close to the BH's Schwarzschild radius. Also it can help us to investigate SMBH in non-active galactic center. Although tidal disruption of stars has been proposed for almost half a century, only until last decade do people realize the importance of such events, after the discovery of a dozens of tidal disruption candidates (Komossa 2002; Komossa & Merritt 2008; Liu *et al.* 2014).

In order to compute the TD event rate, loss cone theory was developed (Frank & Rees 1976; Lightman & Shapiro 1977) and mostly assumed spherical symmetry of the stellar system. However, the host NSC of SMBH may not be spherical object as shown by both observation (Feldmeier *et al.* 2014; Schödel *et al.* 2014) and simulation (Antonini *et al.* 2012), which motivates us to study TD event in non-spherical systems. For this

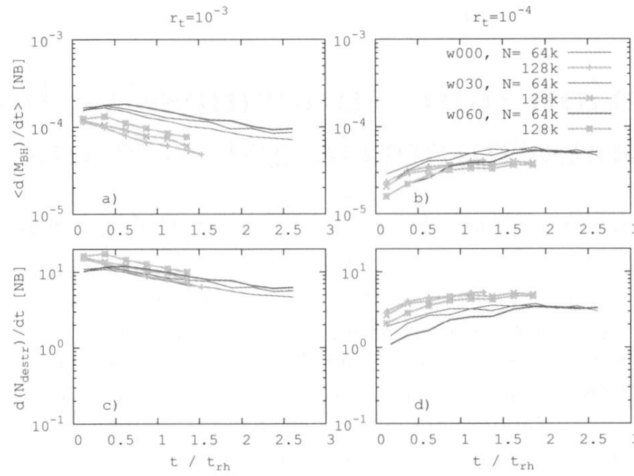

Figure 1. x axis is time in unit of initial half-mass relaxation time. y axis for panel a) and b) is the averaged mass accretion rate in given time range (i.e. $1/4\ t_{rh}$); y axis for panel c) and d) is the number accretion rate. Panel a) and c) show the result for $r_t = 10^{-3}$. Panel b) and d) show the result for $r_t = 10^{-3}$. Line thickness indicate different rotating parameters.

purpose, we performed direct N-body simulations with model stellar clusters constructed by rotating King model (Ernst *et al.* 2007) and put a massive particle in the center which representing the SMBH. A fixed (and artificially enhanced) tidal radius is set for the SMBH ($r_t = 10^{-3}, 10^{-4}$ in model unit). We choose different rotating parameters ($\omega_0 = 0.0, 0.3, 0.6$) to study the effect of rotation on TD. In the following sections we briefly report our results.

2. Tidal disruption rate

Fig. 1 summarizes the tidal disruption rate (TDR) results taken from simulations with different rotating parameters, particle numbers and tidal radius. TDR curves in left column represent large r_t models. These models quickly enter empty loss cone regime. And faster rotation results in higher TDR. However, TDR curves in right column show an opposite trend: faster rotation results in lower TDR. These systems are still in full loss cone regime and BH has large Brownian motion which may cause this discrepancy. And TDR curves of different models converge at the late stage. After entering empty loss cone regime, their behaviors are consistent with that of large r_t models. The results of $\omega_0 = 0.0$ models resemble those of Plummer model (Zhong *et al.* 2014).

3. Loss cone shape in axisymmetric potential

The enhancement of TDR could caused by faster relaxation process in rotating systems, or could be the consequence of an enlarged loss cone. In order to address this question, we perform test particle experiments to investigate the loss cone in axisymmetric potential.

In axisymmetric potential, the total angular momentum J of a test particle is not conserved. As a result, star outside of the loss cone ($J > J_{lc} \approx \sqrt{2GM_\bullet r_t}$) has chance to drift into the loss cone and star with $J < J_{lc}$ can drift out. So the boundary of loss cone in J dimension may be a few times larger than classical J_{lc} (Magorrian & Tremaine 1999). J_z is still conserved and the condition $J_z < J_{lc}$ shall always be satisfied. At the time of disruption, $J < J_{lc}$ is required. We perform the experiment in phase space coordinated by energy E, angular momentum J and its z component J_z. Given a combination of

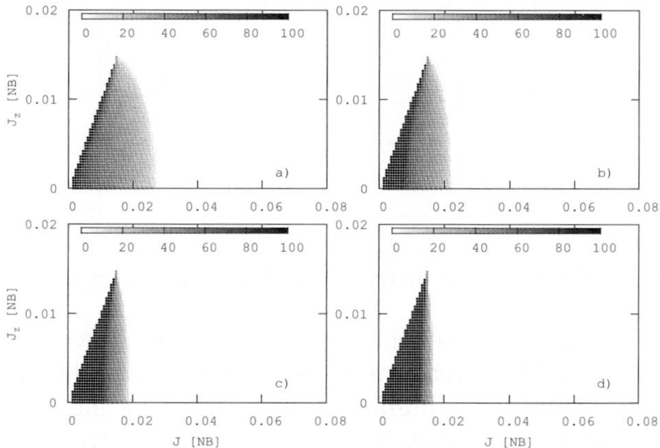

Figure 2. x axis is module of angular momentum in N-body unit. y axis is z component of angular momentum. Panel a) correspond to $E = -1.3$, b) $E = -1.5$, c) $E = -1.7$, d) $E = -1.9$. Gray scale indicate the filling factor P in percentage.

(E, J, J_z), we put the test particle at its apocenter position and integrate its orbit for one orbital period in the potential generated with Self-Consistent Field (SCF, Hernquist & Ostriker 1992) code, based on snapshots from the direct N-body simulation. If the particle reaches BH's tidal radius we mark it as loss cone star, otherwise it is out of loss cone. We also found that the fate of stars with same (E, J, J_z) might be different, depending on the zenith angle θ of their apocenter. We incorporate these dependence into a parameter P which means among all stars with same (E, J, J_z) only a fraction P of them is inside loss cone. The results are shown in Fig. 2.

The whole plane in each panel comprises 3 regions along J axis: 1) inner region where P equals 1, particles with these (E, J, J_z) can definitely hit the BH within one dynamical time scale; 2) transition region where P is non-zero but less than 1; 3) outer region where $P = 0$. Transition region shrinks from high energy to low energy case, which reflect the fact that variation of J becomes smaller as energy decreases. Because in low energy case, test particles are close to the BH, the potential is more spherical. While in high energy case, test particles can move to the outer region where the axisymmetric stellar potential dominate. We calculate the effective area of loss cone for different values of E and find that in the energy range where most of the disrupted stars come from, the effective area of loss cone is larger comparing to spherical case, thus can account for the enhancement of TDR.

4. Origin of disrupted stars

Loss cone theory indicate that most of the disrupted stars should originate from the place around the critical radius, which is roughly the same as influence radius of the SMBH. We measure the last apocenter position for the disrupted stars and study their distribution in both r and θ dimension. The r distribution is similar to those obtained in spherical case (Fig. 3 left panel, also see Zhong et $al.$ 2014). However, θ distribution show double peaks around the equatorial plane which is unexpected (Fig. 3, middle panel). After doing orbit classification for the disrupted stars, we find that this feature is caused by the different orbital types of the disrupted stars (Fig. 3, right panel).

Inside BH's influence radius r_h, there are two types of regular orbits, namely short axis tube (SAT) and saucer. The apocenter of SAT orbit can cross the equatorial plane and

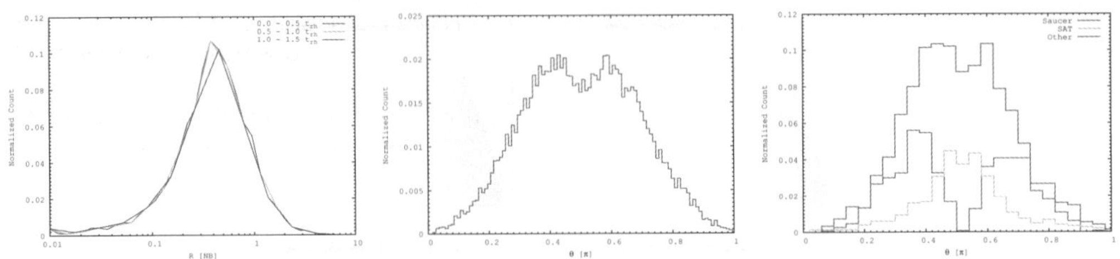

Figure 3. Normalized r and θ distribution of last apocenter of disrupted stars.

minimum J is acquired on the equatorial plane, so the last apocenter of disrupted stars moving on SAT orbit should concentrate to equatorial plane. The apocenter of saucer orbit avoid equatorial plane, the minimum J is acquired at the region above or below it. Thus the saucer orbit is responsible for the double peaks. Outside r_h, star orbits are mostly chaotic, thus θ distribution do not have the double-peak feature.

5. Conclusion

In our N-body simulations we find that BHs in axisymmetric system exhibit higher TD rate than spherical system. This enhancement is caused by an enlarged loss cone in phase space. Most of the disrupted stars are coming from place far from the central BH. The distribution of their angular position show double-peak feature, which can be explained by the different orbit types of the disrupted stars.

References

Antonini, F., Capuzzo-Dolcetta, R., Mastrobuono-Battisti, A. & Merritt, D. 2012 *ApJ* 750, 111

Böker, T., Laine, S., van der Marel, R. P., Sarzi, M., Rix, H.-W., Ho, L. C., & Shields, J. C. 2002, *AJ*, 123, 1389

Böker, T., Sarzi, M., McLaughlin, D. E., van der Marel, R. P., Rix, H.-W., Ho, L. C., & Shields, J. C. 2004, *AJ*, 127, 105

Ernst, A., Glaschke, P., Fiestas, J., Just, A. & Spurzem, R. 2007 *MNRAS* 377, 465

Evans, C. R. & Kochanek, C. S. 1989, *ApJL*, 346, L13

Feldmeier, A., Neumayer, N., Seth, A., Schödel, R., Lützgendorf, N., de Zeeuw, P. T., Kissler-Patig, M., Nishiyama, S. & Walcher, C. J. 2014 *A&A* 570, A2

Frank, J. & Rees, M. J. 1976 *MNRAS* 176, 633

Hernquist, L. & Ostriker, J. P. 1992 *ApJ* 386, 375

Komossa, S. 2002 *Reviews in Modern Astronomy*, 15, 27

Komossa, S. & Merritt, D. 2008 *ApJL* 683, L21

Lightman, A. P. & Shapiro, S. L. 1976 *ApJ* 211, 244

Liu, F. K., Li, S. & Komossa, S. 2014 *ApJ* 786, 103

Magorrian, J. & Tremaine, S. 1999 *MNRAS* 309, 447

Rees, M. J. 1988, *Nature*, 333, 523

Schödel, R., Feldmeier, A., Kunneriath, D., Stolovy, S., Neumayer, N., Amaro-Seoane, P. & Nishiyama, S. 2014 *A&A* 566, A47

Zhong, S., Berczik, P. & Spurzem, R. 2014 *ApJ* 792, 137

Star clusters and black holes in galaxies across cosmic time
Proceedings IAU Symposium No. 312, 2014
Y. Meiron, S. Li, F.-K. Liu & R. Spurzem, eds.

© International Astronomical Union 2016
doi:10.1017/S1743921315007644

The effect of gaseous accretion disk on dynamics of the stellar cluster in AGN

Bekdaulet Shukirgaliyev

Fesenkov Astrophysical Institute Observatory 23, 050020 Almaty, Kazakhstan
email: `bekdaulet@aphi.kz`

Abstract. There is a supermassive black hole, a gaseous accretion disk and compact star cluster in the center of active galactic nuclei, as known today. So the activity of AGN can be represented as the result of interaction of these three subsystems. In this work we investigate the dynamical interaction of a central star cluster surrounding a supermassive black hole and a central accretion disk. The dissipative force acting on stars in the disk leads to an asymmetry in the phase space distribution of the central star cluster due to the rotating accretion disk. In our work we present some results of Stardisk model, where we see some changes in density and phase space of central star cluster due to influence of rotating gaseous accretion disk.

Keywords. galaxies: nuclei, galaxies: active, accretion, accretion discs, stellar dynamics

1. Introduction

Physical nature of AGN is far from full understanding because of big distances and relatively small sizes of energy-producing regions. Therefore development of AGN theory still remains one of the main problems in astrophysics (Beckmann & Shrader 2013, Wu *et al.* 2015).

According to the dominant model, the phenomenon of AGN is explained by the accretion of matter onto a supermassive black hole in the center of galaxies (according to modern data (Kormendy & Ho 2013), there are supermassive black holes with mass from a few billion to several trillions of solar masses in the center of most galaxies). In the process of accretion potential and kinetic energy of substances is effectively converted into radiation energy, which may explain the stable and very powerful radiation from a very small region, observed in AGNs. Since the angular momentum of accreting substances is retained, it forms a disk, and the system becomes axisymmetric.

However, there are typically spherically symmetric compact stellar clusters around central black hole in the central region of galaxies. The previous work (Just *et al.* 2012, Vilkoviskij *et al.* 2013) investigated how star-disk interactions effect the evolution of AGN using simpler model of accretion disk and numerical simulations. In particular, it was found that the dissipative star-disk interactions can significantly increase the rate of accretion of stars onto SMBH as stars transfer portion of the energy to gas and their orbits are stacked in the plane of accretion disk, and as a result they also accreting onto the central black hole (Just *et al.* 2012). However, to assess the impact of the accretion disk in the orbital and phase characteristics of the stars, one needs to explore a more realistic model of the disk.

The investigated AGN model includes three subsystems: a compact star cluster (CSC), the accretion disk (AD), and the central supermassive black hole (SMBH). A star cluster is simulated by direct integration of the individual stars interaction with each other (direct N-body simulations) and with a gas disk and black hole. Gaseous accretion disk is defined phenomenologically with density distribution constant in time, and has a

Keplerian rotation curve. Black hole is also defined phenomenologically as a Newtonian potential. If the particle comes inside the region with radius less than R_{accr} (radius of accretion), then we consider that it is accreted onto the supermassive black hole. Once it happened, the particle disappears and its mass is added to the SMBH. We used modified version of phiGRAPE code (Harfst *et al.* 2007)for our simulations, which uses parallel computing technology of NVIDIA CUDA and MPI. The stellar component of the system is defined as a Plummer density profile initially. A more detailed description of the numerical model can be found in Just *et al.* 2012, Vilkoviskij *et al.* 2013.

In Kazakhstan the first computer cluster specialized for N-body simulations was created in 2008 within the frame of the joint Kazakh-German "STARDISC" project which combined forces of groups in Heidelberg University and Fesenkov Astrophysical Institute. This work presents results of investigations of evolution and physics problems in AGN are calculated in this computer cluster as well as in computer cluster of Astronomisches Rechen-Institute in Germany.

2. Accretion disk model

Let us consider the model of the accretion disk. We take as a basis a three-dimensional, axisymmetric stationary disk, which is characterized by differential rotation with the local angular velocity. The radial profile of the surface density is specified as

$$\Sigma(R) = \Sigma_d \left(\frac{R}{R_d} \right)^{-\alpha}, \tag{2.1}$$

here $\alpha = \frac{3}{4}, R^2 = x^2 + y^2, R_d$ is disk radius and Σ_d is surface density value at $R = R_d$. The value $\alpha = 3/4$ corresponds to the outer boundary of the disk model of the Novikov & Thorne 1973. Disk mass is equal to

$$M_d = 2\pi \int_0^{R_d} \Sigma(R) R dR = \frac{2\pi}{2 - \alpha} \Sigma_d R_d^2. \tag{2.2}$$

For the numerical integration of the equations of motion the force acting on the particle should be smooth and continuous function, so one needs enter the exponential factor that ensures a smooth (but fast enough) decreasing of gas density at the edge of the disk (Just *et al.* 2012).

$$\Sigma(R) = \Sigma_d \left(\frac{R}{R_d} \right)^{-\alpha} e^{-\beta_s \left(\frac{R}{R_d} \right)^s}. \tag{2.3}$$

For the resulting expression to correspond to the equation 2.2 we selected

$$\beta_s = \Gamma \left(1 + \frac{2 - \alpha}{s} \right),$$

wherein $\Gamma(x)$ - the gamma function. Let us take $s = 4$, then $beta_s = 0.70$ for $\alpha = 3/4$. In this case, if $R = R_d$ the surface density is equal to $\Sigma(R_d) = 0.49\Sigma_d$ (Just *et al.* 2012). To numerically model the described model we select isothermal density profile, defined as

$$\rho_g(R, z) = \frac{\Sigma(R)}{\sqrt{2\pi}h_z} exp \left(-\frac{z^2}{2h_z^2} \right). \tag{2.4}$$

Previous works (Just *et al.* 2012, Vilkoviskij *et al.* 2013) used constant thickness accretion disk model

$$h_z = hR_d. \tag{2.5}$$

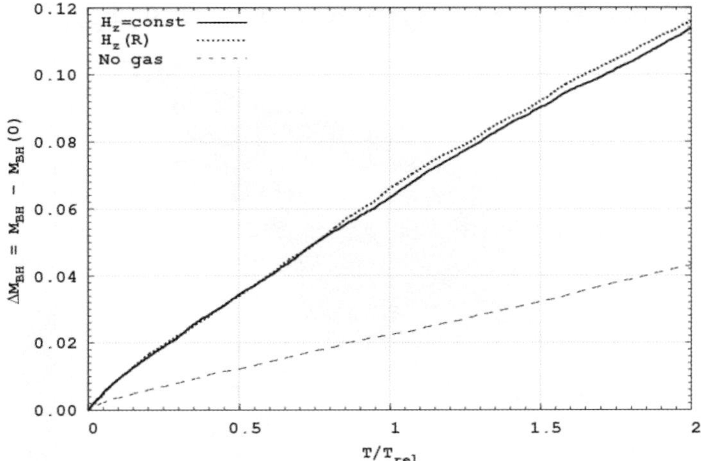

Figure 1. The black hole mass growth as a function of time for three model runs - the old model of disk (solid line), with the new model (dotted line) and a control run without gas disk (dotted line). The mass of black hole is in dimensionless N-body units, the unit of time specified in the relaxation time of the system.

If the relationship 2.5 is substituted into the expression 2.4, we obtain the expression for the density of disk with a constant height:

$$\rho(R, z) = \frac{2 - \alpha}{2\pi\sqrt{2\pi}} \frac{M_d}{hR_d^3} \left(\frac{R}{R_d}\right)^{-\alpha} e^{-\beta_s \left(\frac{R}{R_d}\right)^s} e^{-\frac{z^2}{2h^2 R_d^2}}. \tag{2.6}$$

In this work we consider a new model, which is a modification of the first model of accretion disk with the introduction of a linear function to increase the disk half thickness in the inner part. This modification is based on the physical properties of the inner accretion disk, which is described by the approach of Shakura & Sunyaev 1973

$$h_z = hR_d \left(\frac{R}{R_{crit}}\right). \tag{2.7}$$

The transition point from linearly increasing to constant thickness is in the point where the disk is vertically self-gravitating at $R = R_{crit}$. In our simulations $R_{crit} = 0.0257314$ in N-body dimensionless unit system (Hénon 1971).

The properties of the accretion disc are fixed by the reduced mass with analytical density distribution according to equation 2.6 with the values $\alpha = 3/4, s = 4$ and $h = 10^{-3}$. The disk has Keplerian rotation in the potential of SMBH neglecting the gravitational influence of the disk and pressure gradients within the disk (Just *et al.* 2012).

3. Results

To ensure that the new model of disk does not change the global dynamics of the system and thereby does not contradict the results we obtained earlier, we compared the black hole mass growth rate due to the accretion of stars (Fig. 1).

As can be seen in Fig. 1 the evolution of the black hole during two relaxation times for both accretion disk models is identical. At the same graph control run without gaseous disk is shown. In this case, the black hole mass growth is due to only the capture of those stars, which accidentally happened to be within the accretion radius. We also investigate the impact of accretion disk on the dynamical characteristics of stellar cluster. In

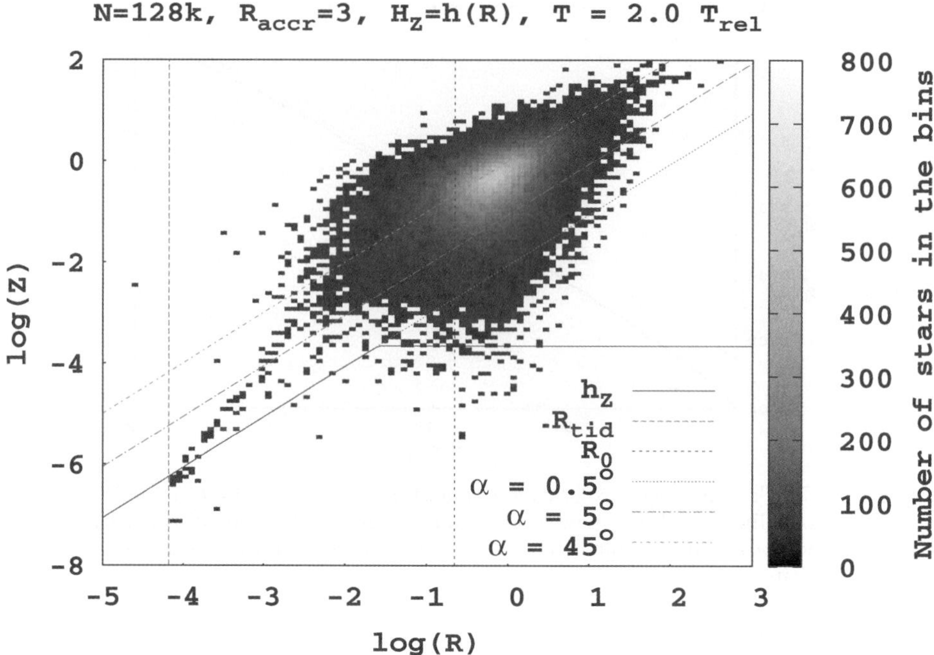

Figure 2. Stellar distribution in cylindrical coordinate system. Here stellar disk formed in the inner part of the system shown as a tail to the center of the system.

particular, we analysed orbital parameters of accreted particles. We found that these two disk models give different results in orbital parameters distribution of captured particles. Specifically, the new disk model allows stars to survive in the accretion disk for longer times than in old disk model. So we get much more particles accreted in near-circular orbits ($e \approx 0$) in the disk plane ($i < 0.5°$). Although both disk models initiate the same growth rate of the central SMBH (Fig. 1), the gaseous disk with a constant thickness captures more counter-rotating stars, in contrast to the disk with a variable thickness.

If we decrease the accretion radius value in the previous model of the disk, the gas density increases in the central part, but its thickness remains constant and it rotates around the center on Keplerian orbits. This leads to deceleration of many stars on orbits counter-rotating with the disk in the inner part of the system, including stars in nearly perpendicular orbits to the disk plane. In case of an improved disk model, super-dense gas occurs practically only in the equatorial plane near the central black hole, and that allows many stars in the central part to evolve towards the direction of rotation of the disk. Stars evolved in the positive direction of rotation relatively to the direction of the disk rotation could increase the contribution of the positive Z-component of the angular momentum in the central part of the cluster.

We found that the disk model with varying thickness leads to formation of relatively stable disk from stars in the inner part of the system. It captures stars which are going to be accreted and allows them to live longer in the disk. One can see formation of such disk in (Fig. 2). The central star cluster changes its structure in the inner part and forms stellar disk in the early stage of simulation ($T \approx 0.01 T_{rel}$). This structure of star cluster retained during two relaxation times (up to the end of the simulations).

Thus, we see that the resulting phenomenological model of gas accretion disk is physically adequate and the studied density profile can be used in the future, including modeling of the gas disk directly using the methods of hydrodynamics. In the future, we plan to improve the numerical model of AGNs in order to allow additional processes (collisions of stars in the central part, stellar evolution, primordial binaries, binary black holes, etc.), and to perform one-to-one simulations, i.e. with the number of stars equal to that in real clusters (about one million stars).

Acknowledgements

I would like to thank my collaborators G. Kennedy, Y. Meiron, P. Berczik, A. Just, T. Panamarev, M. Makukov, D. Yurin, C. Omarov, E. Vilkoviski, and R. Spurzem. We will publish a more detailed presentation of results elsewhere. I also thank International Astronomical Union for its grant which gave me this opportunity to participate in IAU Symposium No. 312 in Beijing in 2014.

References

Beckmann, V. & Shrader, C. 2013, *Active galactic nuclei* (John Wiley & Sons.), 374 p.
Harfst, S., Gualandris, A., Merritt, D., Spurzem, R., Zwart, S. P., & Berczik, P. 2007, *New Astronomy*, 12, 357
Hénon, M 1971, *Astrophys. Space Sci.*, 14, 151
Just, A., Yurin, D., Makukov, M., Berczik, P., Omarov, Ch., Spurzem, R., & Vilkoviskij, E. Y. 2012, *ApJ*, 758, 51
Kormendy J. & Ho L. C. 2013, *Annu. Rev. Astron. Astrophys*, 51, 511
Novikov I. D., Thorne K. S. 1973, in Dewitt C., Dewitt B. S., eds, *Black Holes (Les Astres Occlus) Astrophysics of black holes..*, pp 343-450
Shakura N. I. & Sunyaev R. A. 1973, *A&A*, 24, 337
Vilkoviskij E., Makukov M., Omarov Ch., Panamarev T., Spurzem R., Berczik P., & Just A. 2013, *A&AT*, 28, 151
Wu X.-B., Wang F., Fan X., Yi W., Zuo W., Bian F., Jiang L., McGreer I. D., Wang R., Yang J., Yang Q., Thompson D., & Beletsky Y 2015, *Nature*, 518, 512

Star clusters and black holes in galaxies across cosmic time
Proceedings IAU Symposium No. 312, 2014
Y. Meiron, S. Li, F.-K. Liu & R. Spurzem, eds.
© International Astronomical Union 2016
doi:10.1017/S1743921315007656

The interaction between supermassive black holes and globular clusters

Mario Spera[1,2], Manuel Arca-Sedda[2,3] and Roberto Capuzzo-Dolcetta[2]

[1] INAF-Osservatorio Astronomico di Padova, Vicolo dell'Osservatorio 5, I-35122, Padova, Italy
email: mario.spera@oapd.inaf.it

[2] Sapienza-Universitá di Roma, P.le A. Moro 5, I-00165 Rome, Italy

[3] Universitá di Tor Vergata, Via O. Raimondo 18, I-00173 Rome, Italy

Abstract. Almost all galaxies along the Hubble sequence host a compact massive object (CMO) in their center. The CMO can be either a supermassive black hole (SMBH) or a very dense stellar cluster, also known as nuclear star cluster (NSC). Generally, heavier galaxies (mass $\gtrsim 10^{11}\,\mathrm{M_\odot}$) host a central SMBH while lighter show a central NSC. Intermediate mass hosts, instead, contain both a NSC and a SMBH. One possible formation mechanisms of a NSC relies on the *dry-merger (migratory)* scenario, in which globular clusters (GCs) decay toward the center of the host galaxy and merge. In this framework, the absence of NSCs in high-mass galaxies can be imputed to destruction of the infalling GCs by the intense tidal field of the central SMBH. In this work, we report preliminary results of N-body simulations performed using our high-resolution, direct, code `HiGPUs`, to investigate the effects of a central SMBH on a single GC orbiting around it. By varying either the mass of the SMBH and the mass of the host galaxy, we derived an upper limit to the mass of the central SMBH, and thus to the mass of the host, above which the formation of a NSC is suppressed.

Keywords. galaxies: star clusters, methods: n-body simulations

---

1. Introduction

The innermost regions of galaxies often host a very compact stellar cluster with a typical half-light radius, r_{hl}, of few parsecs, luminosity $\sim 10^7\,\mathrm{L_\odot}$ and total mass $\sim 10^7\,\mathrm{M_\odot}$. These compact structures are known as nuclear star clusters (NSCs) (Böker *et al.* 2004; Côté *et al.* 2006; Turner *et al.* 2012). NSCs are observed in galaxies of all the Hubble types and, sometimes, they co-exist with a central supermassive black hole (SMBH) (Graham & Spitler 2009). For instance, the Milky Way hosts both a NSC with mass $\sim 10^7\,\mathrm{M_\odot}$ and a SMBH of mass $\sim 4 \times 10^6\,\mathrm{M_\odot}$ (Schödel *et al.* 2009).

Two (not exclusive) mechanisms have been proposed for NSC formation:

(*a*) *dissipative*: in this framework the NSC should form thanks to a continuous radial inflow of gas and a subsequent, in situ, star formation (Milosavljević 2004; Bekki 2007);

(*b*) *dissipationless*: this is also known as *dry-merging* scenario in which globular clusters (GCs) decay towards the galactic center via dynamical friction forming and accreting the NSC (Tremaine *et al.* 1975; Capuzzo-Dolcetta 1993; Capuzzo-Dolcetta & Miocchi 2008).

While the in-situ scenario is purely speculative, the dry-merging scenario has been quantitatively investigated and matches observational data and correlations (Arca-Sedda & Capuzzo-Dolcetta 2014; Antonini *et al.* 2012). Anyway, likely the actual formation mechanism can be the result of some combination between the dissipative and the dissipationless processes. Moreover, the presence of a central SMBH could significantly alter the process of formation and growth of the NSC; many authors have shown that the masses of the SMBH, of the galaxy and of the NSC are strongly related. In particular, there is observational evidence of a clear distinction between low-mass galaxies (with mass $\lesssim 10^{10}\,\mathrm{M_\odot}$),whose nuclei are dominated by a NSC, and heavier galaxies, dominated by a SMBH (Scott & Graham 2013).

In this work, and in the framework of the dry-merging scenario, we investigate the interaction between SMBHs and GCs, in order to determine if the mutual influence may play a role in the co-existence of SMBHs and NSCs. We give here preliminary results of a more extended work still in progress (Arca-Sedda *et al.* 2015).

We approach the problem by means of high-precision, direct, N-body simulations following the dynamical evolution of several astrophysical systems composed of a galaxy bulge, a central SMBH and a GC moving on different orbits. To this purpose, we used the N-body code `HiGPUs`(Capuzzo-Dolcetta *et al.* 2013) which, thanks to the hardware acceleration given by graphics processing units, allowed us to use $\sim 1\mathrm{M}$ particles in our simulations, obtaining scientific results with high spatial resolution.

2. Model

Globular cluster The *test* GC in our simulations is built according to a King's mass density profile with central potential parameter $W_0 = 6$, a King's radius $r_\mathrm{k} = 0.24$ pc and a total mass $M_{\mathrm{GC}} = 10^6\,\mathrm{M_\odot}$.

In all the investigated cases, the GC is initially placed at 50 pc from the central SMBH. Therefore, we assume that the GC has already decayed toward the inner galactic region, where the presence of the SMBH may play an important dynamical role. It is worth noting that the choice of a large initial GC mass ($10^6\,\mathrm{M_\odot}$) is motivated by the requirement of orbital shrinking via dynamical friction in less than a Hubble time.

Galaxies and central black holes Since direct N-body simulations cannot handle more than $\sim 10^6$ particles in an efficient way, we decided to model our galaxies by sampling only their central regions. In this work, we focus our attention on elliptical galaxies represented as Dehnen's mass density profiles family (at varying γ, see Dehnen 1993), $\rho_D(r)$, truncated (McMillan & Dehnen 2007) as

$$\rho_{tr}(r) = \rho_D(r)\,\mathrm{sech}\left(\frac{r}{r_\mathrm{cut}}\right). \qquad (2.1)$$

The $\rho_{tr}(r)$ profile falls off as e^{-r/r_cut} for $r \gtrsim r_\mathrm{cut}$ allowing a good representation of the region of interest with a reasonable number of particles. The mass value of the central SMBH is assigned according to the formula by Scott & Graham (2013)

$$\log\left(\frac{M_{\mathrm{BH}}}{\mathrm{M_\odot}}\right) = 1.37\log\left(\frac{M_{\mathrm{gal}}}{10^{11}\,\mathrm{M_\odot}}\right) + 8.06. \qquad (2.2)$$

In this work we spanned the mass ranges $10^{10}\,\mathrm{M_\odot} < M_{\mathrm{gal}} < 3.2 \times 10^{11}\,\mathrm{M_\odot}$ and $5 \times 10^6\,\mathrm{M_\odot} < M_{\mathrm{BH}} < 5 \times 10^8\,\mathrm{M_\odot}$. Table 1 summarizes the parameters adopted in each simulation.

3. Results for GC circular orbits

For each galaxy model, we made three simulations corresponding to a circular, an eccentric ($e \sim 0.75$) and a radial orbit of the GC. In this preliminary work we report the results of the GC moving on a circular orbit.

Table 1. Parameters of the models. M_{gal} and $M_{\mathrm{gal,cut}}$ are the total and truncated (in dependence on r_{cut}) galaxy masses, M_{BH} is the SMBH mass; r_s and γ is the parameter of the galactic Dehnen's profile; N_{gal} and N_{GC} are the numbers of particles used in our simulations for the galaxy and for the GC.

M_{gal} ($\mathrm{M_\odot}$)	M_{BH} ($\mathrm{M_\odot}$)	r_s (kpc)	r_{cut} (pc)	γ	$M_{\mathrm{gal,cut}}$ ($\mathrm{M_\odot}$)	N_{gal}	N_{GC}
10^{10}	5×10^6	0.995	70	0.3	3.4×10^7	1,018,742	29,832
3.2×10^{10}	2×10^7	1.512	70	0.3	4.1×10^7	1,024,025	24,550
10^{11}	10^8	1.917	70	0.2	5.9×10^7	1,031,338	17,237
3.2×10^{11}	5×10^8	2.876	70	0.2	6.8×10^7	1,033,332	15,243

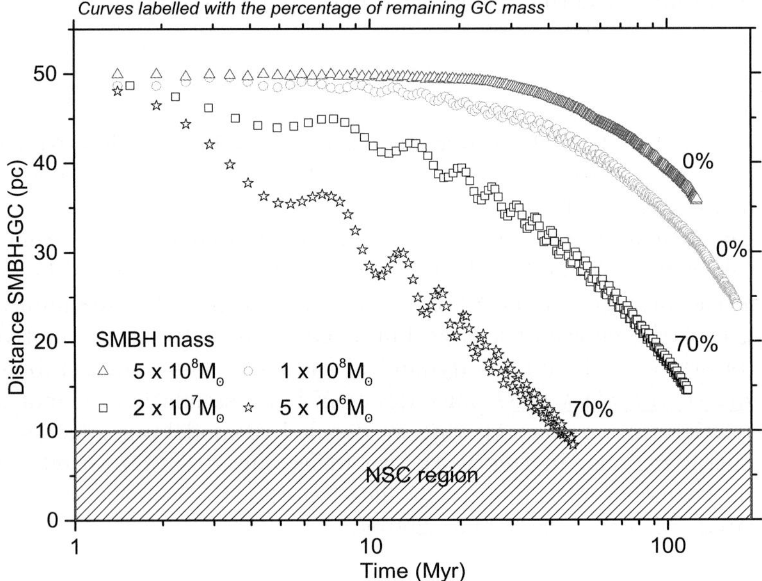

Figure 1. Time evolution of the distance of the decaying GC from the SMBH in the models of Table 1. Each curve is labeled with the percentage to the initial mass that the GC keep bound at the end of each simulation. The shaded area (distances SMBH-GC smaller than 10 pc) marks the region where a NSC may form and grow.

The four different curves in Fig. 1 refer to the time evolution of the GC galactocentric distance in our four models of galaxy. Fig. 1 shows that heavier SMBHs are able to disrupt the GC before it gets to the NSC region, while the incoming GC can survive to the interaction with the central SMBHs if this is relatively light. In fact, in the case of $M_{\mathrm{BH}} = 5 \times 10^6$ $\mathrm{M_\odot}$, the GC can come closer than 10 pc to the SMBH preserving 70% of its initial mass.

This implies that, in this situation, GCs may indeed significantly contribute to the formation and growth of a NSC while heavier SMBHs tends to *protect* the galaxy center, preventing a local mass accumulation and favouring the tidal dissolution of incoming GCs. Figure 1 provides also evidence of a transition regime between *dynamical friction dominated* galaxies (formation and growth of NSC by mergers) and *tidal disruption dominated* galaxies (no NSC). Our, preliminary, simulations suggest the transition lies in the range of SMBHs masses between 2×10^7 $\mathrm{M_\odot}$ and 10^8 $\mathrm{M_\odot}$.

4. Conclusions

We presented preliminary results about the problem of the dearth of NSCs in high mass (elliptical) galaxies. We performed some high-precision, direct N-body simulations to investigate the dynamical fate of a massive GC orbiting the inner region of a galaxy. We showed that a SMBH heavier than $\sim 10^8\,M_\odot$ can efficiently disrupt the infalling GC before it gets to what we called *NSC region*, and therefore it may inhibit the formation process, by subsequent merging events, of a NSC. On the other hand, the incoming GC survives the interaction with lighter BHs and, thus, can contribute to the formation and growth of a NSC. To conclude, our simulation results are a reliable confirmation of the important role played by a massive central black hole on the infalling GCs, as first pointed out by Capuzzo-Dolcetta (1993) and, more recently, by Antonini (2013). Nevertheless, a firmer statement about the topic studied here deserves:

(a) a wider range of initial conditions for both the GC structure and its initial orbital parameters;

(b) an extension of the galaxy models, to determine more precisely the threshold in M_{BH} below which the dry-merging of GCs easily allows the formation of a NSC;

(c) a better N-body sampling of our models, possible by taking advantage of the next generation hardware and software.

Acknowledgements

MS thanks M. Mapelli for useful discussions, and acknowledges financial support from the MIUR through grant FIRB 2012 RBFR12PM1F. MAS acknowledges financial support provided by MIUR through the grant PRIN 2010 LY5N2T 005.

References

Antonini, F. 2013, *ApJ*, 763, 62

Antonini, F., Capuzzo-Dolcetta, R., Mastrobuono-Battisti, A., & Merritt, D. 2012, *ApJ*, 750, 111

Arca-Sedda, M. & Capuzzo-Dolcetta, R. 2014, *MNRAS*, 444, 3738-3755

Arca-Sedda, M., Capuzzo-Dolcetta, R., & Spera, M. 2015, submitted to *MNRAS* (arXiv:1510.01137)

Bekki, K. 2007, *PASA*, 24, 77

Böker, T., Sarzi, M., McLaughlin, D. E., *et al.* 2004, *AJ*, 127, 105

Capuzzo-Dolcetta, R. 1993, *ApJ*, 415, 616

Capuzzo-Dolcetta, R. & Miocchi, P. 2008, *ApJ*, 681, 1136

Capuzzo-Dolcetta, R., Spera, M., & Punzo, D. 2013, JCP, 236, 580

Côté, P., Piatek, S., Ferrarese, L., *et al.* 2006, *ApJS*, 165, 57-94

Dehnen, W. 1993, *MNRAS*, 265, 250

Graham, A. W. & Spitler, L. R. 2009, *MNRAS*, 397, 2148

McMillan, P. J. & Dehnen, W. 2007, *MNRAS*, 378, 541

Milosavljević, M. 2004, *ApJ*, 605, L13

Schödel, R., Merritt, D., & Eckart, A. 2009, *A&A*, 502, 91

Scott, N. & Graham, A. W. 2013, *ApJ*, 763, 76

Tremaine, S. D., Ostriker, J. P., & Spitzer, L., Jr. 1975, *ApJ*, 196, 407

Turner, M. L., Côté, P., Ferrarese, L., *et al.* 2012, *ApJS*, 203, 5

Star clusters and black holes in galaxies across cosmic time
Proceedings IAU Symposium No. 312, 2014
Y. Meiron, S. Li, F.-K. Liu & R. Spurzem, eds.

© International Astronomical Union 2016
doi:10.1017/S1743921315007668

Formation and evolution of nuclear star clusters

Alessandra Mastrobuono-Battisti and Hagai B. Perets

Physics Department, Technion – Israel Institute of Technology, Haifa 3200003, Israel
email: `amastrobuono@physics.technion.ac.il`

Abstract. Nuclear stellar clusters (NSCs) are dense stellar systems known to exist at the center of most of the galaxies. Some of them host a central massive black hole (MBH). They are though to form through in-situ star formation following the infall of gas to the galactic center and/or because of the infall and merger of several stellar clusters. Here we explore the latter scenario by means of detailed self-consistent N-body simulations, proving that a NSC built by the infall and following merger of stellar clusters shows many of the observed features of the Milky Way NSC. We also explore the possibility that the infalling clusters host intermediate mass black holes (IMBHs). Once decayed to the center, the IMBHs act as massive-perturbers accelerating the relaxation of the NSC, filling the loss-cone and boosting the tidal disruption rate of stars up to a value larger than the observational estimates, therefore providing a cumulative constraint on the existence of IMBHs in NSCs. Studying how the properties of the infalling clusters map to the properties of the resulting NSC, we find that, in the IMBHs-free case, the infall mechanism is able to produce many different observational signatures in the form of age segregation.

Keywords. galaxies: bulges – galaxies: kinematics and dynamics – galaxies: nuclei – galaxies: star clusters:general – stars: black holes

1. Introduction

NSCs are dense massive stellar clusters (with typically a few $\times 10^6 - 10^7$ stars), residing in galactic nuclei and most of them hosting a MBH at their center (see Böker 2010, for an overview). They are characterized by effective radii of a few pc and central luminosities up to $\sim 10^7$ L$_\odot$. The formation mechanism of NSCs is still unknown and two are the main competing proposed models: (i) in the in-situ star formation scenario the NSC forms from the gas that migrates to the center of the galaxy (Schödel *et al.* 2008), (ii) in the cluster-infall scenario massive clusters, like globular clusters (GCs) decay to the center via dynamical friction and merge to form a dense nucleus (Tremaine *et al.* 1975; Capuzzo-Dolcetta 1993). These two scenarios are not mutually exclusive, and both can contribute to the formation of NSCs. Here we explore the cluster-infall model by means of detailed N-body simulations both for the Milky Way (MW) and for generic galaxies. The constituent clusters could host IMBHs (Madau & Rees 2001; Ebisuzaki *et al.* 2001; Bromm *et al.* 2003) and thus we also explore this possibility studying the effects of these massive and compact objects on the formation and evolution of NSCs. In Sect. 2 we describe the initial conditions and the simulations, in Sections 3 and 4 we show and discuss the results obtained.

2. Simulations and initial conditions

The models and methods used in our simulations are described in Antonini *et al.* (2012) and in Mastrobuono-Battisti *et al.* (2014), where details of the initial conditions of the infalling GCs, as well as of the galaxy model for the background stellar population can be

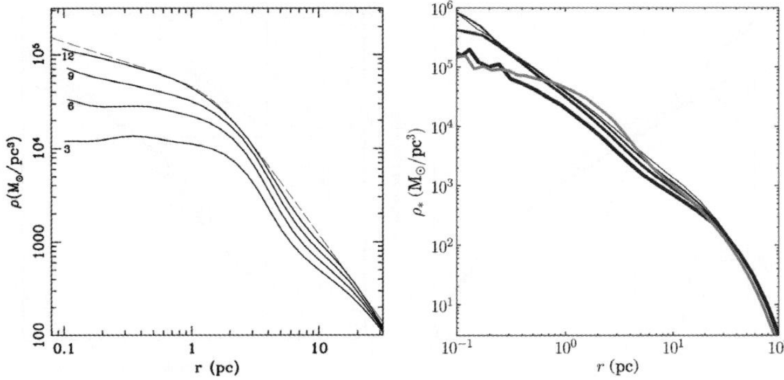

Figure 1. Left panel: Spatial profile of the central NSC after 3, 6, 9 and 12 mergers. The dashed line is the fit to the NSC profile obtained at the end of the last merging event using a broken power law model. Right panel: The same but for the simulation with IMBHs. The thickness of lines decreases with time. The red solid line represents the density profile of the NSC after 12 merging events in the IMBHs-free case. From Antonini *et al.* (2012) and Mastrobuono-Battisti *et al.* (2014).

found. In brief, we followed the decay and the merging of 12 massive dense clusters inside the Galactic bulge by means of fully self-consistent N-body simulations. The bulge of the simulated galaxy has been modeled based on recent observations (Launhardt *et al.* 2002) of the MW, while the individual GC N-body representation has been done starting from a tidally limited, massive and compact King model. A MBH of $4 \times 10^6 M_\odot$ is located at the center of the galaxy. We run simulations both with GCs containing only stars and with GCs hosting a central IMBH of $10^4 M_\odot$. After the first GC decayed to the galactic center, we waited for the NSC to reach a quasi-steady state and we added a second GC to the system. This procedure is iterated until the 12 clusters merged in the inner regions of the galaxy. After the end of all the infalls we followed the relaxation process of the resulting stellar cluster. The total simulation time, rescaled with the relaxation time of the system as described in Mastrobuono-Battisti & Perets (2013), extends to ~ 12 Gyr. We ran our simulations on the GPU partition of the Tamnun cluster at the Technion using $\phi GRAPE$ (Harfst *et al.* 2007), a direct-summation code optimized for computer clusters accelerated by GRAPE boards (Makino 1998) and recently optimized to run on graphic processing units (GPUs, Gaburov *et al.* 2009).

3. Results

The left panel of Fig. 1 shows the spatial profile of the system, in the IMBHs-free case, after the complete merging of 3, 6, 9 and 12 clusters. The density grows with time and, at the end of the last merging event, the core radius of the NSC is 4 times larger than the observed value. The merging phase is followed by a two body relaxation which is predicted to shrink the initial core with time, approaching the Bahcall & Wolf (1976) cusp steady state. The time scale for the process is of the order of the relaxation time at the influence radius of the central MBH, r_infl (Merritt 2010). After the end of the mergings, the system indeed evolves in a self-similar way, keeping almost unaltered the external radial slope (~ -1.8) while its core shrinks up to a final size essentially identical to the size of the core observed at the center of the Milky Way (~ 0.5 pc, see the green line in the left panel of Fig. 2 and Antonini *et al.* 2012). The NSC evolves toward velocity isotropy (and spherical symmetry) and, at the end of the whole simulation, the NSC has only a small bias toward tangential motion. This is consistent with proper-motion data that indicate a

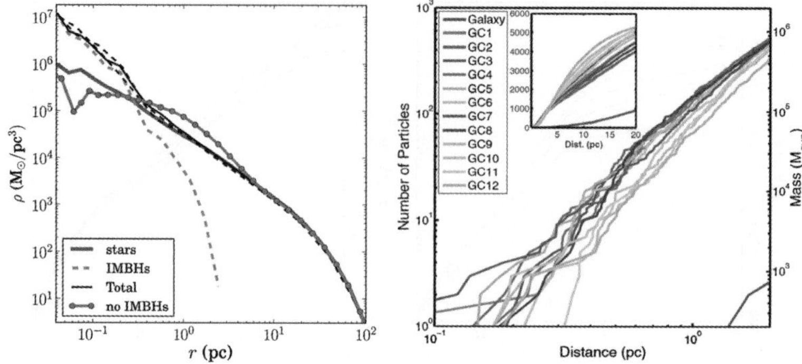

Figure 2. Right panel: The black solid line is the density profile of the NSC at the end of the simulation. The two dashed black lines are the same density profiles obtained from two additional simulations run. The blue solid line shows the spatial profile of the stellar component, while the red dashed line is the final density profile of the IMBHs. The green solid line with bullets shows the density profile of the NSC without IMBHs. Right panel: Radial distribution of the infalling clusters' stellar populations in the case without IMBHs at 12.4 Gyr. Insets show the large scale distribution. From Mastrobuono-Battisti *et al.* (2014) and Perets & Mastrobuono-Battisti (2014).

slight degree of tangential anisotropy (Schödel *et al.* 2014; Merritt 2010). As it is apparent from the right panel of Fig. 1, that shows the density profile of the stellar component of the system after 3, 6, 9 and 12 merging events in the case with IMBHs, the NSC evolution changes significantly when these objects are introduced inside the decaying GCs (Mastrobuono-Battisti *et al.* 2014). After the host GC is destroyed, each IMBH keeps inspiraling to the center leading to a fast relaxation of the NSC stellar population. At the end of the last infall the inner slope of the radial density profile for the whole system is ~ -2.1, steeper then what found in the IMBHs-free case. Indeed, the IMBHs act as massive perturbers, significantly decreasing the relaxation time of the host system (Perets *et al.* 2007). Therefore, at the end of the simulation, both the stellar cusp and steep radial distribution of the IMBHs are fully developed (see left panel of Fig. 2). The IMBHs cause the system to be in the full loss cone regime at any radius, while without them the loss-cone is empty inside the radius of influence of the MBH. We evaluated the integrated tidal disruption event (TDE) rate using analytic considerations combined with the results of the simulations. Accounting only for stars, we find a rate of 1.6×10^{-5} yr^{-1}, while including the IMBHs the rate becomes two order of magnitude higher ($10^{-3}\,\mathrm{yr}^{-1}$). The inferred rate of TDEs from observations is of the order of $10^{-5} - 10^{-4}\,\mathrm{yr}^{-1}$ (see e.g. Khabibullin & Sazonov 2014) i.e. lower than expected for multiple-IMBHs hosting NSCs such as seen in our simulations. We therefore conclude that typical NSCs should not contain IMBHs. Finally, in order to find signatures left by the merging process in the IMBHs-free case, we explored the mapping between the properties of stellar populations in NSCs, and their relation to the initial characteristics of the stellar populations in the progenitor clusters (Perets & Mastrobuono-Battisti 2014). We find that stars in the NSC originating from early-infall clusters are more segregated to the center of the NSC than their stellar counterparts from late-infall clusters. This is true both at the end of the merging process and even after few Gyr of evolution (see right panel of Fig. 2). We also find that the three-dimensional structure of the stellar populations of each of the GCs could significantly differ.

4. Discussion and conclusions

Using detailed N-body simulations we explored the infall-scenario for the formation and evolution of NSCs. We found that the merging of dense stellar clusters produces a system characterized by a central and almost flat core in the density profile and an external slope of -1.8, as actually observed in the MW NSC. Moreover the NSC keeps a slight degree of tangential anisotropy matching the observational data (e.g. Schödel *et al.* 2014). Thus, the process of consecutive infalls and merging in the inner region of the MW of massive dense clusters could provide an explanation for the formation of the Galactic NSC. If present inside the infalling clusters, once decayed to the center, IMBHs are found to scatter each other and the NSC stars, gaining high eccentricities. As a consequence, the NSC develops a cusp already in its early evolutionary stages. The NSC is strongly mass-segregated (Alexander & Hopman 2009) meaning that IMBHs and stars evolve separately and settle on two different profiles characterized by a central cusp with different slopes. The system is in the full loss cone regime and the tidal disruption event (TDE) rates are two orders of magnitude larger than TDE rate estimates from observations, suggesting that typical NSCs, and hence dense stellar clusters, which in the infall scenario are the NSC building blocks, may not contain IMBHs. Moreover, we explored the signatures and the implications of the cluster infall scenario (in the IMBHs-free case), on the structure of NSCs and on their multiple stellar populations. We found that the infall history is reflected in the final structure of the NSC, where stellar populations from earlier infalling clusters are more concentrated in the central parts of the NSC compared to late ones. This dynamical age segregation process can potentially leave behind a signature in the form of an age and/or metallicity radial gradient in the NSC stellar population. In addition, any primordial/early mass segregation in the infalling GCs is mapped into a mass segregated population in the galactic nucleus, potentially leading to efficient mass segregation in NSCs.

References

Alexander, T., & Hopman, C. 2009, *ApJ*, 697, 1861
Antonini, F., Capuzzo-Dolcetta, R., Mastrobuono-Battisti, A., & Merritt, D. 2012, *ApJ*, 750, 111
Bahcall, J. N., & Wolf, R. A. 1976, *ApJ*, 209, 214
Böker, T. 2010, in IAUS, Vol. 266, IAUS, ed. R. de Grijs & J. R. D. Lépine, 58–63
Bromm, V., Yoshida, N., & Hernquist, L. 2003, *ApJL*, 596, L135
Capuzzo-Dolcetta, R. 1993, *ApJ*, 415, 616
Ebisuzaki, T., Makino, J., & Tsuru, T. G., *et al.* 2001, *ApJL*, 562, L19
Gaburov, E., Harfst, S., & Portegies Zwart, S. 2009, *NewAstron*, 14, 630
Harfst, S., Gualandris, A., & Merritt, D., *et al.* 2007, *NewAstron*, 12, 357
Khabibullin, I., & Sazonov, S. 2014, *MNRAS*, 444, 1041
Launhardt, R., Zylka, R., & Mezger, P. G. 2002, *A&A*, 384, 112
Madau, P., & Rees, M. J. 2001, *ApJL*, 551, L27
Makino, J. 1998, *Highlights of Astronomy*, 11, 597
Mastrobuono-Battisti, A., & Perets, H. B. 2013, *ApJ*, 779, 85
Mastrobuono-Battisti, A., Perets, H. B., & Loeb, A. 2014, *ApJ*, 796, 40
Merritt, D. 2010, *ApJ*, 718, 739
Perets, H. B., Hopman, C., & Alexander, T. 2007, *ApJ*, 656, 709
Perets, H. B., & Mastrobuono-Battisti, A. 2014, *ApJL*, 784, L44
Schödel, R., Feldmeier, A., Neumayer, N., Meyer, L., & Yelda, S. 2014, *ArXiv e-prints*
Schödel, R., Merritt, D., & Eckart, A. 2008, *Journal of Physics Conference Series*, 131, 012044
Tremaine, S. D., Ostriker, J. P., & Spitzer, Jr., L. 1975, *ApJ*, 196, 407

Star clusters and black holes in galaxies across cosmic time
Proceedings IAU Symposium No. 312, 2014
Y. Meiron, S. Li, F.-K. Liu & R. Spurzem, eds.
© International Astronomical Union 2016
doi:10.1017/S174392131500767X

Sculpting the central parsec of our Galaxy

Xian Chen

Max-Planck Institute for Gravitational Physics, Am Mühlenberg 1, Potsdam, Germany
email: xian.chen@aei.mpg.de

Abstract. Recent observations have revealed various structures within the gravitational influence of Sgr A* – the massive black hole in the Galactic center. These structures apparently defy the fundamental principles of star formation and stellar dynamics. On one hand, the red giants display a flat density profile, contrary to the cuspy one predicted by conventional stellar relaxation. On the other, Wolf-Rayet and OB stars are observed where in-situ star formation should have been prohibited by the strong tidal force from Sgr A*, and their spatial and phase-space distributions also contradict our understanding of stellar dynamics. To explain each of these inconsistencies, many scenarios have been proposed, which render the model increasingly complicated. Here, we suggest that the sub-parsec stellar disk surrounding Sgr A*, which was recently discovered, can reconcile all the above inconsistencies. We show that during the fragmenting past of this disk, the star-forming clumps could efficiently deplete red giants by repeatedly colliding with them. We also show that because of the torque exerted by the disk, stars within the central arcsec from Sgr A* would quickly mix in the angular-momentum space, which naturally explains the observed distributions of Wolf-Rayet and OB stars. Our results imply that Sgr A* was fueled by gas and stars several millions years ago and could have been an energetic AGN. We discuss future observations that can further testify our model.

Keywords. black hole, Galactic center, dynamics

There are at least three dilemmas that theorist face when they are confronted with the rich phenomena in the Galactic Center (GC) (see Genzel *et al.* 2010 for a review):

(*a*) **The missing red giant problem**: the lack of a cusp in the density distribution of the red giants, which contradicts a longstanding hypothesis that old stars in the deep potential well of a supermassive black hole can develop a cusp under the dynamical process of two-body relaxation.

(*b*) **The paradox of youth**: the existence of a dynamically (nearly) relaxed B-star cluster (S-cluster) in the central 1 arcsec (0.04 pc) region around Sgr A*, where neither in-situ star formation nor fast stellar relaxation is thought possible.

(*c*) **The inverse mass segregation**: a counterintuitive stellar distribution such that the less-massive B-stars are populating the central 1 arcsec region but the more-massive Wolf-Rayet and O-stars are exclusively further out.

Conventionally, these problems were thought to be unrelated and hence many separate solutions have been devised. This is not ideal as it renders the dynamical model of the GC progressively more complex.

Motivated by the recent discovery of a "mini disk", which consists of hundreds of massive young stars revolving closely around Sgr A*, we study the dynamical impact of this disk on the other structures. We find that the interplay between the mini disk and the other stellar components provides a single, unified resolution to all the previous inconsistencies between theory and observation:

(*a*) Following analytical arguments, we found that during the star-forming phase of the disk, the gaseous clumps, which must have existed to give birth to the massive stars, could efficiently strip the envelopes off the red giants through repeatedly collisions. These clumps are the culprits of the missing red giants (Amaro-Seoane & Chen 2014).

Predictions: (1) The cores released from the destructed RGs are still populating the GC. (2) The surviving RGs lie preferentially on the inclined orbits relative to the mini disk, otherwise, they cannot avoid large number of collisions and should have been depleted.

(*b*) Considering the possibility that the disk in the past was more massive and extended closer to Sgr A, we showed that it will induce a rapidly evolving region in the GC where a star could quickly migrate in the eccentricity space. This theoretical discovery provides a mechanism that fully resolves the paradox of youth (Chen & Amaro-Seoane 2014).

Predictions: (1) The gravitational torque of the disk will induce a super-thermal distribution in the eccentricities of S-stars. (2) The old stellar population are subject to the same disk torque and will also exhibit a super-thermal distribution.

(*c*) Now the phenomenon of the inverse mass segregation also can be understood: the orbits of the inner Wolf-Rayet and O stars, within the rapidly evolving region, could be perturbed sufficiently such that they are placed on highly eccentric orbits, leading to their tidal disruption by Sgr A* (Chen & Amaro-Seoane 2014).

Predictions: (1) The surviving Wolf-Rayet and O stars should lie exclusively outside the rapidly evolving region. (2) S-stars in a narrow radial range of $0.5 \sim 0.8$ arcsec should also be depleted due to tidal disruption.

Our results point to a picture that several million years ago Sgr A* was surrounded by an accretion disk and our GC was an active galactic nucleus (AGN); this AGN was powered by gas accretion as well as by tidal disruptions of WR/O stars.

References

Amaro-Seoane, P. & Chen, X. 2014, *ApJ*, 781, L18
Chen, X. & Amaro-Seoane, P. 2014, *ApJ*, 786, L14
Genzel, R., Eisenhauer, F., & Gillessen, S. 2010, *RvMP*, 82, 3121

Star clusters and black holes in galaxies across cosmic time
Proceedings IAU Symposium No. 312, 2014
Y. Meiron, S. Li, F.-K. Liu & R. Spurzem, eds.

© International Astronomical Union 2016
doi:10.1017/S1743921315007681

Orientation of galactic bulge planetary nebulae toward the Galactic center

Ashkbiz Danehkar[1] and Quentin A. Parker[1,2]

[1]Department of Physics and Astronomy, Macquarie University, Sydney, NSW 2109, Australia
email: `ashkbiz.danehkar@students.mq.edu.au`

[2]Australian Astronomical Observatory, PO Box 915, North Ryde, NSW 1670, Australia

Abstract. We have used the Wide Field Spectrograph on the Australian National University 2.3-m telescope to perform the integral field spectroscopy for a sample of the Galactic planetary nebulae. The spatially resolved velocity distributions of the Hα emission line were used to determine the kinematic features and nebular orientations. Our findings show that some bulge planetary nebulae toward the Galactic center have a particular orientation.

Keywords. Galaxy: bulge – planetary nebulae: general – Galaxy: kinematics and dynamics

1. Introduction

The majority of planetary nebulae (PNe) show predominantly axisymmetric morphologies, i.e. elliptical and bipolar. There is a strong argument in favor of the axisymmetric morphologies of PNe being related to the angular momentum of their central engines (e.g., Miszalski *et al.* 2009; Nordhaus *et al.* 2010). Recently, it has been found that the nebular inclinations are closely related to the orbital inclinations of their close-binary central stars (see e.g., Jones *et al.* 2012; Tyndall *et al.* 2012).

PN as a test particle can be used to trace the dynamics of the Galaxy. The major axes of bipolar PNe are assumed to be perfectly aligned with the angular momentum vectors of their central stars, so a population of bipolar Galactic PNe can be used to determine the angular momentum distribution of the local stars within the Galaxy. Recent studies showed that some bipolar PNe within the bulge near the Galactic center have a homogeneous orientation (Weidmann & Díaz 2008; Rees & Zijlstra 2013).

2. IFU observations of galactic planetary nebulae

In this study, we aim to disentangle 3D gaseous structures of some Galactic PNe by means of integral field spectroscopy. We used the Wide Field Spectrograph (WiFeS; Dopita *et al.* 2010) on the 2.3-m ANU telescope at Siding Spring Observatory. WiFeS is an image-slicing integral field unit (IFU) feeding a double-beam spectrograph. It has a field-of-view of $25'' \times 38''$ and a spatial resolution of $1''$. Our observations were carried out with a spectral resolution of $R \sim 7000$. The spectra were reduced using the IRAF pipeline WIFES (described in detail by Danehkar *et al.* 2013).

We obtain the kinematic features by applying a Gaussian curve fit to each spaxel of IFU datacube (see Fig. 1). The observed kinematic maps are then compared to 3-D kinematic models in order to find the orientations of the nebular symmetric axes onto the plane of the sky, i.e. the position angle (PA) measured from the north toward the east in the equatorial coordinate system (see Danehkar *et al.* 2014). The PA is transferred into the Galactic position angle (GPA) in the Galactic coordinate system. Fig. 1 (bottom) shows orientations of the PNe based on their symmetric axes projected onto the sky plane.

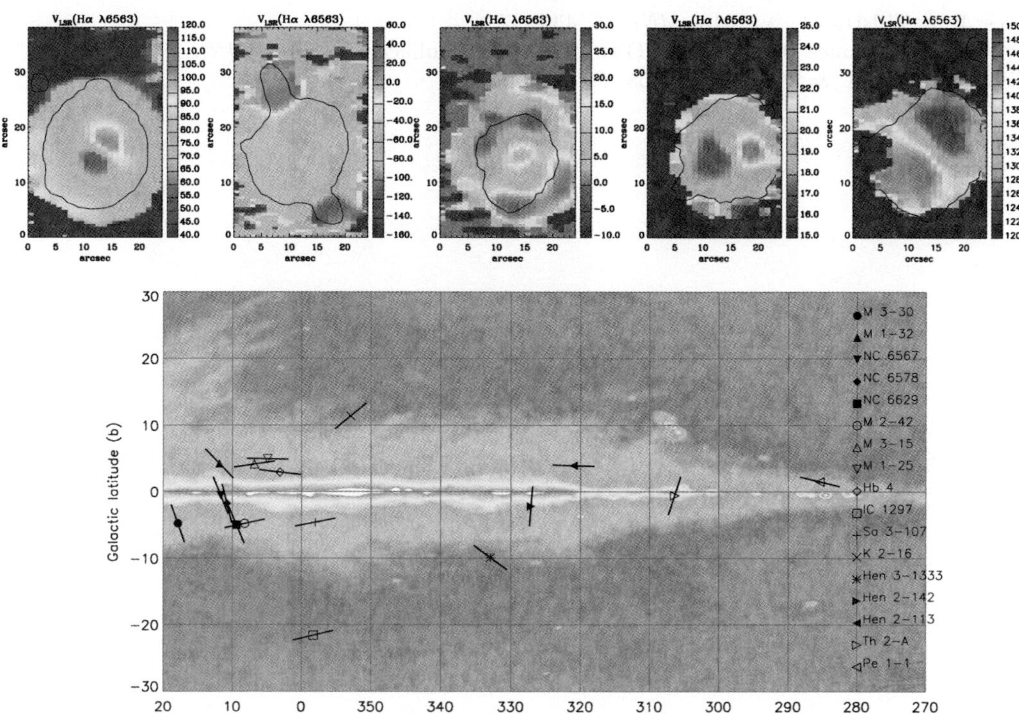

Figure 1. Top panels: From left to right, IFU kinematic maps of M 3-30, Hb 4, IC 1297, NGC 6578, and NGC 6567, respectively. North is up and east is toward the left-hand side. Bottom panel: Orientations of PNe in Galactic coordinates. The lines display the nebular symmetric axes. The background is the radio continuum (408 MHz) brightness from Haslam *et al.* (1982).

The spatial orientations of the Galactic PNe have been studied by few authors (Phillips 1997; Corradi *et al.* 1998; Weidmann & Díaz 2008; Rees & Zijlstra 2013). Although the orientations of nearby Galactic PNe are found to be completely arbitrary in the Galactic disc, some distant PNe seem to have a particular orientation within the Galactic bulge, as seen in Fig. 1 (in agreement with Weidmann & Díaz 2008; Rees & Zijlstra 2013). Galactic bulge PNe are affected by the bulge dynamics and the supermassive black hole located at the Galactic center, and are required to be observed and investigated further.

Acknowledgments

AD acknowledges travel grants from IAU and Australian Institute of Physics.

References

Corradi, R. L. M., Aznar, R., & Mampaso, A. 1998, *MNRAS*, 297, 617
Danehkar A., Parker Q. A., & Ercolano B., 2013, *MNRAS*, 434, 1513
Danehkar A., Todt H., Ercolano B., & Kniazev A. Y., 2014, *MNRAS*, 439, 3605
Dopita M. *et al.*, 2010, *Ap&SS*, 327, 245
Haslam, C. G. T., Salter, C. J., Stoffel, H., & Wilson, W. E. 1982, *A&AS*, 47, 1
Jones D. *et al.*, 2012, *MNRAS*, 420, 2271
Miszalski B., Acker A., Parker Q. A., & Moffat A. F. J., 2009, *A&A*, 505, 249
Nordhaus J., Spiegel D. S., Ibgui L., Goodman J., & Burrows A., 2010, *MNRAS*, 408, 631
Phillips, J. P. 1997, *A&A*, 325, 755

Rees, B. & Zijlstra, A. A. 2013, *MNRAS*, 435, 975
Tyndall A. A., Jones D., Lloyd M., O'Brien T. J., & Pollacco D., 2012, *MNRAS*, 422, 1804
Weidmann, W. A. & Díaz, R. J. 2008, *PASP*, 120, 380

Star clusters and black holes in galaxies across cosmic time
Proceedings IAU Symposium No. 312, 2014
Y. Meiron, S. Li, F.-K. Liu & R. Spurzem, eds.

© International Astronomical Union 2016
doi:10.1017/S1743921315007693

Eddington capture sphere around luminous relativistic stars

Maciek Wielgus

Nicolaus Copernicus Astronomical Center, PL-00-716, Warszawa, Poland
email: `maciek.wielgus@gmail.com`

Abstract. We discuss the interplay of gravity and radiation in a static, spherically symmetric spacetime. Because of the spacetime curvature, balance between radiation pressure from spherical star and effective force of gravity may be established in a particular distance from the star surface, on so-called Eddington capture sphere. This is in contrast with the Newtonian scenario, for which Eddington luminosity of the radiation assures gravity-radiation balance at any radius. We explore properties of this relativistic equilibrium and the dynamics of test particles under radiation influence in the strong gravity regime.

Keywords. Test particle, radiation, general relativity, neutron stars

1. Introduction

A peculiar relativistic equilibrium of gravity and radiation was first indicated by Phinney (1987). Abramowicz *et al.* (1990) gave analytic expression for the radiation stress-energy tensor of static spherical central source, emitting radiation isotropically and homogeneously in its rest frame. Hence, their formulas apply to the case of Schwarzschild spacetime and so do the results presented in this paper. In recent years several groups pursued this subject in more details, investigating its aspects such as the drag forces dynamical importance, Bini *et al.* (2009), Oh *et al.* (2011), Stahl *et al.* (2012), possible astrophysical applications for bursts Stahl *et al.* (2013), Mishra & Kluzniak (2014) and quasi-oscillatory behavior, Wielgus *et al.* (2012).

In this paper we recall the derivation of the equilibrium condition and investigate trajectories of test particles under radiation influence in curved Schwarzschild spacetime.

2. Analytic formulation

We begin with the Schwarzschild spacetime metric,

$$\mathrm{d}s^2 = -\xi\mathrm{d}t^2 + \xi^{-1}\mathrm{d}r^2 + r^2\mathrm{d}\theta^2 + r^2\sin^2\theta\mathrm{d}\phi^2\,, \qquad (2.1)$$

and denote $\xi = 1 - 2/r$. We utilize geometric units $G = c = 1$ and also put $M = 1$, since the equations scale with mass of the central object. Assuming that the central mass (neutron star) is emitting radiation isotropically and homogeneously in its rest frame, the stress-energy tensor components in the static observer's orthonormal frame can be found by integrating specific intensity over the observer's local sky. The necessary calculations were performed by Abramowicz *et al.* (1990), yielding

$$T^{(t)(t)} = 2\pi I(r)(1 - \cos\alpha_0)\,, \qquad (2.2)$$

$$T^{(t)(r)} = \pi I(r)\sin^2\alpha_0\,, \qquad (2.3)$$

$$T^{(r)(r)} = \frac{2}{3}\pi I(r)(1 - \cos^3 \alpha_0) \,, \tag{2.4}$$

$$T^{(\theta)(\theta)} = T^{(\phi)(\phi)} = \frac{1}{3}\pi I(r)(2 - 3\cos\alpha_0 + \cos^3 \alpha_0) \,. \tag{2.5}$$

Static observer's orthonormal frame is denoted by indices in brackets. In Schwarzschild spacetime it is elementary to calculate coordinate frame components from orthonormal frame components by simple rescaling. $I(r)$ is a specific intensity, R denotes radius of the star and $\alpha_0(r)$ is the star viewing angle,

$$I(r) = I(R)\left(\frac{1 - 2/R}{\xi}\right)^2 \,, \tag{2.6}$$

$$\alpha_0(r) = \arcsin\left[\frac{R}{r}\left(\frac{\xi}{1 - 2/R}\right)^{1/2}\right] \,. \tag{2.7}$$

Interestingly, similar analytic results were recently obtained assuming the observer located *inside* the luminous sphere in Schwarzschild spacetime. This case was derived by Wielgus & Abramowicz (2015) and discussed in details by Wielgus *et al.* (2014), in this paper we focus on the more standard scenario of observer located outside of the luminous sphere. Knowing the radiation stress-energy tensor, we may evaluate the radiation flux, being

$$F^i = -h^i{}_j T^{jk} u_k \,, \tag{2.8}$$

where $h^i{}_j = \delta^i{}_j + u^i u_j$ is the projection tensor and the corresponding four-force f^i can be compared to the four-acceleration a^i, giving

$$\frac{f^i}{m} = \frac{\sigma}{m}F^i = a^i = u^k \nabla_k u^i = \frac{d^2 x^i}{ds^2} + \Gamma^i_{jk} u^j u^k \,. \tag{2.9}$$

Test particle effective cross section is denoted with σ and its mass is denoted with m (typically Thomson cross-section of electron and mass of a proton). Evaluating the Christoffel symbols, we find the equations of motion as functions for the proper time τ

$$\frac{d^2 r}{d\tau^2} = (r - 3)\left(\frac{d\phi}{d\tau}\right)^2 - \frac{1}{r^2} + \frac{\sigma}{m}\xi T^{rt} u^t - \frac{\sigma}{m}\left(\xi^{-1}T^{rr} + T^{ij}u_i u_j\right)\frac{dr}{d\tau} \,, \tag{2.10}$$

$$\frac{d^2 \phi}{d\tau^2} = -\frac{2}{r}\frac{dr}{d\tau}\frac{d\phi}{d\tau} - \frac{\sigma}{m}\left(r^2 T^{\phi\phi} + T^{ik}u_i u_k\right)\frac{d\phi}{d\tau} \,. \tag{2.11}$$

We identify four terms on the right hand side of Eq. 2.10 as, respectively: centrifugal acceleration, gravitational acceleration, radiation pressure, radial radiation drag. We also identify two terms on the right hand side of Eq. 2.11 as Coriolis acceleration and radiation azimuthal drag (Poynting-Robertson drag).
Putting all velocities to zero, we find the following critical point condition

$$\frac{1}{r^2} = \frac{\sigma}{m}\xi^{1/2}T^{rt} \,. \tag{2.12}$$

Utilizing Eq. 2.3 and the classic Newtonian definition of the Eddington luminosity, $L_{\mathrm{Edd}} = 4\pi m/\sigma$, we find the critical point radial location

$$r = R_{\mathrm{Edd}} = \frac{2}{1 - \beta^2} \,, \tag{2.13}$$

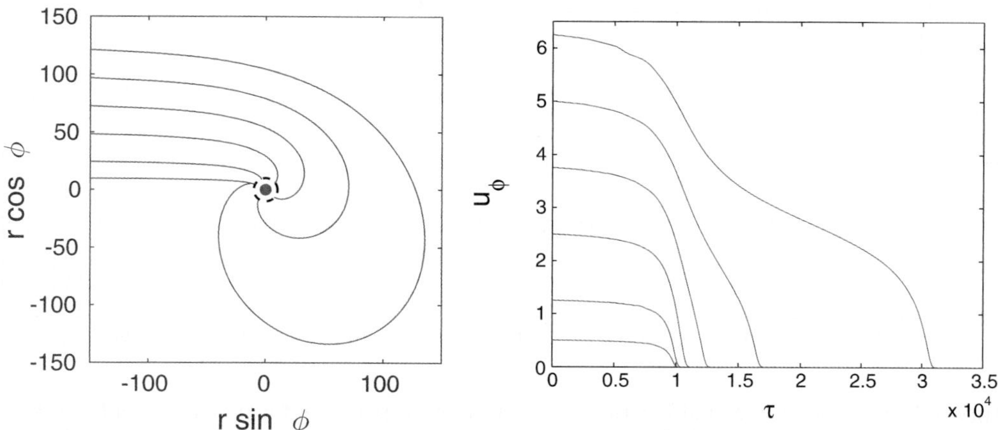

Figure 1. Left: trajectories of particles. Right: angular momentum u_ϕ variation along the trajectories.

where $\beta = L_\infty / L_{\mathrm{Edd}}$ and luminosity L_∞ is

$$L_\infty = \lim_{r \to \infty} 4\pi r^2 T^{tr} . \qquad (2.14)$$

Clearly, the ECS exists above the star surface for the range of β parameter between $(1 - 2/R)^{1/2}$ and $\beta = 1$. The equilibrium of the Eddington Capture Sphere, at $r = R_{\mathrm{Edd}}$ is of a stable kind, net force always pointing towards the equilibrium. Radial and azimuthal drags work as efficient damping forces.

3. Numerically calculated trajectories

We give two numerical illustrations of the ECS related behavior of the test particles. To find the related trajectories, one needs to numerically integrate Eqs. 2.10-2.11. Following Stahl *et al.* (2012) we consider a stream of particles moving towards the luminous star (Hoyle-Lyttleton accretion). We observe that the typical behavior of the particles is to become captured by the ECS in a static equilibrium state, after drag forces remove most of the angular momentum. In the considered example, Fig. 1, we have $R_{\mathrm{Edd}} = 15M$ and particles' velocity at infinity is equal to 0.05 c, 6 trajectories of different impact parameter $(10M, 25M, 50M, 75M, 100M, 125M)$ are shown. The right panel of Fig. 1 indicates, for the corresponding left panel trajectories, how effective is the angular momentum removing effect of drag forces in the strong radiation field. Because of the strong radiation and ECS presence, particles for significantly wider range of impact parameter can be captured. On the other hand, they remain levitating on the ECS rather than falling onto the star surface. This may result in cutting off the accretion-fueled radiation emission and hence destruction of the ECS, from which particles may infall and power up emission of radiation again. Such an oscillatory behavior was investigated by Wielgus *et al.* (2012).

The second numerical illustration concerns the stability of the static equilibrium on the ECS. We consider ECS located at $R_{\mathrm{Edd}} = 10M$ and test particles initially static on the ECS. Then we perturb the radial velocity, Fig. 2 left panel, or the azimuthal velocity, Fig. 2 right panel. The specific values of the perturbed velocities are $0.01c, 0.05c, 0.1c, 0.2c$ for the radial velocities and $0.01c, 0.05c, 0.1c, 0.2c, 0.3c$ for the azimuthal velocities. Clearly, a very large initial velocity is necessary for the particle to escape the ECS and for all considered values the particle returned to the ECS.

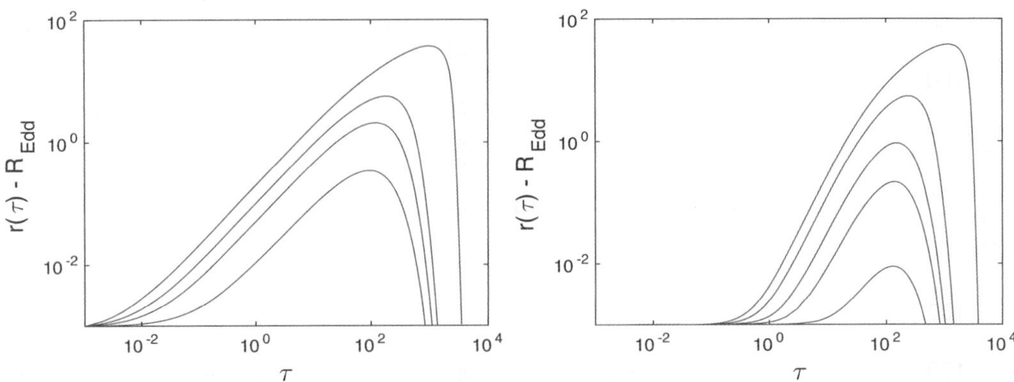

Figure 2. Left: particle radial location as function of proper time for perturbed radial velocity. Right: particle radial location as function of proper time for perturbed azimuthal velocity.

4. Implications

Spacetime around luminous relativistic star is divided into two regions, separated by the *Eddington capture sphere*. In the inner region radiation is effectively super-Eddingtonian, while outside of the ECS it is effectively sub-Eddingtonian. The radiation-gravity balance is only achieved on the ECS. The ECS is an effect derived from the very first principles, and almost certainly (i.e., if general theory of relativity is correct) appears in Nature. More relevant question is whether there is some astrophysical relevance of this mechanism. It seems that at least in two contexts answer to this question should be positive. First is the quasioscillatory behavior related to the subsequent creation and disintegration of the ECS, Wielgus *et al.* (2012). The other is related to the dynamics of matter ejection from inner region of the superluminous system. Here the presence of the ECS and related radiation drag forces alter the ejected matter motion, e.g., Stahl *et al.* (2013).

5. Acknowledgments

The author thanks M. Abramowicz and W. Kluźniak for fruitful discussions. This research was supported by the Polish NCN grant UMO-2013/08/A/ST9/00795.

References

Abramowicz, M. A., Ellis, G. F. R., & Lanza, A. 1990, *ApJ*, 361, 470

Bini, D., Jantzen, R. T., & Stella, L. 2009, *Class. Q. Grav.*, 26, 055009

Mishra, B. & Kluzniak, W. 2013, *A&A*, 566, A62

Oh, J. S., Kim, H., & Lee, H. M., 2011, *New Astron.*, 16, 3, 183

Phinney, E. S. 1987, in *Superluminal Radio Sources*, eds. J. A. Zensus, & T. J. Pearson (Cambridge: Cambridge University Press), 12

Stahl, A., Wielgus, M., Abramowicz, M. A., Kluzniak, W., & Yu, W. 2012, *A&A*, 546, A54

Stahl, A., Kluzniak, W., Wielgus, M., & Abramowicz, M. 2013, *A&A*, 555, A114

Wielgus, M., Stahl, A., Abramowicz, M. A., & Kluzniak, W. 2012, *A&A*, 545, A123

Wielgus, M., Ellis, G. F. R., Vincent, F. H., & Abramowicz, M. A. 2014, *Phys. Rev. D*, 90, 124024

Wielgus, M., & Abramowicz, M., 2015, *Proc. RAGtime*, arXiv:1501.01540

Star clusters and black holes in galaxies across cosmic time
Proceedings IAU Symposium No. 312, 2014
Y. Meiron, S. Li, F.-K. Liu & R. Spurzem, eds.
© International Astronomical Union 2016
doi:10.1017/S174392131500770X

Shadow of a Kerr-like black hole

Farruh Atamurotov

Institute of Nuclear Physics, Ulughbek, Tashkent 100214, Uzbekistan
email: `farruh@astrin.uz`

Abstract. The shadow of a Kerr-like black hole has been considered and it was shown that in addition to the specific angular momentum a, deformation parameter of Kerr-like space-time essentially deforms the shape of the black hole shadow. For a given value of the black hole spin parameter a, the presence of a deformation parameter ϵ reduces the shadow and enlarges its deformation with respect to the one in the Kerr space-time.

Keywords. Kerr-like space-time, Photon motion, Shadow of black hole.

1. Introduction

Recently, such a model was suggested by Johannsen & Psalti (2009) who considered deviations from the Schwarzschild and Kerr solutions and found a regular outside the event horizon space-time, which reduces to the Kerr one, when the deformation parameters vanish. The Johannsen and Psalti's metric is not a vacuum solution of the Einstein equations, but is obtained in a kind of perturbative way in order to include various possible deviations from the Kerr solution in alternative theories of gravity.

A black hole is not visible, it may be observable nonetheless since it may create a shadow if it is in front of a bright background. The apparent shape of an extremely rotating black hole has been first studied by Bardeen (1973). Recently, shadow of black hole was investigated by several authors (e.g. Atamurotov *et al.* 2013).

2. Photon motion around Kerr-like black hole

The deformed Kerr-like metric which describes a stationary axisymmetric, and asymptotically flat vacuum space-time, in the standard Boyer-Lindquist coordinates, can be expressed as Johannsen & Psalti (2009)

$$
\begin{aligned}
\mathrm{d}s^2 \;=\; & -\left(1 - \frac{2Mr}{\Sigma^2}\right)(1+h)\mathrm{d}t^2 + \frac{\Sigma^2(1+h)}{\Delta + a^2 h \sin^2\theta}\mathrm{d}r^2 + \\
& \Sigma^2 \mathrm{d}\theta^2 - \frac{4aMr\sin^2\theta}{\Sigma^2}(1+h)\mathrm{d}t\mathrm{d}\phi + \\
& \sin^2\theta\left[\Sigma^2 + \frac{a^2(\Sigma^2 + 2Mr)\sin^2\theta}{\Sigma^2}(1+h)\right]\mathrm{d}\phi^2,
\end{aligned}
\tag{2.1}
$$

where

$$
\begin{aligned}
\Sigma^2 &= r^2 + a^2\cos^2\theta, \\
\Delta &= r^2 - 2Mr + a^2. \\
h &= \frac{\epsilon M^3 r}{\Sigma^4},
\end{aligned}
$$

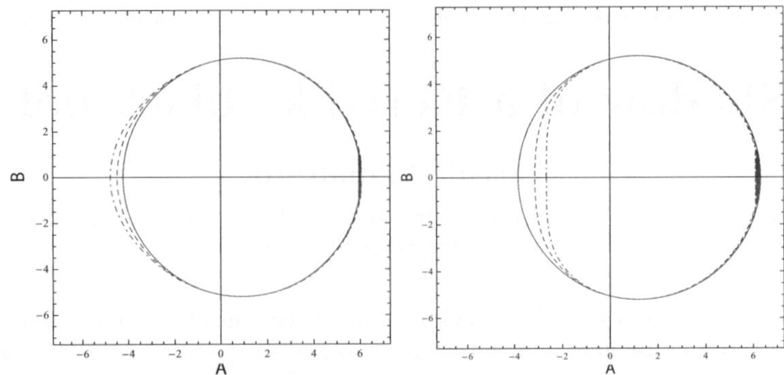

Figure 1. Shadow cast by a Kerr-like black hole situated at the origin of coordinates with inclination angle $\theta = \pi/2$, having a rotation parameter a and a deformation parameter ϵ. Left panel: for $a/M = 0.45$, $\epsilon = 0$ (solid line), $\epsilon = -5$ (dashed line), $\epsilon = -10$ and right panel: $a/M = 0.6$ and $\epsilon = 0$ (solid line), $\epsilon = 5$ (dashed line), $\epsilon = 10$ (dashed-dotted line).

Using the celestial coordinates one can easily describe the shadow (see for example Vázquez & Esteban 2004):

$$\alpha = \lim_{r_0 \to \infty} \left(-r_0^2 \sin\theta_0 \frac{\mathrm{d}\phi}{\mathrm{d}r} \right), \qquad (2.2)$$

and

$$\beta = \lim_{r_0 \to \infty} r_0^2 \frac{\mathrm{d}\theta}{\mathrm{d}r}, \qquad (2.3)$$

In Fig. 1, we show the contour of the shadows of black holes with rotation parameters and for several values of the deformation parameter ϵ.

3. Conclusion

We have analyzed how the shadow of the black hole is distorted by the presence of the deformation parameter ϵ (Atamurotov *et al.* 2013). We have shown that with increasing deformation parameter (Atamurotov *et al.* 2013), the radius of the shadow of the black hole decreases. This phenomena also related to the fact that the increase of deformation parameter forces photon orbits to come closer which corresponds to the decrease of gravitational force acting on photons. In the case of negative ϵ the deflected photons with particular value of impact parameter could be absorbed by central object while they could escape in pure Kerr black hole case with the same impact parameter. The increase of distortion parameter δ_s with the increase of module of deformation parameter ϵ corresponds to deviation of the shape of shadows from pure circle (Atamurotov *et al.* 2013). The deformed rotating black hole's shadow is also going to be deformed independently on sign of deformation parameter.

References

Atamurotov, F., Abdujabbarov, A., & Ahmedov, B. 2013, *Phys. Rev. D*, 88, 064004.

Bardeen, J. M., *Black holes: École D'été de Physique Théorique*, Les Houches 1972, 215–239, Gordon and Breach Science Publishers, New York, 1973.

Johannsen, T. & Psaltis, D. 2011, *Phys. Rev. D*, 83, 124015.

Vázquez S. & Esteban, E. 2004, *Nuovo Cim. 119B*, 489.

Star clusters and black holes in galaxies across cosmic time
Proceedings IAU Symposium No. 312, 2014
Y. Meiron, S. Li, F.-K. Liu & R. Spurzem, eds.

© International Astronomical Union 2016
doi:10.1017/S1743921315007711

The *Fermi* bubbles inflated by winds from the accretion flow in Sgr A*

Guobin Mou[1,2]

[1]Shanghai Astronomical Observatory, Chinese Academy of Sciences, 80 Nandan Rd, Shanghai 200030, China
email: gbmou@shao.ac.cn

[2] University of Chinese Academy of Sciences, 19A Yuquan Road, Beijing 100049, China

Abstract. By performing three-dimensional hydrodynamical simulations, we show that the *Fermi* bubbles could be inflated by winds launched from the "past" hot accretion flow in Sgr A*. The parameters of the accretion flow required in the model are consistent with those obtained independently from other observational constraints. The wind parameters are taken from small scale MHD numerical simulations of hot accretion flows.

Keywords. accretion, black hole physics, jets, hydrodynamics

A pair of giant gamma-ray bubbles which extend 50 degrees above and below the Galactic plane with a width of 40 degrees are revealed by the *Fermi* Gamma-ray Space Telescope (Su *et al.* 2010). The formation mechanism of the bubbles is still under debate.

Many observations have strongly indicated that the activity of the supermassive black hole located in the Galactic center, Sgr A*, is likely much stronger than the present time (Totani 2006), and the *Fermi* bubbles may be the result of this activity. Specifically, the previous independent quantitative studies to the past activity show that while Sgr A* was also in a hot accretion regime, the accretion rate should be $3 \sim 4$ orders of magnitude higher than the present value and last for 10^7 yr.

One of the most important features of hot accretion flow is the existence of strong wind (Yuan *et al.* 2012; see review in Yuan & Narayan 2014). The main properties of winds such as mass flux and velocity have been obtained from the analysis to the MHD numerical simulation data (Yuan *et al.* 2012).

Based on these knowledge and constraints, we have performed three-dimensional hydrodynamical numerical simulations to study the formation of the *Fermi* bubbles. The properties of wind such as the mass flux and velocity are not free parameters but are obtained from previous works on MHD numerical simulations of hot accretion flows once the mass accretion rate is given. The power of wind is $P_w = 2 \times 10^{41}$ erg s^{-1}. The supersonic winds blow into the ISM in the galactic halo, and produce shock. The contact discontinuity separating the shocked ISM and the shocked winds is the boundary of the *Fermi* bubbles. We find that this process can well explain the observed morphology of the *Fermi* bubbles. The active phases is required to last for about 10 million years and the later quiescent state should last for no more than 0.2 million years. Disc-like and massive Central Molecular Zone (CMZ), which are located along the galactic plane around Sgr A*, changes the wind orientation to be approximately towards Galactic poles. This is the reason why we obtain a narrow waist of the *Fermi* bubbles near the galactic plane. Viscosity suppresses the Rayleigh-Taylor (RT) instability and Kelvin-Helmholtz (KH) instability, inducing a smooth edge of the bubble. The observed ROSAT X-ray features can be interpreted by the shocked interstellar medium (ISM) and the interaction region between wind and CMZ gas. We find that the temperature in the shocked ISM at high

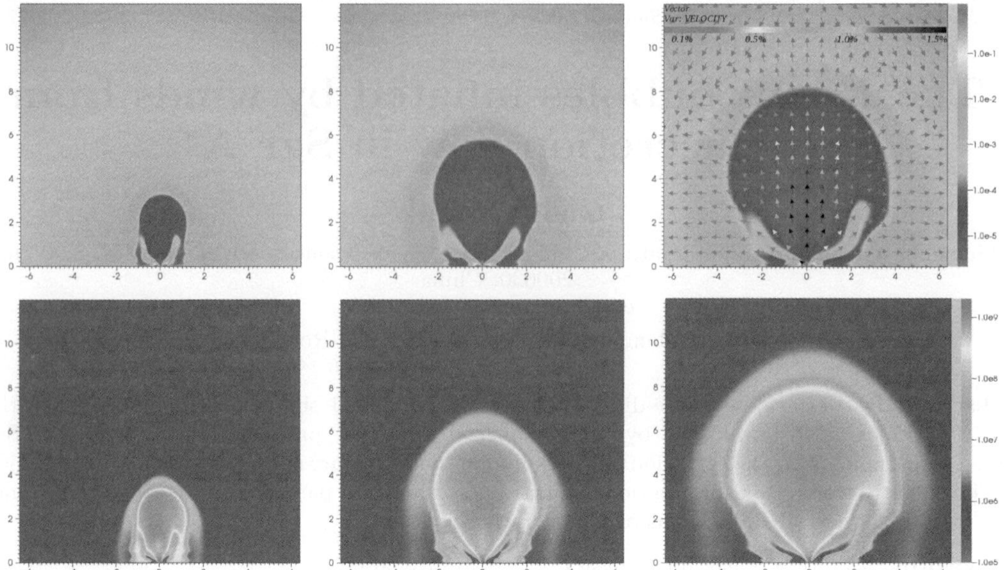

Figure 1. Evolution of electron number density (top) and temperature (bottom) distributions in the X-Z slice. Taken from Mou *et al.* (2014). From left to right, the plots correspond to $t = 4$, 8, and 12.3 Myr, respectively. The velocity field is added in the top right panel with values in units of the light speed.

latitude ($\geqslant +40°$) is about 0.4 keV, and the thermal pressure is 1.2×10^{12} dyn/cm^2. These values are in very good consistency with the results of 0.3 keV and 2×10^{12} dyn/cm^2 obtained in the recent Suzaku observations (Kataoka *et al.* 2013). We leave the study of the production of γ-ray photons and the explanation of the spectrum to our next work (Mou *et al.* 2015, in preparation.).

References

Kataoka, J., Tahara, M., Totani, T., *et al.* 2013, *ApJ*, 779, 57
Mou, G., Yuan,F., Bu, D., Sun, M. & Su, M., *ApJ*, 790, 109
Su, M., Slatyer, T. R., & Finkbeiner, D. P. 2010, *ApJ*, 724, 1044
Totani, T. 2006, *PASJ*, 58, 965
Yuan, F., Bu, D., & Wu, M. 2012, *ApJ*, 761, 130
Yuan, F. & Narayan, R. 2014, *ARA&A*, 52, 529

Star clusters and black holes in galaxies across cosmic time
Proceedings IAU Symposium No. 312, 2014
Y. Meiron, S. Li, F.-K. Liu & R. Spurzem, eds.
© International Astronomical Union 2016
doi:10.1017/S1743921315007723

Radiative efficiency of hot accretion flow and the radio/X-ray correlation in X-ray binaries

Fu-Guo Xie

Key Laboratory for Research in Galaxies and Cosmology, Shanghai Astronomical Observatory,
Chinese Academy of Sciences, 80 Nandan Road, Shanghai 200030, China
email: fgxie@shao.ac.cn

Abstract. Significant progresses have been made since the discovery of hot accretion flow, a theory successfully applied to the low-luminosity active galactic nuclei (LLAGNs) and black hole (BH) X-ray binaries (BHBs) in their hard states. Motivated by these updates, we re-investigate the radiative efficiency of hot accretion flow. We find that, the brightest regime of hot accretion flow shows a distinctive property, i.e. it has a constant efficiency independent of accretion rates, similar to the standard thin disk. For less bright regime, the efficiency has a steep positive correlation with the accretion rate, while for faint regime typical of advection-dominated accretion flow, the correlation is shadower. This result can naturally explain the observed two distinctive correlations between radio and X-ray luminosities in black hole X-ray binaries. The key difference in systems with distinctive correlations could be the viscous parameter, which determines the critical luminosity of different accretion modes.

Keywords. accretion, accretion discs – black hole physics – X-ray: binaries

1. Introduction

The accretion flow around compact objects like black holes can emit a significant amount of energy out. One way to characterise this significance is through radiative efficiency, a ratio between the radiative power L to the rest-mass energy, i.e. $\epsilon \equiv L/\dot{M}c^2$, where $\dot{M}$ is the mass accretion rate of the system. Depending on their temperature, the accretion flows are categorized into hot and cold series. The radiative efficiency is fairly clear for the cold ones, i.e. it is nearly a constant with typical value 10% for the standard Shakura-Sunyaev disk (Shakura & Sunyaev (1973), hereafter SSD), and slightly lower for the Slim disk (Abramowicz et $al.$ (1988)) due to energy advection.

Two important progresses have been achieved in the hot accretion theory (see Xie & Yuan (2012) and Yuan & Narayan (2014) for evidences.), a theory that has been successfully applied in the LLAGNs and the hard states of BHBs, see Yuan & Narayan (2014). First, a large fraction of the accreting material will not be accreted into the BH, but rather be repelled out in the form of outflow (Yuan et $al.$ (2012a), Yuan et $al.$ (2012b)). Secondly, the turbulent dissipation can heat the electrons directly, which leads to an enhancement in the radiative efficiency. Hot accretion flow include both the advection-dominated accretion flow (ADAFs; Narayan & Yi (1994)) at low luminosities and the luminous hot accretion flow (LHAF; Yuan (2001)) at high luminosities. At even higher accretion rates, thermal instability leads the hot accretion flow to a clumpy configuration, i.e. numerous cold clumps embedded within the hot gas. Eventually it will be a disc-corona structure for bright AGNs and BHBs in soft states (e.g. Yang et $al.$ (2015)).

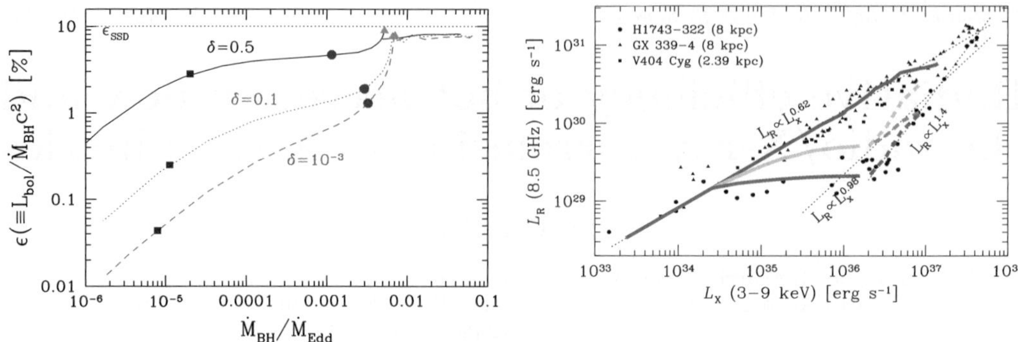

Figure 1. *Left* Radiative efficiency of hot accretion flows. Parameter δ indicates the fraction of viscous heating that directly heats the electrons. The green filled circles separate different modes of hot accretion flows. Adapted from Xie & Yuan (2012). *Right* Radio/X-ray luminosity relationship in black hole X-ray binaries. The red and green curves are for $\alpha = 0.06$, while the blue curves are for $\alpha = 0.6$.

2. Radiative efficiency and radio/X-ray correlation

In ADAFs, most of the viscously liberated energy will be stored as the gas internal energy and advected into BH. Consequently the efficiency is relatively low, which positively correlates with the accretion rate. Besides, more faction of direct viscous heating to the electrons will lead to higher radiative efficiency. The moderate luminosity (LHAF) regime has a very steep $\epsilon - \dot{M}$ correlation. Finally, at the brightest regime where the accretion flow is two-phase, the efficiency is nearly a constant, similar to that of SSD. These results are shown in the left panel of Fig. 1.

One application of the radiative efficiency is to understand the radio/X-ray luminosity correlation observed in black hole X-ray binaries. In X-ray binaries, both the conventional $p \sim 0.6$ (Corbel *et al.* (2013); in the form $L_{\rm R} \propto L_{\rm X}^p$, where $L_{\rm R}$ and $L_{\rm X}$ are respectively the radio and the X-ray luminosities.) and the new hybrid (Coriat *et al.* (2011)) correlations are observed (cf. right panel of Fig. 1, Yuan & Xie (in preparation).).

In the hot accretion flow – jet model, the radio emission comes from the jet, while the X-ray emission comes from the hot accretion flow. With a fixed function of mass ejection rate into the jet versus the accretion rate, we can reproduce the $p \sim 0.6$ correlation at low luminosities. Because the enhancement of efficiency for the bright ADAF and LHAF branches, the X-ray luminosity increases much faster than the radio luminosity in this regime, leads to a much flatter radio/X-ray correlation. At the bright two-phase accretion flow branch, the efficiency is nearly a constant, which reproduces the $p \sim 1.3$ correlation.

References

Abramowicz, M. A., Czerny B., Lasota, J. P., & Szuszkiewicz, E. 1988, *ApJ*, 332, 646
Corbel, S., *et al.* 2013, *MNRAS*, 428, 2500
Coriat, M., *et al.* 2011, *MNRAS*, 414, 677
Narayan, R. & Yi, I. 1994, *ApJ*, 428, L13
Shakura, N. I. & Sunyaev, R. A. 1973, *A&A*, 24, 337
Xie, F. G. & Yuan, F., 2012, *MNRAS*, 427, 1580
Yang, Q. X., Xie, F. G., Yuan, F., Zdziarsksi, A. A., Ho, L. C., & Yu, Z., 2015, *MNRAS*, 447, 1692
Yuan, F. 2001, *MNRAS*, 324, 119
Yuan, F., Bu, D., & Wu, M. 2012b, *ApJ*, 761, 130
Yuan, F., Wu, M., & Bu, D. 2012a, *ApJ*, 761, 129
Yuan, F. & Narayan, R. 2014, *Ann. Rev. Astron. Astroph.*, 52, 529

Star clusters and black holes in galaxies across cosmic time
Proceedings IAU Symposium No. 312, 2014
Y. Meiron, S. Li, F.-K. Liu & R. Spurzem, eds.
© International Astronomical Union 2016
doi:10.1017/S1743921315007735

P-mode oscillation on slim discs

Li Xue and Ju-Fu Lu

Department of Astronomy and Institute of Theoretical Physics and Astrophysics,
Xiamen University, Xiamen, Fujian 361005, China
email: `lixue@xmu.edu.cn`

Abstract. We numerically investigate the thermally unstable accretion discs around spinning black holes with different spins. We adopted an additional evolutionary viscosity equation to replace the standard alpha-prescription based on the results of two MHD simulations. We find an interesting oscillation when accretion switches to slim disc mode. The oscillation arises from the sonic point of accretion flow and propagates outwards. We mimic the bolometric light-curve and find a series of harmonics on its power spectrum. The frequency ratio of those harmonics is a regular integer series. The lowest frequency of the harmonics is identical to the prediction of trapped p-mode in QPO theory.

Keywords. accretion, accretion disks, gravitation, relativistic processes, stars: black holes, X-rays: bursts

---

1. Introduction

High-frequency quasi-periodic oscillations (HFQPOs) have been observed in some black hole (BH) X-ray binaries, which only appear in "steep power law" state in high luminosity ($L > 0.1 L_{\rm Edd}$) and are in range of 40 to 450Hz. These frequencies are comparable to the orbital frequency of innermost stable circular orbit (ISCO) of a stellar-mass BH. It is believed that the mechanism behind them are closely related to the dynamics of inner regions of BH accretion discs (see Remillard & McClintock 2006, Kato *et al.* 2008, Belloni *et al.* 2012, for reviews).

In Xue *et al.* (2011), we have described a code and relevant equations for the axisymmetric relativistic accretion flows around spinning black holes. The viscosity in that code is described by the standard alpha-prescription (Shakura & Sunyaev 1973). In this work, we update the code with adopting an additional evolutionary viscosity equation to replace the alpha-prescription based on two MHD simulations, Hirose *et al.* (2009) and Penna *et al.* (2013). We run the new version code for several thermally unstable accretion models and we find an interesting oscillation when the accretion switches to slim disc mode within the inner disc region. We argue that this oscillation is the trapped p-mode, which is excited by the sonic-point instability on a transonic accretion flow across ISCO (Kato 1978). Indeed, our new code represent a time-dependent accretion model that is able to explain the observational HFQPOs.

2. Results

In fact, the most essential feature in our models is the evolution of viscosity. It reasonably introduces a time-delay between the viscous stress and total pressure, but this delay vanishes in the alpha-prescription. Our colleagues Lin *et al.* (2011) and Ciesielski *et al.* (2012) have focused on the effects of this time-delay on thermal instability and they conclude that it can not remarkably diminish or remove the instability for typical

141

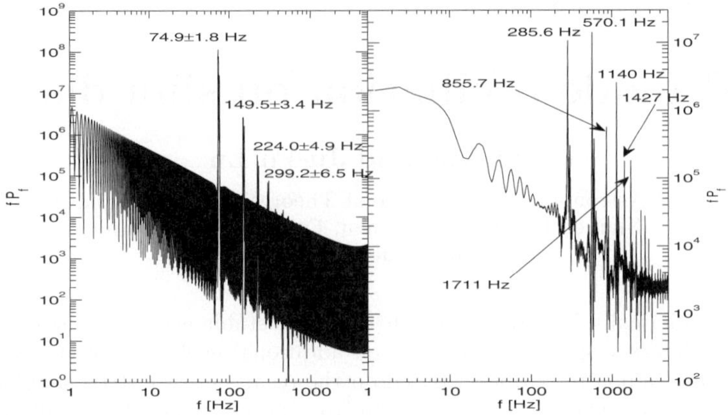

Figure 1. PSDs of light-curves. The left panel is for the disc around a non-spinning black hole ($a^* = 0$, $M = 10M_\odot$, $n = 1$). The right panel is for the disc around a fast-spinning black hole ($a^* = 0.947$, $M = 7.02M_\odot$, $n = 4$).

delay lengths. Indeed, in this work, we find and investigate the other effect of this delay on the oscillation of accretion discs (please concern our paper in *ApJLett* for the detail).

Here, we only show the power spectral densities (PSDs) of oscillating light-curves for two typical models (around non-spinning and fast-spinning black holes) in Figure 1. Their oscillations both arise in high luminosity state ($L \gtrsim 0.2L_{\rm Edd}$). The fundamental frequency (the lowest frequency of harmonics) is ~ 74.9Hz for non-spinning model and ~ 285.6Hz for fast-spinning model, which are both identical to the predicted 71.3Hz and 300Hz (see Kato *et al.* 2008). It is a remarkable feature, in Figure 1, that there are many harmonics on PSDs and the frequency ratio is a regular integer series 1 : 2 : 3.... This interesting feature is potentially useful for explain the observational QPO pairs, but the analogue observation with our numerical results is not straight and convenient. It needs to carefully consider the gravitational red-shift and bending on the emitted photon. However, for convenience, we only construct the light-curves by directly integrating the radiation cooling flux at each radius in this work.

Acknowledgements

This work was supported by the National Natural Science Foundation of China under grants 11233006 and 11373002

References

Belloni, T. M., Sanna, A., & Méndez, M. 2012, *MNRAS*, 426, 1701

Ciesielski, A., Wielgus, M., Kluźniak, W., Sądowski, A., Abramowicz, M., Lasota, J.-P., & Rebusco, P. 2012, *A&A*, 538, 148

Hirose, S., Krolik, J. H., & Blaes, O. 2009, *ApJ*, 691, 16

Kato, S. 1978, *MNRAS*, 185, 629

Kato, S., Fukue, J., & Mineshige, S. 2008, *Black-Hole Accretion Disks: Towards a New Paradigm* (Kyoto: Kyoto Univ. Press), p. 422

Lin, D.-B., Gu, W.-M., & Lu, J.-F. 2011, *MNRAS*, 415, 2319

Penna, R. F., Sądowski, A., Kulkarni, A. K., & Narayan, R. 2013, *MNRAS*, 428, 2255

Remillard, R. A. & McClintock, J. E. 2006, *ARAA*, 44, 49

Shakura, N. I. & Sunyaev, R. A. 1973, *A&A*, 24, 337

Xue, L., Sądowski, A., Abramowicz, M. A., & Lu, J.-F. 2011, *ApJS*, 195, 7

Star clusters and black holes in galaxies across cosmic time
Proceedings IAU Symposium No. 312, 2014
Y. Meiron, S. Li, F.-K. Liu & R. Spurzem, eds.

© International Astronomical Union 2016
doi:10.1017/S1743921315007747

The black hole spins of quasars

Bei You[1,2] and Xinwu Cao[1]

[1]Key Laboratory for Research in Galaxies and Cosmology, Shanghai Astronomical
Observatory, Chinese Academy of Sciences, 80 Nandan Road, Shanghai, 200030, China
email: youbeiyb@gmail.com

[2]University of Chinese Academy of Sciences, 19A Yuquanlu, Beijing 100049, China

Abstract. We present the estimates of the black hole spins of five quasars. The peaks of the spectra of the accretion discs surrounding massive black holes in quasars are in the far-UV or soft X-ray band, which are usually not observed. However, in the disc corona model, the soft photons from the disc are Comptonized to high energy in the hot corona, and the hard X-ray spectra (luminosity and spectral shape) contain the information of the incident spectra from the disc. The values of black hole spin parameter a are inferred from the spectral fitting, which spread over a large range, ~ -0.94 to 0.998.

Keywords. galaxies: nuclei, galaxies: kinematics and dynamics, methods: numerical

1. Introduction

The black hole play an important role on the energetic radiation (Kormendy & Richstone 1995) and the production of the relativistic jets/outflows (Blandford & Znajek 1977) for both X-ray binaries and active galactic nuclei (AGN). The relativistic jets/outflows are important for the evolution of the host galaxy including the star formation (Fabian 2012). Therefore better understanding of these processes requires the measurements of the black hole spins.

One reliable method to constrain the black hole spins of X-ray binaries is the continuum fitting (Zhang*et al.* 1997). In principle, this method can be used to constrain the black hole spins for AGNs. However, the application of this method in AGNs meets a major difficulty, i.e., the spectral peaks of accretion discs surrounding massive black holes with $\sim 10^8 M_\odot$ are in the far-UV or soft X-ray band, which are unobservable for most AGNs.

In the accretion disk corona model, a portion of the soft photons of the spectral peaks originating from the disc undergo Comptonization in the hot corona, and the scattered photons are dominantly in the hard X-ray band (You *et al.* 2012). This implies that the observed hard X-ray spectra (luminosity and spectral shape) contain the information of the incident spectra from the cold disc. It seems possible to constrain the values of black hole spin parameter a in AGNs with a general relativistic accretion disc corona model, even if the spectral peaks of the accretion discs are not observed.

2. Results

Using the relativistic accretion disc corona model (see You *et al.* 2012; You 2014), we fit the observed optical/UV "Big Blue Bump" (BBB) at $\sim 10^{15} - 10^{16}$ Hz and the power-law hard X-ray continuum spectra of five quasars from Shang*et al.* (2011). The left tail of the BBBs can be observed in infared/optical/UV wavebands, which provide useful information of the BBBs.

There are four free parameters in our model calculations, i.e., the mass accretion rate $\dot{m}$, the spin parameter a, the power fraction f dissipated in the corona, and the viewing

Table 1. The fitting parameters of the sources

Object	$\log M$	f	a	$\dot{m}$	θ	$T_e\,(\mathrm{keV})$
PG 1322+659	8.29	$0.47^{+0.10}_{-0.12}$	$-0.94^{+0.14}_{-0.06}$	$0.388^{+0.019}_{-0.020}$	$25.7^{+3.5}_{-4.7}$	$265.3^{+32.1}_{-32.1}$
PG 1115+407	8.18	$0.41^{+0.06}_{-0.03}$	$-0.55^{+0.15}_{-0.15}$	$0.420^{+0.020}_{-0.005}$	$10.0^{+7.3}_{-3.1}$	$241.2^{+65.5}_{-65.5}$
4C 10.06	9.00	$0.60^{+0.15}_{-0.10}$	$0.7^{+0.05}_{-0.05}$	$0.415^{+0.020}_{-0.017}$	$39.4^{+2.2}_{-2.0}$	$194.7^{+15.7}_{-15.7}$
4C 39.25	9.33	$0.47^{+0.03}_{-0.02}$	$0.998^{+0.000}_{-0.004}$	$0.344^{+0.008}_{-0.011}$	$42.5^{+0.9}_{-1.4}$	$158.3^{+4.4}_{-4.4}$
OS 562	8.92	0.33	$0.96^{+0.03}_{-0.04}$	$0.505^{+0.027}_{-0.020}$	$13.0^{+7.2}_{-9.6}$	$142.5^{+8.4}_{-8.4}$

Note. The black hole mass M is taken from Tang *et al.* (2012). T_e is the maximum value of the electron temperature of the corona, which is calculated with the best fitted values of free parameters a and $\dot{m}$.

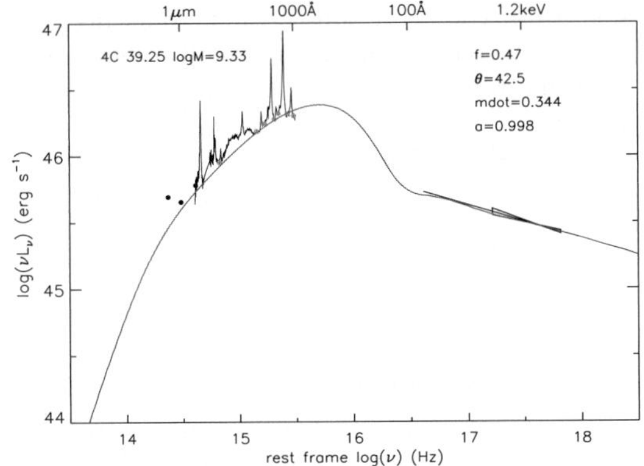

Figure 1. The spectral fitting results for 4C 39.25. The radio and near infrared data are not used in our accretion disc corona model fittings. The photometric points marked as blue filled circles are used for fitting. The solid black line represents the power law X-ray spectrum in $2-10$ keV. The red line represents the best fitting result. The black hole masses M of these five quasars are given in Tang *et al.* (2012), which are estimated with the broad-line width and the scaling relationships. The spectral fitting results for other sources can be found in You (2014).

angle θ with respect to disc axis. We find that the fraction f is mainly determined by the X-ray spectral index Γ, which is almost independent of the values of other parameters. With the derived value of f from Γ, $\dot{m}$ and θ are constrained with the luminosity and spectral shape in optical waveband (the red tail of the BBB). These two parameters are almost independent of the black hole spin parameter a, because the emission in this waveband is dominantly from the outer region of the disc. Finally, the black hole spin parameter a is constrained mainly with the observed hard X-ray luminosity.

Acknowledgements

This work is supported by the NSFC (grants 11173043, 11121062, 11233006, 11073020, and 11373056), the Fundamental Research Funds for the Central Universities (WK2030220004), the CAS/SAFEA International Partnership Program for Creative Research Teams (KJCX2-YW-T23), and Shanghai Municipality.

References

Blandford R. D. & Znajek R. L., 1977, *MNRAS*, 179, 433
Fabian, A. C. 2012, *ARAA*, 50, 455
Kormendy J. & Richstone D., 1995, *ARA&A*, 33, 581
Shang Z., *et al.*, 2011, *ApJS*, 196, 2
Tang B., Shang Z., Gu Q., Brotherton M. S., & Runnoe J. C., 2012, *ApJS*, 201, 38
You, B., Cao, X., & Yuan, Y.-F. 2012, *ApJ*, 761, 109
You, B., Cao, X., & Yuan, Y.-F. 2014, arXiv:1401.5659
Zhang S. N., Cui W., & Chen W., 1997, *ApJ*, 482, L155

PART TWO
Galactic and extragalactic globular clusters

Star clusters and black holes in galaxies across cosmic time
Proceedings IAU Symposium No. 312, 2014
Y. Meiron, S. Li, F.-K. Liu & R. Spurzem, eds.
© International Astronomical Union 2016
doi:10.1017/S1743921315007759

Are globular clusters the natural outcome of regular high-redshift star formation?

J. M. Diederik Kruijssen

Max-Planck Institut für Astrophysik, Karl-Schwarzschild-Straße 1, 85748, Garching, Germany
email: `kruijssen@mpa-garching.mpg.de`

Abstract. We summarise the recent progress in understanding the formation and evolution of globular clusters (GCs) in the context of galaxy formation and evolution. It is discussed that an end-to-end model for GC formation and evolution should capture four different phases: (1) star and cluster formation in the high-pressure interstellar medium of high-redshift galaxies, (2) cluster disruption by tidal shocks in the gas-rich host galaxy disc, (3) cluster migration into the galaxy halo, and (4) the final evaporation-dominated evolution of GCs until the present day. Previous models have mainly focussed on phase 4. We present and discuss a simple model that includes each of these four steps – its key difference with respect to previous work is the simultaneous addition of the high-redshift formation and early evolution of young GCs, as well as their migration into galaxy haloes. The new model provides an excellent match to the observed GC mass spectrum and specific frequency, as well as the relations of GCs to the host dark matter halo mass and supermassive black hole mass. These results show (1) that the properties of present-day GCs are reproduced by assuming that they are the natural outcome of regular high-redshift star formation (i.e. they form according to same physical processes that govern massive cluster formation in the local Universe), and (2) that models only including GC evaporation strongly underestimate their integrated mass loss over a Hubble time.

Keywords. star formation, galaxy evolution, galaxy formation, globular clusters, stellar dynamics

1. Introduction

Approximately 50% of all stars in the Universe formed at redshifts $z > 1$, with a peak in the cosmic star formation history at $z = 2$–3 (Madau & Dickinson 2014). Understanding the cosmological assembly of stellar mass in galaxies therefore requires a census of the conditions under which star formation proceeded in high-redshift systems. Modern observational facilities enable detailed studies on the super-kpc scales of spatially-resolved galaxies at $z > 1$ (Hodge *et al.* 2012; Tacconi *et al.* 2013), revealing gas pressures 3–4 orders of magnitude higher than in the local Universe (Swinbank *et al.* 2011; Kruijssen & Longmore 2013). Despite these efforts, the cloud-scale (~ 100 pc) physics of high-redshift star formation remain out of reach.

Globular clusters (GCs) have the potential to be used as tools for tracing extreme star formation physics in their high-redshift birth environments. However, the resulting insight is only relevant in the context of galaxy formation if GCs did not form through some 'special' mode of star formation, but instead are the natural products of 'regular' high-redshift star formation, i.e. they result from the same physical processes that govern star formation in the local Universe. Given the extreme conditions seen in $z > 1$ systems (e.g. high gas pressures and densities), the observed old ages, high masses, high velocity dispersions, and low metallicities of GCs indeed seem to be consistent with the expected star formation products (see Figure 1 and Elmegreen & Efremov 1997; Shapiro *et al.* 2010;

Figure 1. *Left*: The Galactic GC M80, with a mass of $M \sim 10^{5.6}$ $M_\odot$ and a radius of $R \sim 1.8$ pc. Image credit: NASA and The Hubble Heritage Team (STScI/AURA). *Right*: The most massive young massive cluster in the nearby starburst NGC 1569, with a mass of $M \sim 10^6$ $M_\odot$ and a radius of $R \sim 1.6$ pc. Image credit: NASA, ESA, the Hubble Heritage Team (STScI/AURA), and A. Aloisi (STScI/ESA). The properties of both clusters are very similar, showing that special conditions exclusive to the early Universe are not necessary for the formation of GC-like clusters.

Kruijssen 2014). It is therefore reasonable to evaluate if GCs do indeed result from the dominant mode of star formation at $z > 1$.

To establish whether GCs are indeed the natural outcome of regular, high-redshift star formation and eventually can be used as fossils to trace the underlying physics, it is necessary to obtain an end-to-end description for GC formation and evolution. Only then, it is possible to correct for a Hubble time of GC evolution and 'rewind' the current GC population to their birth sites. In this paper, we summarise our two most recent contributions to solving this problem.

(i) In Kruijssen (2014), we review GC formation in the context of galaxy formation and evolution.

(ii) In Kruijssen (2015), we present a simple, end-to-end model for GC formation and evolution that combines our current knowledge of massive cluster formation, high-redshift star formation, present-day GC populations, and star cluster disruption.

Previous GC formation and evolution models typically rely on GC evaporation to explain the observed properties of present-day GC populations (most notably their mass spectrum, see e.g. Fall & Zhang 2001; Prieto & Gnedin 2008; Kruijssen & Portegies Zwart 2009; Li & Gnedin 2014). However, these models often fail to reproduce other observables (e.g. the specific frequency or the radial invariance within galaxies of the GC mass spectrum). The key new step made in the Kruijssen (2015) model is that it accounts for the 'pre-processing' of the GC population by impulsive tidal shocks in their natal galaxy discs, which leads to their rapid tidal disruption well before evaporation becomes important (also see Elmegreen 2010), as well as the subsequent redistribution of GCs into galaxy haloes by hierarchical galaxy formation. By introducing this extra phase in the evolutionary history of the GC population, the model produces present-day GC populations that are consistent with those observed, e.g. in terms of their mass spectrum, specific frequencies, and colours, as well as their relations to the host dark matter halo mass and supermassive black hole mass. The conclusion is that GCs are consistent with being the products of regular high-redshift star formation.

2. Ingredients for an end-to-end GC formation and evolution model

An end-to-end model for GC formation and evolution should describe the formation of GCs within their high-redshift natal environment, their early evolution within the host galaxy disc, their migration into the galaxy haloes, and their subsequent evolution until the present day. Here, we summarise the physics covered by the Kruijssen (2015) model.

2.1. *Cluster formation*

Studies of stellar cluster formation in the local Universe show that clusters form through the hierarchical collapse of density peaks in the interstellar medium (ISM) (Efremov & Elmegreen 1998; Longmore *et al.* 2014; Rathborne *et al.* 2015). Whether the resulting stellar system becomes a gravitationally bound stellar cluster or an unbound association depends on the star formation efficiency (SFE) – at low SFEs, the removal of the residual gas by feedback unbinds the stellar system (e.g. Hills 1980; Lada *et al.* 1984; Goodwin & Bastian 2006). Because the SFE increases with the number of free-fall times completed before residual gas removal, the highest density peaks end up being gas-poor and remain gravitationally bound (Kruijssen *et al.* 2012a). By integrating the resulting bound fraction of young stars over the density spectrum of the ISM, one can formulate a model to predict the fraction of all star formation in a galaxy that results in bound stellar clusters (Kruijssen 2012), i.e. the cluster formation efficiency (CFE or Γ, see Bastian 2008). In this model, the CFE ranges from $\Gamma \sim 1\%$ in low-pressure galaxies ($P/k < 10^4$ K cm^{-3}) to $\Gamma \sim 50\%$ in high-pressure galaxies ($P/k > 10^6$ K cm^{-3}), which is quantitatively consistent with recent observations (Goddard *et al.* 2010; Adamo *et al.* 2011; Silva-Villa *et al.* 2013). We thus see that the high-pressure conditions seen in high-redshift galaxies promote the formation of bound stellar clusters.

After determining which fraction of the star formation rate results in bound clusters, this *cluster formation rate* must be distributed over some range of cluster masses. Observations (Portegies Zwart *et al.* 2010) and theory (Elmegreen & Falgarone 1996) show that the initial cluster mass function (ICMF) follows a power law $dN/dM \propto M^\alpha$ with index $\alpha = -2$. At the low-mass end, the ICMF continues down to the detection limit. Based on studies of cluster formation in the solar neighbourhood, a commonly-adopted lower mass limit is $M_{\mathrm{min}} = 100$ M$_\odot$ (Lada & Lada 2003). At the high-mass end, the ICMF is limited by the Toomre (1964) mass, i.e. the largest mass scale for gravitational instability in a differentially rotating disc (Hughes *et al.* 2013; Kruijssen 2014). Because the Toomre mass increases with the ambient gas pressure, the high-pressure conditions seen in high-redshift galaxies enable the formation of clusters much more massive ($M_{\mathrm{max}} \sim 10^7$ M$_\odot$) than those forming in low-redshift discs ($M_{\mathrm{max}} \sim 10^5$ M$_\odot$, see e.g. Larsen 2009).

2.2. *Disruption phase 1: early evolution and migration*

The vast majority of stars in the Universe formed within gaseous galaxy discs rather than irregular dwarf galaxies or mergers (Genzel *et al.* 2010; Rodighiero *et al.* 2011). We can therefore assume that after their formation, young stellar clusters reside within the gas-rich disc of their host galaxy. Within such discs, the frequent encounters with giant molecular clouds (or analogously with the massive clumps seen in high-redshift galaxies) lead to the rapid disruption of young clusters by impulsive tidal shocks (Spitzer 1987; Gieles *et al.* 2006). This disruption agent dominates over the more gradual mass loss by evaporation (Kruijssen *et al.* 2011). in Kruijssen (2015), we show that including this *rapid-disruption phase* in the evolutionary history of GCs is crucial for reproducing their present-day properties. If the rapid-disruption phase in the host galaxy disc is excluded, the total dynamical mass loss over a Hubble time is severely underestimated.

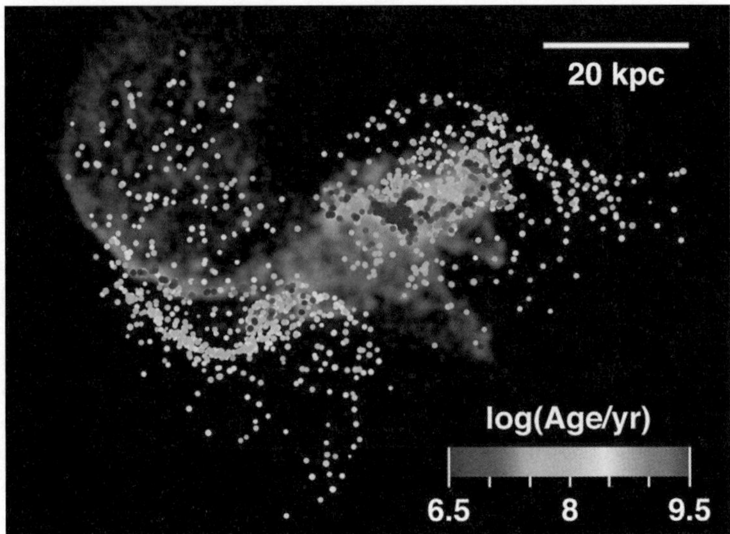

Figure 2. Snapshot of galaxy merger model `1m11` from Kruijssen *et al.* (2012b), which includes a sub-grid model for the formation and evolution of the stellar cluster population. Shown is a merger of two Milky Way-mass galaxies at the time of their first encounter. Coloured dots represent stellar clusters, colour-coded by their ages as indicated by the legend. The grey scale indicates the gas surface density. The snapshot shows how intermediate-age clusters are escaping into the halo, whereas those clusters that formed during the merger reside in the gas-rich, disruptive environment of the discs. This cluster migration process is also seen in observations of nearby galaxy mergers (Bastian *et al.* 2009).

The rapid disruption of young GCs in the host galaxy disc comes to an end when the clusters migrate out of the volume that is occupied by the (molecular) gas. In dwarf galaxies, the bursty nature of star formation could lead to changes of the gravitational potential that facilitate this type of migration (Pontzen & Governato 2012), but across the galaxy mass range the consistently most plausible migration agent seems to be the galaxy mergers associated with hierarchical galaxy formation (White & Frenk 1991; Kravtsov & Gnedin 2005), which take place on $\sim$ Gyr time-scales (Kruijssen 2015). This type of migration can take two different forms. If the GCs' host galaxy undergoes a major merger, the redistribution of material naturally migrates the young GCs into the halo of the merger remnant (Kruijssen *et al.* 2012b). If the GCs' host galaxy is cannibalised by a much (i.e. by a factor of > 3) more massive galaxy, the tidal stripping of the system also migrates the GCs into the halo of the more massive galaxy. The common denominator is that as long as the host galaxy merges with a galaxy of a similar or larger mass, the rapid-disruption phase comes to an end.

In Kruijssen *et al.* (2011, 2012b), we presented numerical simulations of galaxy discs and galaxy mergers with a sub-grid model for the formation and evolution of the stellar cluster population. The cluster populations modelled in these simulations (see Figure 2) provide a good match to the observed age distributions of clusters in M83 (Adamo & Bastian 2015) and in the Antennae galaxies (Kruijssen 2011), and quantitatively show that the survival chances of stellar clusters indeed increase dramatically by migration into the gas-poor galaxy halo during major mergers.

2.3. *Disruption phase 2: quiescent evolution*

Once the (by then intermediate-age) GCs have migrated into the host galaxy halo, their long-term survival has become likely, as their further disruption until the present day is dominated by gradual evaporation. Previous GC formation models have focussed on this phase to model the emergence of the observed GC population, and can be divided in roughly two categories.

(i) One ('environmentally independent') family of models accounts for evaporation-driven mass loss using the classical expression by Spitzer (1987), which is independent of the tidal field and is exclusively set by the mass and radius of the GC under consideration (Fall & Zhang 2001; Prieto & Gnedin 2008; McLaughlin & Fall 2008; Li & Gnedin 2014).

(ii) The other ('environmentally dependent') family of models accounts for the results of N-body simulations showing that the evaporation-driven mass loss rate of GCs is exclusively set by the tidal field strength (i.e. the galactic environment) and the GC mass, and is independent of the cluster radius (Vesperini & Heggie 1997; Baumgardt & Makino 2003; Gieles & Baumgardt 2008).

Both model families have their own problems. The first family reproduces the near-universal shape of the GC mass spectrum not including an environmental dependence, but is physically inconsistent with N-body simulations of evaporating clusters as well as with observations of cluster populations, which show clear indications of environmentally-dependent cluster disruption (e.g. Bastian *et al.* 2012). The second family accounts for the environmental variation of the evaporation rate, but as a result cannot explain the near-universality of the GC mass spectrum or the environmental independence of the specific frequency after dividing out the metallicity dependence (Kruijssen 2014).

In the GC formation model described here (Kruijssen 2015), the above problems are alleviated, because most of the GC disruption takes place during their early evolution in the gas-rich host galaxy disc. In this model, the subsequent redistribution of GCs into galaxy haloes erases the correlation between their mass loss history and the present-day environment. Instead, the total mass loss of GCs should correlate with their formation environment, as is corroborated by the observed decrease of the specific frequency with the metallicity at constant galactocentric radius (Harris & Harris 2002; Lamers *et al.* 2015).

The relative unimportance of GC evaporation in galaxy haloes to their present-day statistics does not imply the process is uninteresting. For instance, this quiescent phase in the history of GCs is crucial in setting their structural properties (Gieles *et al.* 2011) as well as their stellar mass functions (Kruijssen 2009). Both areas still provide a broad range of unanswered questions, which are beyond the scope of this paper.

2.4. *Connection to the host galaxy*

All of the discussed quantities governing the formation, disruption, migration, and evaporation of GCs depend on the properties of the natal galaxy. While present-day GCs are no longer associated with the galaxy in which they formed, their metallicities and ages provide clues to their formation environments. The median age of the Galactic GC population ($\tau \sim 11.5$ Gyr, see Forbes & Bridges 2010) indicates they typically formed at redshift $z \sim 3$. By using the galaxy mass-metallicity relation observed at that redshift (Erb *et al.* 2006; Mannucci *et al.* 2009), the metallicities of present-day GCs can be connected to the masses of the galaxies in which they were born.

After identifying the host galaxy mass and metallicity, we adopt an equilibrium-disc model to describe star and cluster formation, as well as the subsequent rapid cluster disruption and migration. As shown by Krumholz & McKee (2005), such a model is entirely set by the combination of the gas surface density, angular velocity, and Toomre stability parameter (here assumed to be $Q = 1$, indicating marginal stability). The gas surface density is set by the ISM pressure (which, as explained above, follows from the maximum GC mass-scale), and the angular velocity is provided by the observed scaling relation between galaxy mass and rotation rate in $z = 1.1$–3.5 galaxies (Förster Schreiber

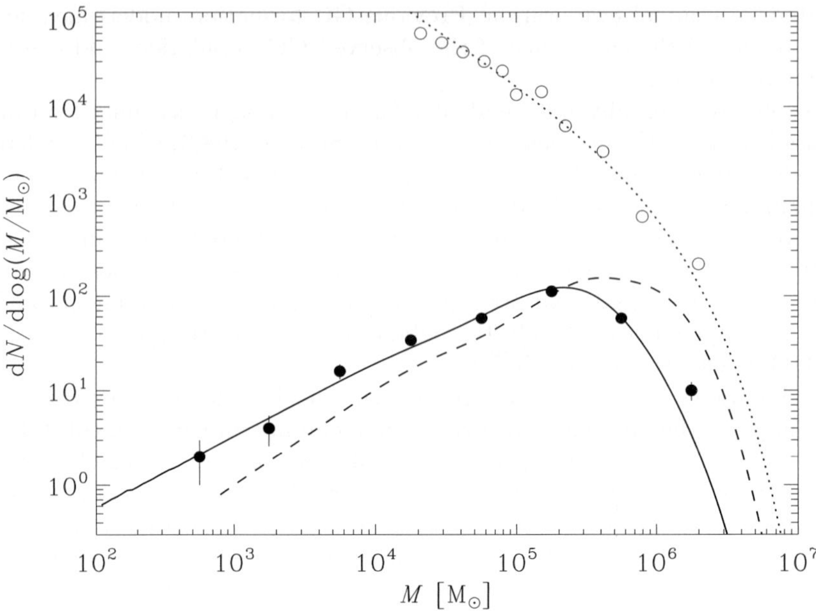

Figure 3. Predicted GC mass functions for a Milky Way-like galaxy model that includes both in-situ GC formation and the accretion of GCs from cannibalised dwarf galaxies. The different lines show the GC mass functions at the times of GC formation (dotted) and GC migration into the gas-poor galaxy halo (dashed), as well as at the present day (solid). The open circles indicate the observed mass spectrum of young clusters in the Antennae galaxies (Zhang & Fall 1999), whereas the solid circles represent the observed mass spectrum of globular clusters in the Milky Way (Harris 1996, 2010 edition). Note that while the model lines reproduce the observational data quite well, they have not been fitted to the observations.

et al. 2009). As a result, we have a working model for star and cluster formation and early evolution as a function of the metallicity of each present-day GC.

In Kruijssen (2015), we show that the ISM pressure (and hence early cluster disruption rate) in the above model weakly *increases* with the galaxy mass (and hence metallicity). In hierarchical galaxy formation models (Springel *et al.* 2005), the rate of mergers with galaxies of at least the GC host galaxy's own mass $M_{\rm host}$ decreases with $M_{\rm host}$. The migration rate into the galaxy halo of young GCs therefore *decreases* with the host galaxy mass (and metallicity). We thus see that the rapid-disruption phase (1) leads to more cluster disruption and (2) lasts longer in massive, high-metallicity galaxies.

3. Model results: the mass spectrum of Galactic globular clusters

The above model can be used to predict the GC mass spectrum and specific frequency, as well as the relations of GCs to the host dark matter halo mass and supermassive black hole mass, as a function of GC metallicity across cosmic time. As discussed in Kruijssen (2015), the model simultaneously reproduces these observed distribution functions and scaling relations. In particular, it is the only model to match the decrease of the specific frequency with metallicity as well as its invariance with the galactocentric radius (Harris & Harris 2002; Lamers *et al.* 2015), which troubled previous models.

We refer the interested reader to Kruijssen (2015) for an in-depth discussion of the model's formulation and results, as well as its broad range of predictions. Instead of going into a similar level of detail here, we briefly discuss a key result of the model.

Figure 3 shows the predicted GC mass spectrum for a metallicity-composite GC population appropriate for the Milky Way. The different lines show the mass spectra at the times of GC formation ($t = 0$) and migration ($t =$ several Gyr), as well as at the present day ($t = 11.5$ Gyr). For comparison, it also includes the observed mass spectrum of young clusters in the Antennae galaxies (Zhang & Fall 1999), as well as the observed mass spectrum of Galactic GCs (Harris 1996, 2010 edition). The figure clearly shows that the adopted initial mass spectrum matches that of young clusters seen forming in the Antennae, and the predicted present-day GC mass spectrum also provides a good match to the observations. We note that the model is not a fit. The good agreement further supports the model's inclusion of the initial rapid-disruption phase in the gas-rich host galaxy disc at the epoch of GC formation.

The discussed model provides a very simple, end-to-end description for GC formation and evolution. It is based on what is currently known of:

(i) the formation of stellar clusters in high-pressure conditions such as galaxy mergers (in the local Universe) or clumpy galaxy discs (i.e. the high-redshift GC formation sites);

(ii) cluster disruption by tidal shocks caused by encounters with giant molecular clouds or other substructure in the high-pressure ISM;

(iii) cluster migration by galaxy mergers during hierarchical galaxy formation;

(iv) gradual cluster disruption by evaporation within galaxy haloes.

While the model only represents a first step towards future, more sophisticated end-to-end models, it is the first to combine all relevant phases of GC formation and evolution across cosmic history. In the process, it provides a surprisingly good match to the observed GC population. With this framework in hand, future studies covering the different phases of the model in more detail (e.g. using numerical simulations) should enable the use of GCs as tracers of high-redshift star formation.

Acknowledgements

I am grateful to the conference organisers for the invitation and financial support through an IAU grant, as well as for organising a very lively and stimulating meeting. In addition, I am indebted to my collaborators for their insights and contributions to the projects discussed here. I particularly would like to thank Angela Adamo, Nate Bastian, Michael Hilker, Henny Lamers, Steve Longmore, Jill Rathborne, and Marina Rejkuba.

References

Adamo A. & Bastian N., 2015, *The lifecycle of clusters in galaxies*. The Birth of Star Clusters, editor S.W. Stahler, Springer, submitted

Adamo A., Östlin G., & Zackrisson E., 2011, *MNRAS*, 417, 1904

Bastian N., 2008, *MNRAS*, 390, 759

Bastian N. *et al.*, 2012, *MNRAS*, 419, 2606

Bastian N., Trancho G., Konstantopoulos I. S., & Miller B. W., 2009, *ApJ*, 701, 607

Baumgardt H. & Makino J., 2003, *MNRAS*, 340, 227

Efremov Y. N. & Elmegreen B. G., 1998, *MNRAS*, 299, 588

Elmegreen B. G., 2010, *ApJ*, 712, L184

Elmegreen B. G. & Efremov Y. N., 1997, *ApJ*, 480, 235

Elmegreen B. G. & Falgarone E., 1996, *ApJ*, 471, 816

Erb D. K., Steidel C. C., Shapley A. E., Pettini M., Reddy N. A., & Adelberger K. L., 2006, *ApJ*, 646, 107

Fall S. M. & Zhang Q., 2001, *ApJ*, 561, 751

Forbes D. A. & Bridges T., 2010, *MNRAS*, 404, 1203

Förster Schreiber N. M. *et al.*, 2009, *ApJ*, 706, 1364

Genzel R. *et al.*, 2010, *MNRAS*, 407, 2091

Gieles M. & Baumgardt H., 2008, *MNRAS*, 389, L28

Gieles M., Heggie D. C., & Zhao H., 2011, *MNRAS*, 413, 2509

Gieles M., Portegies Zwart S. F., Baumgardt H., Athanassoula E., Lamers H. J. G. L. M., Sipior M., & Leenaarts J., 2006, *MNRAS*, 371, 793

Goddard Q. E., Bastian N., & Kennicutt R. C., 2010, *MNRAS*, 405, 857

Goodwin S. P. & Bastian N., 2006, *MNRAS*, 373, 752

Harris W. E., 1996, *AJ*, 112, 1487

Harris W. E. & Harris G. L. H., 2002, *AJ*, 123, 3108

Hills J. G., 1980, *ApJ*, 235, 986

Hodge J. A., Carilli C. L., Walter F., de Blok W. J. G., Riechers D., Daddi E., & Lentati L., 2012, *ApJ*, 760, 11

Hughes A. *et al.*, 2013, *ApJ*, 779, 44

Kravtsov A. V. & Gnedin O. Y., 2005, *ApJ*, 623, 650

Kruijssen J. M. D., 2009, *A&A*, 507, 1409

Kruijssen J. M. D., 2011, PhD thesis, Utrecht University, The Netherlands

Kruijssen J. M. D., 2012, *MNRAS*, 426, 3008

Kruijssen J. M. D., 2014, *Classical and Quantum Gravity*, 31, 244006

Kruijssen J. M. D., 2015, *MNRAS* submitted

Kruijssen J. M. D. & Longmore S. N., 2013, *MNRAS*, 435, 2598

Kruijssen J. M. D., Maschberger T., Moeckel N., Clarke C. J., Bastian N. & Bonnell I. A., 2012a, *MNRAS*, 419, 841

Kruijssen J. M. D., Pelupessy F. I., Lamers H. J. G. L. M., Portegies Zwart S. F., Bastian N., Icke V., 2012b, *MNRAS*, 421, 1927

Kruijssen J. M. D., Pelupessy F. I., Lamers H. J. G. L. M., Portegies Zwart S. F., Icke V., 2011, *MNRAS*, 414, 1339

Kruijssen J. M. D., Portegies Zwart S. F., 2009, *ApJ*, 698, L158

Krumholz M. R., McKee C. F., 2005, *ApJ*, 630, 250

Lada C. J., Lada E. A., 2003, *ARA&A*, 41, 57

Lada C. J., Margulis M., Dearborn D., 1984, *ApJ*, 285, 141

Lamers H. J. G. L. M., Kruijssen J. M. D., Bastian N., Rejkuba M., Hilker M., Kissler-Patig M., 2015, in preparation

Larsen S. S., 2009, *A&A*, 494, 539

Li H., Gnedin O. Y., 2014, *ApJ*, 796, 10

Longmore S. N. et al., 2014, Protostars and Planets VI, 291

Madau P., Dickinson M., 2014, *ARA&A*, 52, 415

Mannucci F. et al., 2009, *MNRAS*, 398, 1915

McLaughlin D. E., Fall S. M., 2008, *ApJ*, 679, 1272

Pontzen A., Governato F., 2012, *MNRAS*, 421, 3464

Portegies Zwart S. F., McMillan S. L. W., Gieles M., 2010, *ARA&A*, 48, 431

Prieto J. L., Gnedin O. Y., 2008, *ApJ*, 689, 919

Rathborne J. M. et al., 2015, *ApJ* in press, arXiv:1501.07368

Rodighiero G. et al., 2011, *ApJ*, 739, L40

Shapiro K. L., Genzel R., Förster Schreiber N. M., 2010, *MNRAS*, 403, L36

Silva-Villa E., Adamo A., Bastian N., 2013, *MNRAS*, 436, L69

Spitzer L., 1987, Dynamical evolution of globular clusters. Princeton, NJ, Princeton University Press, 1987, 191 p.

Springel V. et al., 2005, *Nature*, 435, 629

Swinbank A. M. et al., 2011, *ApJ*, 742, 11

Tacconi L. J. et al., 2013, *ApJ*, 768, 74

Toomre A., 1964, *ApJ*, 139, 1217

Vesperini E., Heggie D. C., 1997, *MNRAS*, 289, 898

White S. D. M., Frenk C. S., 1991, *ApJ*, 379, 52

Zhang Q., Fall S. M., 1999, *ApJ*, 527, L81

Star clusters and black holes in galaxies across cosmic time
Proceedings IAU Symposium No. 312, 2014
Y. Meiron, S. Li, F.-K. Liu & R. Spurzem, eds.

© International Astronomical Union 2016
doi:10.1017/S1743921315007760

Preliminary results from simulations on the sub-galactic structure formation

Kyungwon Chun[1] and Jihye Shin[2]

[1]School of Space Research, Kyung Hee University, Yongin, Gyeonggi, 446-701, Korea
email: kwchun@khu.ac.kr

[2]Kavli Institute for Astronomy and Astrophysics, Peking University, 5 Yiheyuan Road,
Haidian District, Beijing 100871, China

Abstract. We aim to investigate the formation of sub-galactic structure in the Lambda cold dark matter (CDM) cosmology. To accomplish our research goal, we have added various baryonic physics on the existing cosmological hydrodynamic code, `GADGET-2`. We performed two test runs to check our new implementations. We show our preliminary results from these test runs.

Keywords. Galaxy: formation, large-scale structure of universe, globular clusters: general

1. Introduction

Our goal is to investigate the formation of sub-galactic structure in Lambda CDM cosmology. For this, we have modified the `GADGET-2` (Springel 2005) code, a parallel N-body/SPH code, for the more realistic baryonic physics. We calculated radiative heating/cooling rates using `CLOUDY90` package (Ferland *et al.* 1998). Global reionization is considered in the whole simulation volume at redshift z_{re}=8.9 (Haardt & Madau 1996). We assume that the dense gas cloud ($n_H \geqslant 0.014\mathrm{cm}^{-3}$) (Sawala *et al.* 2010) is shielded from the universal UV radiation. Stars form when gas particles satisfy star formation criteria of Saitoh *et al.* (2008). The star particle is considered as a single stellar population with a spectrum of Salpeter mass function (Salpeter 1955). The number of stars that eventually ends up as type II supernovae is calculated using stellar evolutionary tracks of Hurley, Pols, & Tout (2000). Energy, mass, and metals ejected by the supernovae are considered in a way to convey to its neighboring gas particles.

2. Test runs

We performed two test simulations that focus on an evolution of an isolated galaxy and study the cosmological structure formation. Table 1 gives parameters on the simulation. For the isolated galaxy, we were able to confirm that star formation rate (SFR) of isolated galaxy simulation well reproduces the Schmidt-Kennicutt law (Kennicutt 1998) (Fig. 1a). The cosmology is described with the cosmological parameters of $\Omega_M = 0.276$, $\Omega_\Lambda = 0.724$, $\Omega_b = 0.045$, and $h = 0.703$. Halos embedded in the more massive halos are identified using Rockstar halo finder (Behroozi, Wechsler, & Wu 2013). Figure 1b shows that number density profile of subhalo around the main halo of $\sim 10^{10}\,M_\odot$ is described with a softened power law of $\alpha = 3.3$. However, mass function is not consistent with the theoretical prediction of Sheth & Tormen (1999), especially for the low halo mass (Fig. 1c).

Table 1. Parameters of runs

Model	$\mathbf{L_{BOX}}$[1]	$\mathbf{N_{SPH}}$[2]	$\mathbf{m_{SPH}}$[3]	Shielding[4]	$\mathbf{n_{th}}$[5]	$\mathbf{T_{th}}$[6]	$\mathbf{C_*}$[7]
Isolated	×	98,304	$4.2 \times 10^4 M_\odot$	×	$0.1 cm^{-3}$	$15000K$	0.033
Run 1	8Mpc/h	256^3	$3.8 \times 10^5 M_\odot$	○	$0.1 cm^{-3}$	$15000K$	0.033
Run 2	8Mpc/h	256^3	$3.8 \times 10^5 M_\odot$	○	$100 cm^{-3}$	$5000K$	0.033

[1] A side length of a cube box   [2] Number of gas particles   [3] Mass of gas particles
[4] UV Shielding   [5] Threshold number density of star formation
[6] Threshold Temperature of star formation   [7] Star formation efficiency

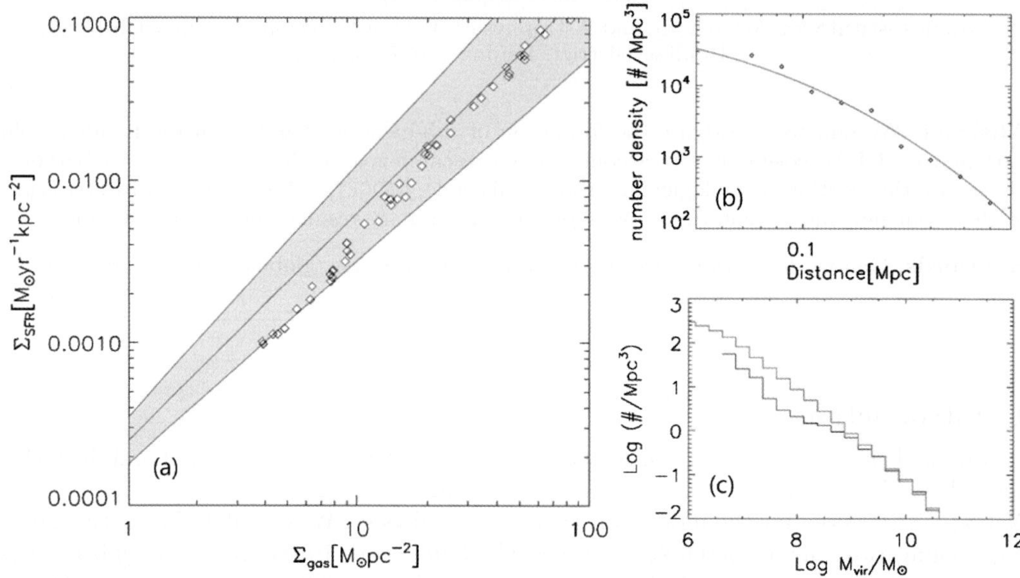

Figure 1. Results of runs. (a) The isolated galaxy simulation represents Schmidt-Kennicutt law. (b) Spatial density profile can be fitted with softened power law ($\alpha = 3.3$). (c) Dark matter mass function with the theoretical mass function(red line) and the simulation result(black line).

3. Future work

We will use more realistic power spectrum and halo-finding algorithm. After confirming our modified `GADGET-2` code is ready for the scientific studies, we will perform a multi-level simulation that resolves sub-galactic structure around the Milky Way-like main halo.

References

Behroozi, P. S., Wechsler, R. H. & Wu, H. Y. 2013a, *ApJ*, 762, 109
Ferland, G. J., Korista, K. T., Verner, D. A., Ferguson, J. W., Kingdon, J. B., & Verner, E. M. 1998, *PASP*, 110, 761
Haardt, F. & Madau, P. 1996, *ApJ*, 461, 20
Hurley, J. R., Pols, O. R., & Tout, C. A. 2000, *MNRAS*, 315, 543
Kennicutt, R. C.,Jr. 1998, *ApJ*, 498,541
Saitoh, T. R., Daisaka, H., Kokubo, E., Makino, J., Okamoto, T., Tomisaka, K., Wada, K., & Yoshida, N. 2008, *PASJ*, 60, 667
Salpeter, E. E. 1955, *ApJ*, 121, 161
Sawala, T., Scannapieco, C., Maio, U., & White, S. 2010, *MNRAS*, 402, 1599
Springel, V. 2005, *MNRAS*, 364, 1105

Star clusters and black holes in galaxies across cosmic time
Proceedings IAU Symposium No. 312, 2014
Y. Meiron, S. Li, F.-K. Liu & R. Spurzem, eds.

© International Astronomical Union 2016
doi:10.1017/S1743921315008078

Globular Clusters in the Local Group

Eva K. Grebel

Astronomisches Rechen-Institut, Zentrum für Astronomie der Universität Heidelberg,
Mönchhofstr. 12–14, 69120 Heidelberg, Germany
email: grebel@ari.uni-heidelberg.de

Abstract. Only twelve of the > 76 Local Group galaxies contain globular clusters, showing a broad range of specific frequencies. Here we summarize the properties of these globular cluster systems. Many host galaxies contain very old globulars, but in some globular cluster formation may have been delayed. An age range of several Gyr is common. Except for the inner regions of the spirals, old globular clusters tend to be metal-poor. Increasingly, light element variations and hints of multiple stellar populations are being found also in extragalactic globulars. There is ample evidence for globular cluster accretion from dwarfs onto massive galaxies, but its magnitude has yet to be quantified. Caution is needed to avoid overinterpreting indirect evidence.

Keywords. Galaxy: globular clusters: general, Galaxy: halo, galaxies: star clusters, Local Group, galaxies: dwarfs, Magellanic Clouds, galaxies: abundances, galaxies: evolution, stellar content

1. Introduction

Star clusters are observed in all but the least massive Local Group galaxies (e.g. Grebel 2002). These clusters span a wide range of ages, metallicities, masses, and radii. Star clusters that can in principle survive for a Hubble time as bound stellar systems are called 'globular clusters'. For a review on globular cluster formation, see, e.g. Kruijssen (2014). Star clusters may dissolve due to their internal dynamical evolution, in particular because of relaxation processes. External influences such as dynamical friction and varying tidal fields (caused, e.g. by disk or bulge passages in their parent galaxy) may also lead to their gradual erosion, emphasizing the role of cluster environment. A cluster's survival depends on its mass and density and on its host galaxy's gravitational potential as well as the cluster's distance from and orbit in its host galaxy (e.g. Gnedin & Ostriker 1997). Considering proper-motion-based orbits, Dinescu *et al.* (1999) find that the destruction rates due to internal two-body relaxation are higher than those due to tidal shocks.

Yet we still call an old star cluster currently undergoing disruption, such as Palomar 5 (Odenkirchen *et al.* 2001, 2002, Rockosi *et al.* 2002), a globular cluster. More generally, we may consider clusters that are plausibly massive and compact enough to survive as bound stellar aggregates for a Hubble time in the absence of destructive external or internal influences to be globular clusters. In fact, it is believed that many of the globular clusters that once formed were since destroyed by external gravitational effects. Indeed, a number of Galactic globular clusters show evidence for early stages of tidal disruption (e.g. Andreuzzi *et al.* 2001, Odenkirchen & Grebel 2004, Grillmair & Johnson 2006, Jordi & Grebel 2010, Chun *et al.* 2010, Niederste-Ostholt *et al.* 2010, Kunder *et al.* 2014).

On the other hand, we do not usually call clusters much younger than ~ 9 Gyr globular clusters. Hence we consider the Small Magellanic Cloud (SMC) to contain only one true globular cluster (Glatt *et al.* 2008a), although many of its prominent intermediate-age clusters are massive enough to pass as likely globular candidates (see Glatt *et al.* 2008b, 2011). *For the purpose of this review, we follow the convention of requiring an old age.*

There are additional properties that seem to be common to possibly all (Galactic) globular clusters, for instance, the lack of dark matter (e.g. Jordi *et al.* 2009; Frank *et al.* 2012; Sollima *et al.* 2012; but see Ibata *et al.* 2013). An important characteristic common to the majority of globular clusters are light element abundance variations (e.g. Kraft 1994; Harbeck *et al.* 2003a, 2003b; Gratton et al. 2004; Kayser *et al.* 2008), and/or the presence of multiple stellar populations (Gratton, Carretta, & Bragaglia 2012). While massive old open clusters may overlap in age and even structural properties with low-mass, younger globular clusters, the absence of light element anticorrelations and of multiple populations appears to be a unique identifying property of open clusters (e.g. de Silva *et al.* 2009; Carretta *et al.* 2010a; Bragaglia *et al.* 2014; McLean *et al.* 2015). Such abundance information, however, is not yet available for most nearby extragalactic clusters and can thus not be used as a defining or distinguishing feature in this review. Conversely, as light element abundance anticorrelations seem to occur *only* in globular clusters, surveying field stars for such anomalies allows one to constrain the contribution of dissolved globular clusters to the Galactic field populations. These studies suggest that at least 17% of the present-day stellar mass observed as Galactic halo field stars has a globular cluster origin (Martell & Grebel 2010; Martell *et al.* 2011).

Apart from light element abundance inhomogeneities, most globular clusters are fairly homogeneous in their heavy elements including iron (Carretta *et al.* 2009) with only very small star-to-star variations (e.g. Ivans *et al.* 1999, 2001; Cohen 2011; Yong *et al.* 2013) of ~ 0.03 to 0.05 dex in [Fe/H]. The few globular clusters that are not mono-metallic include the most massive Galactic globular cluster, ω Centauri, which is believed to be the stripped nucleus of a former dwarf galaxy, the globular cluster at the center of the Sagittarius dwarf spheroidal (dSph) galaxy, M54, as well as a few other globulars (e.g. M2, M22, Terzan 5, NGC 5824). For details, see, for instance, Kraft (1994), Da Costa *et al.* (2009, 2014), Ferraro *et al.* (2009), Marino *et al.* (2009), Johnson & Pilachowski (2010), Carretta *et al.* (2010b), Origlia *et al.* (2013), Yong *et al.* (2014), Massari *et al.* (2014), and Milone *et al.* (2015). For several of these unusual objects a (stripped) nuclear star cluster origin has been proposed to explain their internal range in metallicities. Those globular clusters with heavy element abundance spreads for which orbits have been determined seem to have originated from independent progenitors (Casetti-Dinescu *et al.* 2013).

Globular clusters differ from similarly luminous dwarf galaxies by their typically much more compact sizes with half-light radii of less than 10 pc (although the recently discovered so-called extended globular clusters are more diffuse; see Huxor *et al.* 2005), by their apparent lack of dark matter (e.g. Gilmore *et al.* 2007a), by (usually) being mono-metallic systems (while even low-mass dwarf galaxies show appreciable abundance spreads; e.g. Grebel 1997, 2000; Harbeck *et al.* 2001; Shetrone *et al.* 2001; Grebel *et al.* 2003; Koch *et al.* 2006, 2007a, 2007b; Norris *et al.* 2008; Adén *et al.* 2009, 2011; Kirby *et al.* 2011; Simon *et al.* 2011), and by *not* following a metallicity-luminosity relation. Globular clusters and dSph galaxies lie on different fundamental planes, but globular clusters and ultra-compact dwarf (UCD) galaxies show some overlap (Tollerud *et al.* 2011).

The separation of globular clusters from dSphs in mass-radius-luminosity space is primarily driven by whether objects have the mass-to-light ratios typical for old stellar populations (~ 3, as globular clusters do), or whether they are dark-matter-dominated (Tollerud *et al.* 2011; see also Zaritsky *et al.* 2006 and Misgeld & Hilker 2011). The UCD locus, with which the brightest globular clusters overlap, tilts away from the pure mass-follows-light relation (Tollerud *et al.* 2011). Kinematic studies of UCDs show that some have globular-cluster-like mass-to-light ratios suggesting that they are indeed very massive star clusters (e.g. Mieske *et al.* 2008; Frank *et al.* 2011), while others have higher mass ratios supporting a "galaxian" nature (e.g. stripped nuclei or lower-mass compact ellipticals; see, for instance, Brodie *et al.* 2011 and Forbes *et al.* 2014).

2. Differences and commonalities of the M31, M33, and Galactic globular cluster systems

M31, the dominant spiral galaxy in the Local Group, also has the richest globular cluster system. More than 700 globular cluster candidates have been identified in M31 and its surroundings so far (see Huxor *et al.* 2014; Galleti *et al.* 2004, and the regularly updated Revised Bologna Catalogue†). In contrast, in the Milky Way only about 160 globular clusters are known (see Harris 1996 and the continuing updates of the Harris catalog‡).

The luminosity functions of the globular cluster systems of the Milky Way and M31 have similar median absolute luminosities of $M_V = -7.3$ and -7.6, respectively (Huxor *et al.* 2014). When considering only M31 globular clusters at projected galactocentric distances larger than 30 kpc, the resulting luminosity function shows a secondary maximum at lower luminosities (~ -5.5). This resembles the bimodal values in the sparsely sampled globular cluster luminosity function of Sagittarius (van den Bergh 1998). This as well as the fact that many of the remote globular clusters of M31 are associated with the stellar tidal streams found around this galaxy suggests an accretion origin for the outer halo globular cluster population (Mackey *et al.* 2010, 2013; Huxor *et al.* 2014).

More metal-rich Milky Way and M31 globular clusters tend to be located at smaller galactocentric distances, although there is a range of metallicities at any distance (Wang & Ma 2013). More metal-rich globulars typically have smaller half-light radii. Globular cluster half-light radii and tidal radii increase with galactocentric distance in both galaxies (McLaughlin & van der Marel 2005; Barmby *et al.* 2007; Huxor *et al.* 2011; Wang & Ma 2013) as expected in scenarios of environmentally driven cluster survivability (e.g. Gnedin & Ostriker 1997). Faint globular clusters show a range of half-light radii, whereas the brighter ones tend to be primarily compact (van den Bergh 1996; Huxor et al. 2014).

The most metal-rich globular clusters ([Fe/H] > -0.5 dex) in M31 are mainly located in the bulge region in the innermost 5 kpc. They follow the steeply rising H I rotation curve of the bulge (Lee *et al.* 2008; Galleti *et al.* 2009). Even the more widely distributed metal-poor globular clusters ([Fe/H] < -1.0 dex) exhibit a clear rotational pattern although their velocity dispersion is much larger (Galleti *et al.* 2009). This may be the signature of a rotating, yet pressure-supported halo (Lee *et al.* 2008). A subset of M31 globular clusters shows X-ray emission and follows the rotation of the disk (Lee *et al.* 2008).

The most metal-rich globular clusters ([Fe/H] > -0.8 dex) in the Milky Way are almost exclusively situated within the inner 10 kpc around the Galactic center and are concentrated in the bulge region. They show a flattened distribution and are believed to belong to the bulge and the thick disk (e.g. Mackey & Gilmore 2004a). For a growing number of Galactic globular clusters, orbit determinations are becoming available. Some globular clusters (metal-rich ones but also a few metal-poor ones) were found to belong kinematically to the thick disk or to the bar (Casetti-Dinescu *et al.* 2010). Others show clear classical hot halo kinematics with extended, highly eccentric and in part retrograde orbits (Dauphole *et al.* 1996; Dinescu *et al.* 1999). The so-called "young" halo globular clusters (Zinn *et al.* 1993) with predominantly red horizontal branches, but [Fe/H] < -1 dex stand out as a hot population with mainly high-energy, large-size, and highly eccentric

† http://www.bo.astro.it/M31/
‡ http://physwww.mcmaster.ca/~harris/mwgc.dat

orbits in the outer regions of the Milky Way, consistent with a possible accretion origin (e.g. Dinescu *et al.* 1999). These objects have the widest spatial distribution and are found out to distances of more than 100 kpc (e.g. Mackey & Gilmore 2004a). The radial mass density profile of the surviving old halo globular clusters, on the other hand, follows the expected profile of the cold protogalactic gas (Parmentier & Grebel 2005).

Most bulge, disk, and old halo globular clusters in the Milky Way have core radii < 3 pc, while globular clusters of the "young" halo and in Galactic satellites may exhibit core radii up to ~ 20 pc (Mackey & Gilmore 2004a). This similarity in structural properties along with the eccentric and in part retrograde orbits may support a dwarf galaxy origin for "young halo" globulars. Typical compact globular clusters in the Milky Way and M31 have half-light radii of up to 5 pc, but the diffuse, "extended globular clusters" described by Huxor *et al.* (2005) can reach half-light radii of 30 pc. Such objects are found at galactocentric distances from ~ 15 to ~ 150 kpc around M31 and our Galaxy. Extended globular clusters just seem to sample the upper end of the globular cluster size distribution while behaving like regular, old, metal-poor globulars in all other aspects (Huxor *et al.* 2014). Da Costa *et al.* (2009b) suggested that there may be two modes of cluster formation, one leading to compact and one to extended clusters, with extended globular clusters forming (and surviving) preferentially in lower tidal field environments such as dwarf galaxies (see also Elmegreen 2008). In plots of the radius-luminosity relation, extended globular clusters begin to fill the gap between "normal" globular clusters and ultra-faint dSphs, raising interesting questions about the presence of dark matter or gradual dynamical or tidal dissolution (Misgeld & Hilker 2011; Huxor *et al.* 2011).

The designation "young" for a subset of halo clusters mentioned earlier describes their location in a diagram of horizontal branch type versus metallicity: these globular clusters have a lower metallicity than expected from their horizontal branch type; i.e., one would expect a bluer horizontal branch at their metallicity. Hence a *second* parameter must be responsible for their horizontal branch morphology. When plotting horizontal branch isochrones in the above diagram, these clusters appear to be ~ 1 to 3 Gyr younger than the oldest globular clusters (e.g. Lee *et al.* 1994; Keller *et al.* 2012). However, the locus in the horizontal branch type – metallicity diagram is not only determined by age. Other factors, such as helium abundance or central density, also play a role. Nonetheless, age dating based on deep color-magnitude diagrams (CMDs) often corroborates a younger age (e.g. Stetson *et al.* 1999; Rey *et al.* 2001; Dotter *et al.* 2010; VandenBerg *et al.* 2013).

Keller *et al.* (2012) find that the "young halo" globular clusters at Galactocentric distances larger than 10 kpc lie close to a plane that is very similar to the plane encompassing the Galactic satellites, supporting the suggestion that they were accreted. VandenBerg *et al.* (2013) argue that the age-metallicity relation of Galactic globular clusters splits into two branches, which are offset by 0.6 dex. The more metal-rich branch shows disk-like kinematics, while the more metal-poor one has halo kinematics. Leaman *et al.* (2013) find good agreement between the metal-poor branch and the age-metallicity relation of Local Group irregular galaxies. They propose that a large fraction of the halo globular cluster system was accreted from such galaxies. – In summary, there are many different lines of arguments favoring accretion of part of the Galactic globular cluster system.

Among luminous M31 globulars extended blue horizontal branches are common, suggestive of multiple stellar populations (Perina et al. 2012). Taking the locus in the horizontal branch type – metallicity diagram as a proxy for age, Perina *et al.* (2012) find that M31 globular clusters at projected galactocentric distances of 10 to 100 kpc are ~ 0.4 dex more metal-rich at a given horizontal branch type than their Galactic counterparts. This suggests that they typically formed 1 to 2 Gyr later than even the "young" Galactic halo globulars if helium abundance variations can be neglected – an intriguing systematic difference. For M31's globular clusters there are generally no sufficiently deep CMD data available for an independent, accurate age determination. However, for one globular cluster Hubble Space Telescope (HST) imaging reaching $m_V \sim 30.5$ mag was obtained, which provides data extending below the main-sequence turn-off. The resulting isochrone-based age is 2 to 3 Gyr younger than for typical Galactic globular clusters at that metallicity and horizontal branch morphology (Brown *et al.* 2004).

We have less detailed information about the globular clusters of M33. Surveys of the star cluster system of this smallest Local Group spiral reveal a large number of young and intermediate-age clusters and a distribution of integrated colors similar to that of the Large Magellanic Cloud (LMC) (e.g. Sarajedini & Mancone 2007). Thus far these surveys suffer from spatial and luminosity incompleteness, but also from the difficulty to distinguish genuine star clusters from other extended sources such as background galaxies or asterisms in crowded regions (e.g. Park & Lee 2007; Zloczewski & Kaluzny 2009).

Moreover, age estimates based on resolved CMDs obtained from HST data are available only for a subset of cluster candidates (e.g. Cockcroft *et al.* 2011; Zloczewski & Kaluzny 2009; Park *et al.* 2009; San Roman *et al.* 2009), and these are typically too shallow to permit an accurate age dating of old cluster candidates. Globular cluster candidates have thus been identified mainly via integrated colors, e.g. from their overlap with the locus of Galactic globular clusters in two-color diagrams (e.g. Ma 2012, 2013) and age-dated via their multi-color spectral energy distributions (e.g. Fan & de Grijs 2014). The Fan & de Grijs catalog contains 75 globular cluster candidates older than 10 Gyr and approximately 17 candidates located in the 8–10 Gyr range within the uncertainties of their age estimates. In addition, there are at least six outer halo globular cluster candidates within projected radii of 10 to 50 kpc around M33 (Cockcroft *et al.* 2011). This comparatively low number may indicate that the outer halo of M33 was significantly stripped during past encounters with M31 or that M33 experienced a much more quiescent accretion history (Cockcroft *et al.* 2011). At least one outer halo globular cluster in M33 is an extended cluster similar to the ones found in the outskirts of the Milky Way and M31 (Stonkutė *et al.* 2008).

The "specific frequency" of a galaxy is its number of globular clusters, N_{GC}, normalized by unit parent galaxy luminosity; $S_N = N_{GC} \cdot 10^{0.4(M_V + 15)}$ (Harris & van den Bergh 1981). If all globular cluster candidates in M33 are genuine globulars, then M33's specific frequency is 2 to 2.7, implying a rather rich globular cluster system for a low-mass disk galaxy. The S_N values of M31 and of the Milky Way are > 2.3 and ~ 0.7, respectively. Given their incompleteness and uncertain detections, these numbers are likely to change.

3. Globular clusters in Local Group dwarf spheroidal galaxies

The Local Group contains a large number of metal-poor, gas-deficient dSphs, most of which are satellites of the two massive spirals (see Grebel *et al.* 2003 for details). In recent years, many new, very faint dSph companions of the Milky Way and M31 were discovered (e.g. Zucker *et al.* 2004, 2006a, 2006b, 2007; Belokurov *et al.* 2006; Martin *et al.* 2006; Walsh *et al.* 2007). A comprehensive tabulation of the key properties of Local Group dwarf galaxies including the recent discoveries is given in McConnachie (2012)† and its continued updates. Only the two most luminous dSphs, Sagittarius and Fornax, host globular clusters.

The Sagittarius dSph galaxy, which is currently merging with the Milky Way, contains five high-, four moderate, and two low-confidence globular cluster members (Law

† `https://www.astrosci.ca/users/alan/Nearby_Dwarfs_Database.html`

& Majewski 2010). Assuming that Sagittarius' total luminosity was $M_V = -15$ prior to its disruption, its resulting specific frequency is 5 to 9 depending on how many globular clusters are associated with it (Law & Majewski 2010). The specific frequency of Sagittarius (henceforth Sgr) is much higher than that of more luminous galaxies, where S_N scatters around 1 (Harris *et al.* 2013). Higher values of S_N are typically only found in the most massive elliptical galaxies (such as cD or brightest cluster galaxies) as well as in low-mass, gas-deficient galaxies such as dSphs and nucleated dwarf ellipticals (Harris *et al.* 2013; Grebel 2002).

As noted by Da Costa & Armandroff (1995), Sgr contributes both "old halo" as well as "young halo" globular clusters to the Milky Way – a cautionary point that is important to remember when discussing globular cluster accretion. The Sgr globular cluster candidates follow a fairly narrow age-metallicity relation, which can be well approximated by a simple closed-box model with a continuous star formation rate (Forbes & Bridges 2010).

The spatial coincidence of the very massive, very luminous globular cluster M54 with the center of Sgr indicates that M54 may be the nucleus of this dSph galaxy (Sarajedini & Layden 1995), possibly akin to what is seen in more massive nucleated dwarf ellipticals (e.g. Lisker *et al.* 2006, 2007). However, more recent studies revealed that Sgr has a metal-rich nucleus with a flat velocity dispersion profile that is independent of the metal-poor M54 with its steeply rising velocity dispersion (Monaco *et al.* 2005; Bellazzini *et al.* 2008). This suggests that M54 may have formed elsewhere in Sgr and eventually sank to its center via dynamical friction (Bellazzini *et al.* 2008). We thus see the actual nucleus and M54 in superposition; a coincidence that would be impossible to disentangle in more distant galaxies. M54 has a metallicity of [Fe/H] ~ -1.6 with an extended tail towards higher metallicities, similar to the metallicity spread found in ω Centauri (Carretta *et al.* 2010c). Just like other globular clusters, it shows a Na–O anticorrelation, which is not seen in the stars attributed to the Sgr nucleus (Carretta *et al.* 2010c).

The Fornax dSph galaxy contains five metal-poor globular clusters, four of which are located in its outer regions, while one of them (globular cluster #4 in the naming convention of Hodge 1961) lies close to its center. This fourth globular cluster is more metal-rich than the others (Larsen *et al.* 2012) and approximately 3 Gyr younger (Buonanno et al. 1999), while the other four are indistinguishable in age and as old as the oldest Galactic globular clusters (Buonanno *et al.* 1998). Fornax' globular cluster #1 is the most metal-poor globular cluster known to date ([Fe/H] $= -2.5$ dex, Letarte *et al.* 2006). The specific frequency of Fornax is ~ 22.8, the highest value of any Local Group galaxy.

The metal-poor, ancient, roughly coeval globular clusters #3 and #5 show pronounced differences in their horizontal branch morphologies (Smith *et al.* 1996), again adding a cautionary note to the use of horizontal branch indices as age indicators. If Fornax were to be accreted by the Milky Way, it, too, would contribute both genuinely old as well as "young halo" globular clusters. Nitrogen abundance variations and horizontal branch morphologies in its four old globular clusters suggest the presence of multiple stellar populations, just like in Galactic globulars (D'Antona *et al.* 2013; Larsen *et al.* 2014a).

Oh *et al.* (2000) pointed out that due to dynamical friction it is unexpected to find Fornax' four old globular clusters in the outer regions of this galaxy instead of in the center, forming a merged, very massive nucleus. Oh *et al.* (2000) suggest tidal heating and significant tidal mass loss of Fornax as a possible mechanism preventing the clusters' orbital decay. Goerdt *et al.* (2006) and Cole *et al.* (2012) argue that a cored dark matter halo (as opposed to a cuspy cold dark matter halo) extending to the positions of the globular clusters would likewise prevent this. Strigari *et al.* (2006), on the other hand, challenge the notion of a very large core. Generally though, cored dark matter halos seem to reproduce best the kinematic data of galaxies of a range of different Hubble types (e.g.

Gilmore *et al.* 2007b; Donato *et al.* 2009; Salucci *et al.* 2012). For dSphs like Fornax, a cuspy profile was found to be excluded at high significance (Walker & Peñarrubia 2011).

Several other Galactic dSphs show dynamically cold substructures, which may also share similar metallicities (e.g. Kleyna *et al.* 1998, 2004; Walker *et al.* 2006; Battaglia *et al.* 2011). If real (see the cautionary remarks of Ural *et al.* 2010), these structures may be the remnants of dissolved old star clusters. Just like the old globular clusters in Fornax, the longevity and survival of these potential remnants support cored dark matter halos (e.g. Lora *et al.* 2012a, 2012b, 2013). We note that initially cuspy cold dark matter halos can evolve into cored dark matter profiles due to the redistribution of matter via star formation and feedback (e.g. Mashchenko *et al.* 2006; Pasetto *et al.* 2010).

4. Globular clusters in Local Group dwarf ellipticals

M31 is the only galaxy in the Local Group that has dwarf elliptical (dE) companions (e.g. Grebel 1997; 2001). While its three dE satellites NGC 147, NGC 185, and NGC 205 all host globular clusters, its compact elliptical companion M32 apparently (and unexpectedly, given its luminosity) does not (e.g. van den Bergh 2000; Rudenko *et al.* 2009). NGC 205 and NGC 185 show prominent intermediate-age populations as well as recent star formation, gas, and dust, whereas NGC 147 is largely old and gas-deficient (see Grebel 1999 and references therein). NGC 205 and M32 both contain conspicuous nuclei.

Globular cluster candidates in NGC 205 were first identified by Hubble (1932). Follow-up studies, particularly metallicity and age determinations from integrated spectra and shallow HST CMDs, suggest that Hubble I, II, IV, VI, VII, and VIII are genuine old globular clusters with estimated ages ranging from 7–11 Gyr (± 2 Gyr) and [Fe/H] from -1.1 to -2.0 dex (Da Costa & Mould 1988; Sharina *et al.* 2006; Colucci & Bernstein 2011). The enhanced [α/Fe] ratios in the old globular clusters indicate high star formation rates at early times (Colucci & Bernstein 2011). The resulting S_N of NGC 205 is 1.7.

The first globular cluster candidates in NGC 147 and NGC 185 were discovered by Baade (1944). More candidates were added in the following decades, including recent detections by Sharina & Davoust (2009) in NGC 147 and by Veljanoski *et al.* (2013) in both dEs. The new additions are in part much fainter than the early detections and tend to be located at larger distances from the centers of the galaxies. Contamination by clusters in the disk or spheroid of M31 seen in superposition are an issue. Many of the globular cluster candidates in both dEs have been analyzed using various methods based on integrated spectra, radial velocities, and integrated photometry (e.g. Da Costa & Mould 1988; Sharina *et al.* 2006; Sharina & Davoust 2009; Veljanoski *et al.* 2013). The inferred ages and metallicities differ considerably in some cases (as also for NGC 205). Some of the candidates were found to be too young to be considered bona fide globular clusters, although the age estimates are rather uncertain and can be affected by the surrounding integrated field star light (Veljanoski *et al.* 2013). Regardless of the method, most globular clusters are found to be metal-poor ([Fe/H] < -1.2 dex). Sharina *et al.* (2006) point out that many of the globular cluster candidates in the three dEs overlap with the Galactic "young halo" locus in the metallicity – horizontal-branch-type diagram.

NGC 147 may contain up to ten globular clusters, but two of these (the metal-rich Hodge IV, which exhibits a somewhat discrepant radial velocity, and SD-10) may not be genuine globulars (Sharina & Davoust 2009; Veljanoski *et al.* 2013). NGC 185 hosts seven promising globular cluster candidates. Accounting for the uncertainties, Veljanoski *et al.* (2013) estimated the specific frequencies to be 8 ± 2 for NGC 147 and 5.5 ± 0.5 for NGC 185. These values follow the trend in S_N in other dEs of similar luminosity.

5. Globular clusters in Local Group (dwarf) irregular galaxies

The LMC is the most luminous, most massive irregular galaxy in the Local Group. It contains at least 15 globular clusters older than ~ 10 Gyr (e.g. Mackey & Gilmore 2004b; Baumgardt *et al.* 2013) and one slightly younger candidate (ESO 121-SC03; age ~ 8.9 Gyr; Xin *et al.* 2008). The LMC's specific frequency is 0.6. It also hosts numerous intermediate-age and young populous clusters (e.g. Girardi *et al.* 1995; Hunter *et al.* 2003; Glatt *et al.* 2010), but shows an as yet unexplained lack of clusters with ages of about 5 to 9 Gyr (Olszewski *et al.* 1991). Many more clusters than expected from chance superpositions appear to be double or multiple with similarly old components (Bhatia & Hatzidimitrou 1988; Bhatia *et al.* 1991; Dieball & Grebel 2000; Dieball *et al.* 2002).

The oldest globular clusters in the LMC are indistinguishable in age from the oldest Galactic, Sagittarius, and Fornax globular clusters (see Olsen *et al.* 1998; Johnson *et al.* 1999; Grebel & Gallagher 2004). Intriguingly, the SMC's only globular cluster, NGC 121, is 2–3 Gyr younger (Glatt *et al.* 2008a). Moreover, and in contrast to the LMC (and the Milky Way), the SMC formed massive star clusters throughout its history (e.g. Da Costa & Hatzidimitriou 1998; Glatt *et al.* 2008b), which is unexpected especially if the two Clouds were indeed close to and interacted with each other for most of their lifetime.

Despite their small age range, the LMC globulars cover a variety of horizontal branch morphologies and metallicities (Olszewski et al. 1991; Olsen *et al.* 1998; Mackey & Gimore 2004b; Johnson *et al.* 2006; Mucciarelli *et al.* 2010). High-resolution spectra of individual red giants show that the old globular clusters in the LMC show similar Na–O anticorrelations as Galactic globular clusters (Mucciarelli *et al.* 2009; Mateluna *et al.* 2012). Abundances from integrated spectra reveal an $[\alpha/\text{Fe}]$ spread in old LMC globular clusters of similar metallicity, suggesting that the interstellar medium was not well mixed at the time of their formation (Colucci *et al.* 2012). Several old LMC globular clusters show much lower $[\alpha/\text{Fe}]$ ratios than comparable Galactic globular clusters (Johnson *et al.* 2006; Mateluna *et al.* 2012), which poses an interesting challenge for scenarios where the early accretion of massive satellites contributed substantially to the build-up of the Galactic halo and its globular clusters (e.g. De Lucia & Helmi 2008; Leaman *et al.* 2013). On the other hand, for even more metal-poor old LMC field stars, $[\alpha/\text{Fe}]$ is in very good agreement with Galactic halo stars of the same metallicity (Haschke *et al.* 2012a).

With only one genuine, albeit slightly younger globular cluster it is difficult to analyze the SMC ($S_N \sim 0.1$) in the same manner. In any case, this galaxy shows a range of cluster metallicities at any given age (Glatt *et al.* 2008b; Cignoni *et al.* 2013), which implies that its interstellar gas was not well mixed throughout its history. The SMC's star clusters do not show any preferred rotation (e.g. Da Costa & Hatzidimitriou 1998), consistent with the highly perturbed structure of this interacting dwarf irregular galaxy (e.g. Haschke *et al.* 2012b). In contrast, even the old LMC globular clusters follow disk as opposed to halo kinematics (Schommer *et al.* 1992; Grocholski *et al.* 2006).

Only two of the "isolated" dwarf irregular galaxies in the Local Group host globular clusters. In NGC 6822 eight globular cluster candidates have been identified, including five extended, diffuse clusters and three compact and luminous clusters (Hwang *et al.* 2011; Huxor *et al.* 2013). Assuming that they are all genuine globular clusters, the resulting $S_N \sim 3$. In its central regions, NGC 6822 also contains a number of intermediate-age and young star clusters. The old, extended globular clusters appear to lie roughly along the major axis of the old stellar halo, almost perpendicular to the elongated H I distribution of NGC 6822 (Hwang *et al.* 2011; Huxor *et al.* 2013). Just like in the SMC some of the globular clusters have high ellipticities. Integrated spectroscopy suggests that the globular clusters are metal-poor with $[\text{Fe}/\text{H}]$ from -1.5 to -2.5 and ages between ~ 8 to

Table 1. Globular cluster systems of Local Group galaxies

Galaxy	Type	M_V	N_{GC}	S_N	Remarks
M31	SA(s)b	−21.2	> 700	> 2.3	Rotation, [Fe/H] gradient, on avg. younger; accretion
Milky Way	SBbc	−20.9	∼ 160	∼ 0.7	[Fe/H] gradient, Na–O anticorrelation, accretion
M33	SA(s)cd	−18.9	75–98	2–2.7	Outer halo stripped?
Sgr	dSph(t)	∼ −15.0	5–11	5–9	"Young" & "old halo" GCs; Na–O anticorrelation
For	dSph	−13.1	5	22.8	"Young" & "old halo" GCs; multiple pop.? [Fe/H] grad.
NGC 205	dE5, N	−16.4	∼ 6	1.7	"Young" & "old halo" GCs; [α/Fe] enhanced
NGC 185	dE3	−15.6	∼ 7	∼ 4	"Young" & "old halo" GCs; ∼ solar [α/Fe]
NGC 147	dE5	−15.1	8–10	∼ 7–9	Near-solar [α/Fe]?
LMC	SB(s)m	−18.5	15–16	0.6	Rotation, "young" & "old halo", Na–O, [α/Fe] spread
SMC	SB(s)mp	−17.1	1	0.1	Younger than oldest MW & LMC GCs, high ellipticity
NGC 6822	IB(s)m	−16.0	8	3	Some GCs: high ellipt., one w. solar [α/Fe]. No rotation
WLM	IB(s)m	−14.4	1	1.7	Na–O anticorr.? [α/Fe] enhanced, high ellipticity

Notes: Column 1: galaxy name; column 2: galaxy classification (mainly from the NASA/IPAC Extragalactic Database or NED; `ned.ipac.caltech.edu`); column 3: absolute V-band magnitude (see Grebel 2002, Grebel *et al.* 2003); column 4: number of globular clusters (caution: incompleteness, false detections!); column 5: specific frequency (same warning); column 6: some of the special properties of a galaxy's globular cluster system (see relevant chapters of this review for details and references). MW stands for Milky Way, GC for globular cluster.

14 Gyr (Cohen & Blakeslee 1998; Hwang *et al.* 2014). The old cluster Hubble VII shows solar [Ca/Fe] (Colucci & Bernstein 2011) and is close to the center of NGC 6822, while the other old, metal-poor globular clusters have galactocentric distances > 2 kpc (Hwang *et al.* 2014). They do not rotate with the disk of NGC 6822 (Hwang *et al.* 2014).

In WLM, the faintest dIrr in the Local Group known to contain a globular cluster, only one such object is known, resulting in a relatively high specific frequency of 1.7. WLM's globular cluster is unusual in several respects: It is very luminous, highly eccentric yet very compact (Stephens *et al.* 2006), and lies in the outer regions of WLM at a projected distance of about 400 – 500 pc from the main body of the dIrr. Only one other, but very young luminous star cluster has been found in WLM so far, located in its central regions, and there are no known populous intermediate-age star clusters (Hodge *et al.* 1999).

A deep HST CMD and integrated spectra suggest that the globular cluster in WLM is old (> 12 Gyr) and metal-poor with inferred metallicities ranging from ∼ −1.5 to ∼ −2 dex (see Hodge *et al.* 1999; Stephens *et al.* 2006; Colucci & Bernstein 2011; Larsen *et al.* 2014b). Stephens *et al.* (2006) find a fairly high central velocity dispersion of ∼ 10 km s^{-1} for the globular cluster and no evidence of rotation. They suggest that a high velocity dispersion anisotropy is the most likely reason for the cluster's high ellipticity. Larsen *et al.* (2014b) show that WLM's globular cluster has similar abundance patterns as the old, metal-poor globular clusters in the Milky Way and in Fornax, including α element enhancement (see also Colucci *et al.* 2011). While integrated spectra do not allow one to measure light element abundance anticorrelations, Larsen *et al.* interpret the cluster's high [Na/Fe] ratio as a possible indication of a second generation of star formation.

6. Concluding remarks

Out of the more than 76 galaxies of the Local Group, only twelve host one or more globular clusters. The globular cluster systems exhibit a bewildering range of richness and specific frequencies, while the lack of globular clusters in other galaxies of similar type and luminosity is not yet understood. Many globular clusters were only discovered in recent years though. Their census is likely to continue to increase in the coming years, particularly with respect to faint, extended clusters. Most host galaxies contain very old globular clusters (as well as some that are up to a few Gyr younger), but in some galaxies (such as the SMC and perhaps M31) all globular clusters seem to be systematically younger. The reasons for such a possibly delayed cluster formation are still unclear.

There is ample direct and indirect evidence of globular cluster accretion from dwarf galaxies onto M31 and the Milky Way, but its magnitude remains to be quantified. As dwarf galaxies may contribute both "young" and "old" halo globular clusters, as well as globulars with or without [α/Fe] enhancement, one needs to beware of too simplistic conclusions. Globulars in dwarf galaxies often show large core radii and/or high ellipticities. The preferred occurrence of globular clusters with large core radii in the outer halos of massive galaxies may imply an accretion origin and/or suggest that more extended clusters form and survive more easily in environments with lower tidal fields. Na–O anticorrelations or extended horizontal branches in globular clusters in several dwarf galaxies indicate that multiple stellar populations may be a global property of globular clusters. Finally, globular clusters can help to constrain the dark halo properties of their hosts.

References

Adén, D., et al. 2009, A&A, 506, 1147

Adén, D., et al. 2011, A&A, 525, A153

Andreuzzi, G., et al. 2001, A&A, 372, 851

Baade, W. 1944, ApJ, 100, 147

Barmby, P., McLaughlin, D. E., Harris, W. E., Harris, G. L. H., & Forbes, D. A. 2007, AJ, 133, 2764

Battaglia, G., Tolstoy, E., Helmi, A., Irwin, M., Parisi, P., Hill, V., & Jablonka, P. 2011, MNRAS, 411, 1013

Baumgardt, H., Parmentier, G., Anders, P., & Grebel, E. K. 2013, MNRAS, 430, 676

Bellazzini, M., et al. 2008, AJ, 136, 1147

Belokurov, V., et al. 2006, ApJL, 647, L111

Bhatia, R. K. & Hatzidimitriou, D. 1988, MNRAS, 230, 215

Bhatia, R. K., Read, M. A., Hatzidimitriou, D., & Tritton, S. 1991, A&AS, 87, 335

Bica, E., Bonatto, C., Dutra, C. M., & Santos, J. F. C. 2008, MNRAS, 389, 678

Brodie, J. P., Romanowsky, A. J., Strader, J., & Forbes, D. A. 2011, AJ, 142, 199

Brown, T. M., et al. 2004, ApJL, 613, L125

Buonanno, R., et al. 1998, ApJL, 501, L33

Buonanno, R., et al. 1999, AJ, 118, 1671

Carretta, E., Bragaglia, A., Gratton, R., D'Orazi, V., & Lucatello, S. 2009, A&A, 508, 695

Carretta, E., et al. 2010a, A&A, 516, A55

Carretta, E., et al. 2010b, ApJL, 714, L7

Carretta, E., et al. 2010c, A&A, 520, A95

Casetti-Dinescu, D. I., Girard, T. M., Korchagin, V. I., van Altena, W. F., & López, C. E. 2010, AJ, 140, 1282

Casetti-Dinescu, D. I., et al. 2013, AJ, 146, 33

Chun, S.-H., et al. 2010, AJ, 139, 606

Cignoni, M., et al. 2013, ApJ, 775, 83

Cockcroft, R., et al. 2011, ApJ, 730, 112

Cohen, J. G. 2011, ApJL, 740, L38

Cohen, J. G. & Blakeslee, J. P. 1998, AJ, 115, 2356

Cole, D. R., Dehnen, W., Read, J. I., & Wilkinson, M. I. 2012, MNRAS, 426, 601

Colucci, J. E. & Bernstein, R. A. 2011, in A Universe of Dwarf Galaxies, EAS Publications Series 48, eds. M. Koleva, P. Prugniel, & I. Vauglin (EDP Sciences: Paris), 275

Colucci, J. E., Bernstein, R. A., Cameron, S. A., & McWilliam, A. 2012, ApJ, 746, 29

Da Costa, G. S. & Mould, J. R. 1988, ApJ, 334, 159

Da Costa, G. S. & Armandroff, T. E. 1995, AJ, 109, 2533

Da Costa, G. S. & Hatzidimitriou, D. 1998, AJ, 115, 1934

Da Costa, G. S., Held, E. V., Saviane, I., & Gullieuszik, M. 2009a, *ApJ*, 705, 1481

Da Costa, G. S., Grebel, E. K., Jerjen, H., Rejkuba, M., & Sharina, M. E. 2009b, *AJ*, 137, 4361

Da Costa, G. S., Held, E. V., & Saviane, I. 2014, *MNRAS*, 438, 3507

D'Antona, F., et al. 2013, *MNRAS*, 434, 1138

Dauphole, B., Geffert, M., & Colin, J., et al. 1996, *A&A*, 313, 119

De Lucia, G. & Helmi, A. 2008, *MNRAS*, 391, 14

de Silva, G. M., Gibson, B. K., Lattanzio, J., & Asplund, M. 2009, *A&A*, 500, L25

Dieball, A. & Grebel, E. K. 2000, *A&A*, 358, 897

Dieball, A., Müller, H., & Grebel, E. K. 2002, *A&A*, 391, 547

Dinescu, D. I., Girard, T. M., & van Altena, W. F. 1999, *AJ*, 117, 1792

Donato, F., et al. 2009, *MNRAS*, 397, 1169

Dotter, A., et al. 2010, *ApJ*, 708, 698

Elmegreen, B. G. 2008, *ApJ*, 672, 1006

Fan, Z. & de Grijs, R. 2014, *ApJS*, 211, 22

Ferraro, F. R., et al. 2009, *Nature*, 462, 483

Forbes, D. A. & Bridges, T. 2010, *MNRAS*, 404, 1203

Forbes, D. A., et al. 2014, *MNRAS*, 444, 2993

Frank, M. J., et al. 2011, *MNRAS*, 414, L70

Frank, M. J., et al. 2012, *MNRAS*, 423, 2917

Galleti, S., Federici, L., Bellazzini, M., Fusi Pecci, F., & Macrina, S. 2004, *A&A*, 416, 917

Galleti, S., Bellazzini, M., Buzzoni, A., Federici, L., & Fusi Pecci, F. 2009, *A&A*, 508, 1285

Gilmore, G., et al. 2007a, *Nuclear Physics B Proceedings Supplements*, 173, 15

Gilmore, G., et al. 2007, *ApJ*, 663, 948

Girardi, L., Chiosi, C., Bertelli, G., & Bressan, A. 1995, *A&A*, 298, 87

Glatt, K., et al. 2008a, *AJ*, 135, 1106

Glatt, K., et al. 2008b, *AJ*, 136, 1703

Glatt, K., et al. 2009, *AJ*, 138, 1403

Glatt, K., Grebel, E. K., & Koch, A. 2010, *A&A*, 517, A50

Glatt, K., et al. 2011, *AJ*, 142, 36

Gnedin, O. Y. & Ostriker, J. P. 1997, *ApJ*, 474, 223

Goerdt, T., Moore, B., Read, J. I., Stadel, J., & Zemp, M. 2006, *MNRAS*, 368, 1073

Gratton, R., Sneden, C., & Carretta, E. 2004, *ARA&A*, 42, 385

Gratton, R. G., Carretta, E., & Bragaglia, A. 2012, *A&ARv*, 20, 50

Grebel, E. K. 1997, *Reviews in Modern Astronomy*, 10, 29

Grebel, E. K. 1999, in The Stellar Content of Local Group Galaxies, *IAU Symp.* 192, eds. P. Whitelock & R. Cannon (San Francisco: ASP), 17

Grebel, E. K. 2000, in Star Formation from the Small to the Large Scale, 33rd ESLAB Symposium, *ESA SP* 445, eds. F. Favata, A.A. Kaas, & A. Wilson (Noordwijk: ESA), p. 87

Grebel, E. K. 2001, *Astrophysics and Space Science Supplement*, 277, 231

Grebel, E. K. 2002, in Extragalactic Star Clusters, *IAU Symp.* 207, eds. D. Geisler, E. K. Grebel, & D. Minniti (San Francisco: ASP), 94

Grebel, E. K., Gallagher, J. S., & III, Harbeck, D. 2003, *AJ*, 125, 1926

Grebel, E. K. & Gallagher, J. S. 2004, *ApJL*, 610, L89

Grillmair, C. J., & Johnson, R. 2006, *ApJL*, 639, L17

Grocholski, A. J., Cole, A. A., Sarajedini, A., Geisler, D., & Smith, V. V. 2006, *AJ*, 132, 1630

Harbeck, D., et al. 2001, *AJ*, 122, 3092

Harbeck, D., Smith, G. H., & Grebel, E. K. 2003a, *AJ*, 125, 197

Harbeck, D., Smith, G. H., & Grebel, E. K. 2003b, *A&A*, 409, 553

Harris, W. E. 1996, *AJ*, 112, 1487

Harris, W. E. & van den Bergh, S. 1981, *AJ*, 86, 1627

Harris, W. E., Harris, G. L. H., & Alessi, M. 2013, *ApJ*, 772, 82

Haschke, R., et al. 2012a, *AJ*, 144, 88

Haschke, R., Grebel, E. K., & Duffau, S. 2012b, *AJ*, 144, 107

Hodge, P. W. 1961, *AJ*, 66, 83

Hodge, P. W., Dolphin, A. E., Smith, T. R., & Mateo, M. 1999, *ApJ*, 521, 577

Hubble, E. 1932, *ApJ*, 76, 44

Hunter, D. A., Elmegreen, B. G., Dupuy, T. J., & Mortonson, M. 2003, *AJ*, 126, 1836

Huxor, A. P., *et al.* 2005, *MNRAS*, 360, 1007

Huxor, A. P., *et al.* 2011, *MNRAS*, 414, 770

Huxor, A. P., Ferguson, A. M. N., Veljanoski, J., Mackey, A. D., & Tanvir, N. R. 2013, *MNRAS*, 429, 1039

Huxor, A. P., *et al.* 2014, *MNRAS*, 442, 2165

Hwang, N., *et al.* 2011, *ApJ*, 738, 58

Hwang, N., *et al.* 2014, *ApJ*, 783, 49

Ibata, R., *et al.* 2013, *MNRAS*, 428, 3648

Ivans, I. I., *et al.* 1999, *AJ*, 118, 1273

Ivans, I. I., *et al.* 2001, *AJ*, 122, 1438

Johnson, J. A., Bolte, M., Stetson, P. B., Hesser, J. E., & Somerville, R. S. 1999, *ApJ*, 527, 199

Johnson, J. A., Ivans, I. I., & Stetson, P. B. 2006, *ApJ*, 640, 801

Johnson, C. I. & Pilachowski, C. A. 2010, *ApJ*, 722, 1373

Jordi, K., *et al.* 2009, *AJ*, 137, 4586

Jordi, K. & Grebel, E. K. 2010, *A&A*, 522, A71

Kayser, A., Hilker, M., Grebel, E. K., & Willemsen, P. G. 2008, *A&A*, 486, 437

Keller, S. C., Mackey, D., & Da Costa, G. S. 2012, *ApJ*, 744, 57

Kirby, E. N., Lanfranchi, G. A., Simon, J. D., Cohen, J. G., & Guhathakurta, P. 2011, *ApJ*, 727, 78

Kleyna, J. T., Geller, M. J., Kenyon, S. J., Kutz, M. J., & Thorstensen, J. R. 1998, *ApJ*, 115, 2359

Kleyna, J. T., Wilkinson, M. I., Evans, N. W., & Gilmore G. 2004, *MNRASL*, 354, L66

Koch, A., *et al.* 2006, *AJ*, 131, 895

Koch, A., *et al.* 2007a, *ApJ*, 657, 241

Koch, A., *et al.* 2007b, *AJ*, 133, 270

Kraft, R. P. 1994, *PASP*, 106, 553

Kruijssen, J. M. D. 2014, *Classical and Quantum Gravity*, 31, 244006

Kunder, A., *et al.* 2014, *A&A*, 572, A30

Larsen, S. S., Brodie, J. P., & Strader, J. 2012, *A&A*, 546, A53

Larsen, S. S., Brodie, J. P., Grundahl, F., & Strader, J. 2014a, *ApJ*, 797, 15

Larsen, S. S., Brodie, J. P., Forbes, D. A., & Strader, J. 2014b, *A&A*, 565, A98

Law, D. R. & Majewski, S. R. 2010, *ApJ*, 718, 1128

Leaman, R., VandenBerg, D. A., & Mendel, J. T. 2013, *MNRAS*, 436, 122

Lee, Y.-W., Demarque, P., & Zinn, R. 1994, *ApJ*, 423, 248

Lee, M. G., *et al.* 2008, *ApJ*, 674, 886

Letarte, B., *et al.* 2006, *A&A*, 453, 547

Lisker, T., Glatt, K., Westera, P., & Grebel, E. K. 2006, *AJ*, 132, 2432

Lisker, T., Grebel, E. K., Binggeli, B., & Glatt, K. 2007, *ApJ*, 660, 1186

Lora, V., Magaña, J., Bernal, A., Sánchez-Salcedo, F. J., & Grebel, E. K. 2012a, *JCAP*, 2, 011

Lora, V., Just, A., Sánchez-Salcedo, F. J., & Grebel, E. K. 2012, *ApJ*, 757, 87

Lora, V., Grebel, E. K., Sánchez-Salcedo, F. J., & Just, A. 2013, *ApJ*, 777, 65

Ma, J. 2012, *AJ*, 144, 41

Ma, J. 2013, *AJ*, 145, 88

Mackey, A. D. & Gilmore, G. F. 2004a, *MNRAS*, 355, 504

Mackey, A. D. & Gilmore, G. F. 2004b, *MNRAS*, 352, 153

Mackey, A. D., Wilkinson, M. I., Davies, M. B., & Gilmore, G. F. 2008, *MNRAS*, 386, 65

Mackey, A. D., *et al.* 2010, *ApJL*, 717, L11

Mackey, A. D., *et al.* 2013, *MNRAS*, 429, 281

Marino, A. F., *et al.* 2009, *A&A*, 505, 1099

Martell, S. L. & Grebel, E. K. 2010, *A&A*, 519, A14

Martell, S. L., Smolinski, J. P., Beers, T. C., & Grebel, E. K. 2011, *A&A*, 534, A136

Martin, N. F., *et al.* 2006, *MNRAS*, 371, 1983

Mashchenko, S., Couchman, H. M. P., & Wadsley, J. 2006, *Nature*, 442, 539

Massari, D., *et al.* 2014, *ApJ*, 795, 22

Mateluna, R., *et al.* 2012, *A&A*, 548, A82

McConnachie, A. W. 2012, *AJ*, 144, 4

McLaughlin, D. E. & van der Marel, R. P. 2005, *ApJS*, 161, 304

McLean, B. T., De Silva, G. M., & Lattanzio, J. 2015, *MNRAS*, 446, 3556

Mieske, S., *et al.* 2008, *A&A*, 487, 921

Milone, A. P., *et al.* 2015, *MNRAS*, 447, 931

Misgeld, I. & Hilker, M. 2011, *MNRAS*, 414, 3699

Monaco, L., Bellazzini, M., Ferraro, F. R., & Pancino, E. 2005, *MNRAS*, 356, 1396

Mucciarelli, A., Origlia, L., Ferraro, F. R., & Pancino, E. 2009, *ApJL*, 695, L134

Mucciarelli, A., Origlia, L., & Ferraro, F. R. 2010, *ApJ*, 717, 277

Niederste-Ostholt, M., *et al.* 2010, *MNRAS*, 408, L66

Norris, J. E., *et al.* 2008, *ApJL*, 689, L113

Odenkirchen, M., *et al.* 2001, *ApJL*, 548, L165

Odenkirchen, M., *et al.* 2002, *AJ*, 124, 1497

Odenkirchen, M., & Grebel, E. K. 2004, in Satellites and Tidal Streams, *ASP Conf. Ser.* 327, eds. F. Prada & D. Martínez-Delgado (San Francisco: ASP), 284

Oh, K. S., Lin, D. N. C., & Richer, H. B. 2000, *ApJ*, 531, 727

Olsen, K. A. G., Hodge, P. W., & Mateo, M., *et al.* 1998, *MNRAS*, 300, 665

Olszewski, E. W., Schommer, R. A., Suntzeff, N. B., & Harris, H. C. 1991, *AJ*, 101, 515

Origlia, L., *et al.* 2013, *ApJL*, 779, L5

Rey, S.-C., Yoon, S.-J., Lee, Y.-W., Chaboyer, B., & Sarajedini, A. 2001, *AJ*, 122, 3219

Rudenko, P., Worthey, G., & Mateo, M. 2009, *AJ*, 138, 1985

Park, W.-K. & Lee, M. G. 2007, *AJ*, 134, 2168

Park, W.-K., Park, H. S., & Lee, M. G. 2009, *ApJ*, 700, 103

Parmentier, G., & Grebel, E. K. 2005, *MNRAS*, 359, 615

Pasetto, S., Grebel, E. K., Berczik, P., Spurzem, R., & Dehnen, W. 2010, *A&A*, 514, A47

Perina, S., *et al.* 2012, *A&A*, 546, A31

Rockosi, C. M., *et al.* 2002, *AJ*, 124, 349

Salucci, P., *et al.* 2012, *MNRAS*, 420, 2034

San Roman, I., Sarajedini, A., Garnett, D. R., & Holtzman, J. A. 2009, *ApJ*, 699, 839

Sarajedini, A., & Layden, A. C. 1995, *AJ*, 109, 1086

Sarajedini, A. & Mancone, C. L. 2007, *AJ*, 134, 447

Schommer, R. A., Suntzeff, N. B., Olszewski, E. W., & Harris, H. C. 1992, *AJ*, 103, 447

Sharina, M. E., Afanasiev, V. L., & Puzia, T. H. 2006, *MNRAS*, 372, 1259

Sharina, M. & Davoust, E. 2009, *A&A*, 497, 65

Shetrone, M. D., Côté, P., & Sargent, W. L. W. 2001, *ApJ*, 548, 592

Simon, J. D., *et al.* 2011, *ApJ*, 733, 46

Smith, E. O., Neill, J. D., Mighell, K. J., & Rich, R. M. 1996, *AJ*, 111, 1596

Sollima, A., *et al.* 2012, *ApJ*, 744, 196

Stetson, P. B., *et al.* 1999, *AJ*, 117, 247

Stephens, A. W., Catelan, M., & Contreras, R. P. 2006, *AJ*, 131, 1426

Stonkutė, *et al.* 2008, *AJ*, 135, 1482

Strigari, L. E., *et al.* 2006, *ApJ*, 652, 306

Tollerud, E. J., Bullock, J. S., Graves, G. J., & Wolf, J. 2011, *ApJ*, 726, 108

Ural, U., *et al.* 2010, *MNRAS*, 402, 1357

van den Bergh, S. 1996, *AJ*, 112, 2634

van den Bergh, S. 1998, *ApJL*, 505, L127

van den Bergh, S. 2000, The Galaxies of the Local Group, *Cambridge Astrophysics Series* 35 (Cambridge: Cambridge Univ. Press)

VandenBerg, D. A., Brogaard, K., Leaman, R., & Casagrande, L. 2013, *ApJ*, 775, 134

Veljanoski, J., *et al.* 2013, *MNRAS*, 435, 3654

Walker, M. G., Mateo, M., Olszewski, E. W., Pal, J. K., Sen, B., & Woodroofe, M. 2006, *ApJL*, 642, L41

Walker, M. G. & Peñarrubia, J. 2011, *ApJ*, 742, 20

Walsh, S. M., Jerjen, H., & Willman, B. 2007, *ApJL*, 662, L83

Wang, S. & Ma, J. 2013, *AJ*, 146, 20

Xin, Y., Deng, L., de Grijs, R., Mackey, A. D., & Han, Z. 2008, *MNRAS*, 384, 410

Yong, D., *et al.* 2013, *MNRAS*, 434, 3542

Yong, D., *et al.* 2014, *MNRAS*, 441, 3396

Zaritsky, D., Gonzalez, A. H., & Zabludoff, A. I. 2006, *ApJ*, 638, 725

Zinn, R. 1993, in The Globular Cluster-Galaxy Connection, *ASP Conf. Ser.* 48, eds. G. H. Smith & J. P. Brodie (ASP: San Francisco), p. 38

Zloczewski, K. & Kaluzny, J. 2009, *Acta Astronomica*, 59, 47

Zucker, D. B., *et al.* 2004, *ApJL*, 612, L121

Zucker, D. B., *et al.* 2006a, *ApJL*, 643, L103

Zucker, D. B., *et al.* 2006b, *ApJL*, 650, L41

Zucker, D. B., *et al.* 2007, *ApJL*, 659, L21

Star clusters and black holes in galaxies across cosmic time
Proceedings IAU Symposium No. 312, 2014
Y. Meiron, S. Li, F.-K. Liu & R. Spurzem, eds.

© International Astronomical Union 2016
doi:10.1017/S1743921315007772

Exotic populations in globular clusters: blue stragglers as tracers of the internal dynamical evolution of stellar systems

Francesco R. Ferraro

Department of Physics and Astronomy, University of Bologna, Viale Berti Pichat 6/2, 40127-Bologna, Italy
email: `francesco.ferraro3@unibo.it`

Abstract. In this paper I present an overview of the main observational properties of a special class of exotic objects (the so-called Blue Straggler Stars, BSSs) in Galactic Globular Clusters (GCs). The BSS specific frequency and their radial distribution are discussed in the framework of using this stellar population as probe of GC internal dynamics. In particular, the shape of the BSS radial distribution has been found to be a powerful tracer of the dynamical evolution of stellar systems, thus allowing the definition of an empirical "clock" able to measure the dynamical age of stellar aggregates from pure observational properties.

Keywords. globular clusters, blue stragglers, dynamics

1. Introduction

Globular clusters (GCs) are the only cosmic structures that within a time-scale shorter than the age of the Universe undergo nearly all the physical processes known in stellar dynamics (Meylan & Heggie 1997). Gravitational interactions and collisions among single stars and/or binaries are expected to be quite frequent, especially in high density environments (e.g. Hut *et al.* 1992) characterizing the inner regions of GCs. Such a pronounced dynamical activity can generate populations of exotic objects, like X-ray binaries, millisecond pulsars and blue straggler stars (BSSs; see, e.g., Paresce *et al.* 1992, Bailyn 1995, Bellazzini *et al.* 1995, Ferraro *et al.* 2001, Ransom *et al.* 2005, Pooley & Hut 2006, Ferraro *et al.* 2009a, Lanzoni *et al.* 2010).

Blue straggler stars (BSSs) are commonly defined as stars brighter and bluer than the main-sequence (MS) turnoff in the host stellar cluster. They are thought to be central H-burning stars, more massive than the MS turnoff stars (Shara *et al.* 1997, Gilliland *et al.* 1998, De Marco *et al.* 2004, Fiorentino *et al.* 2014). In stellar systems with no evidence of recent star formation, their origin cannot be explained in the framework of normal single-star evolution. Two main formation channels are currently favored: (1) mass transfer (MT) in binary systems (McCrea 1964) possibly up to the complete coalescence of the two stars, and (2) stellar collisions (Hills & Day 1976). Both these processes can potentially bring new hydrogen into the core and therefore "rejuvenate" a star to its MS stage (e.g., Lombardi *et al.* 1995, Lombardi *et al.* 2002; Chen & Han 2009).

MT in binaries might be the dominant formation channels in all environments (e.g., Knigge *et al.* 2009; Leigh *et al.* 2013), and most likely it is so in low-density GCs, open clusters and the Galactic field (Ferraro *et al.* 2006a; Sollima *et al.* 2008; Mathieu *et al.* 2009; Gosnell *et al.* 2014; Preston & Sneden 2000). Collisions are believed to be important especially in dense environments, such as the cores of globular clusters (GCs; Bailyn 1992; Ferraro *et al.* 1993, Ferraro *et al.* 1997, Ferraro *et al.* 2003a) and even the centre of some open clusters (Leonard & Linnell 1992; Glebbeek *et al.* 2008).

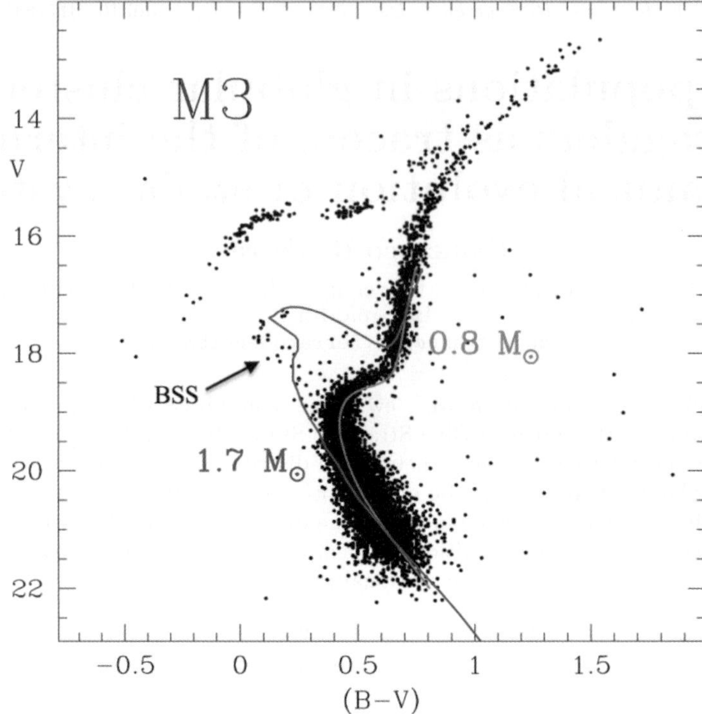

Figure 1. Optical CMD of the globular cluster M3, with the location of BSSs indicated by the arrow. The theoretical track corresponding to $0.8M_\odot$ well reproduces the main evolutionary sequences of the cluster, while BSSs populate a region of the CMD where core hydrogen-burning stars of $\sim 1.7M_\odot$ are expected. From Buonanno *et al.* (1994)

Because of their large number of member stars, GCs are the ideal environment for BSS studies. Nominally all the GCs observed so far have been found to harbor a significant number of BSSs (Piotto *et al.* 2004; Leigh *et al.* 2007). Moreover, the stellar density in GCs varies dramatically from the central regions to the outskirts, and since BSSs in different environments (low- versus high-density) could have different origins (e.g., Fusi Pecci *et al.* 1992; Ferraro *et al.* 1995, Davies *et al.* 2004), these stellar systems allow to investigate both formation channels simultaneously. However a clear distinction is hampered by the internal dynamical evolution of the parent cluster (Ferraro *et al.* 2012). In fact, having masses larger than normal cluster stars, BSSs are affected by dynamical friction, a process that drives the objects more massive than the average toward the cluster centre, over a timescale which primarily depends on the local mass density (e.g., Alessandrini *et al.* 2014). Hence, as the time goes on, heavy objects (like BSSs) orbiting at larger and larger distances from the cluster centre are expected to drift toward the core: as a consequence, the radial distribution of BSSs develops a central peak and a dip, and the region devoid of these stars progressively propagates outward.

Ferraro *et al.* (2012) used this argument to define the so-called "dynamical clock", an empirical tool able to measure the dynamical age of a stellar system from the shape of its BSS radial distribution. This appears indeed to provide a coherent interpretation of the variety of BSS radial distributions observed so far: GCs with a flat BSS radial distribution (Ferraro *et al.* 2006b; Dalessandro *et al.* 2008a; Beccari *et al.* 2011) are dynamically young systems, GCs with bimodal distributions (e.g., Ferraro *et al.* 1993, Ferraro *et al.* 2004; Lanzoni *et al.* 2007a; Dalessandro *et al.* 2008b; Beccari *et al.* 2013, and

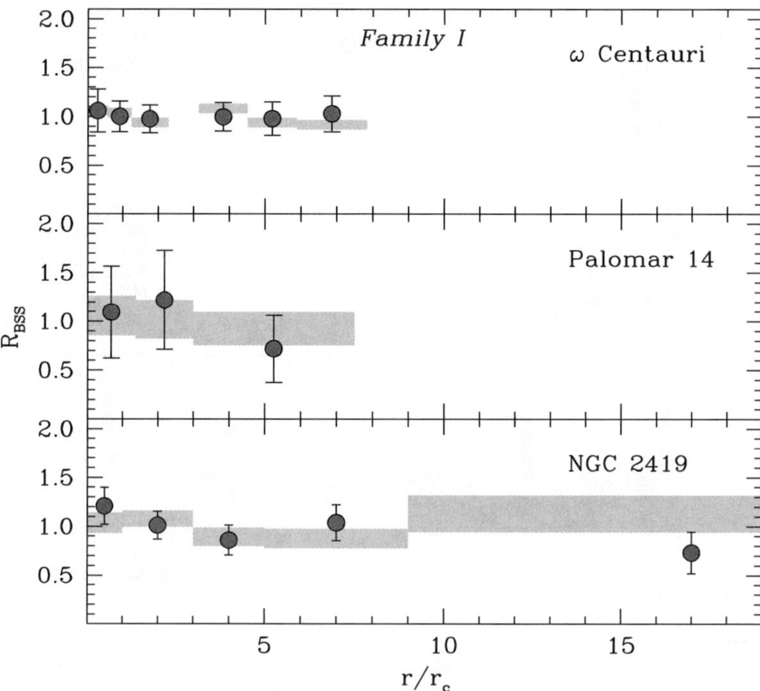

Figure 2. BSS radial distribution observed in ω Centauri, Palomar 14 and NGC 2419, with the large filled circles marking the values of the double normalized specific frequency R_{BSS}. The distribution of the double normalized ratio measured for RGB or HB stars is also shown for comparison (grey strips). The BSS radial distribution is flat and totally consistent with that of the reference population, thus indicating a low degree of dynamical evolution for these three GCs (*Family I*). From Ferraro *et al.* (2012).

references therein) are dynamically intermediate-age systems (their actual dynamical age being determined by the distance of the dip of the distribution from the cluster centre), and GCs with a single-peaked BSS distribution (Ferraro *et al.* 1999a; Lanzoni *et al.* 2007b; Contreras Ramos *et al.* 2012; Dalessandro *et al.* 2013) are dynamically old systems.

2. Setting the dynamical clock for stellar systems

In order to perform meaningful comparisons among different clusters the first step is the definition of an appropriate BSS specific frequency. Ferraro *et al.* (1993) introduced the "double normalized ratio", defined as:

$$R_{\mathrm{BSS}} = \frac{(N_{\mathrm{BSS}}/N_{\mathrm{BSS}}^{\mathrm{tot}})}{(L^{\mathrm{sampled}}/L_{tot}^{\mathrm{sampled}})},\qquad(2.1)$$

where N_{BSS} is the number of BSSs counted in a given cluster region, $N_{\mathrm{BSS}}^{\mathrm{tot}}$ is the total number of BSSs observed, and $L^{\mathrm{sampled}}/L_{tot}^{\mathrm{sampled}}$ is the fraction of light sampled in the same region, with respect to the total measured luminosity. The same ratio can be defined for any post-MS population. Theoretical arguments (Renzini & Fusi Pecci 1988) demonstrate that the double normalized ratio is equal to unity for any population (such as red giant branch and horizontal branch stars, RGB and HB respectively) whose radial distribution follows that of the cluster luminosity. Ferraro *et al.* (2012) presented a comparison of the BSS radial distribution of 21 GCs with very different structural properties

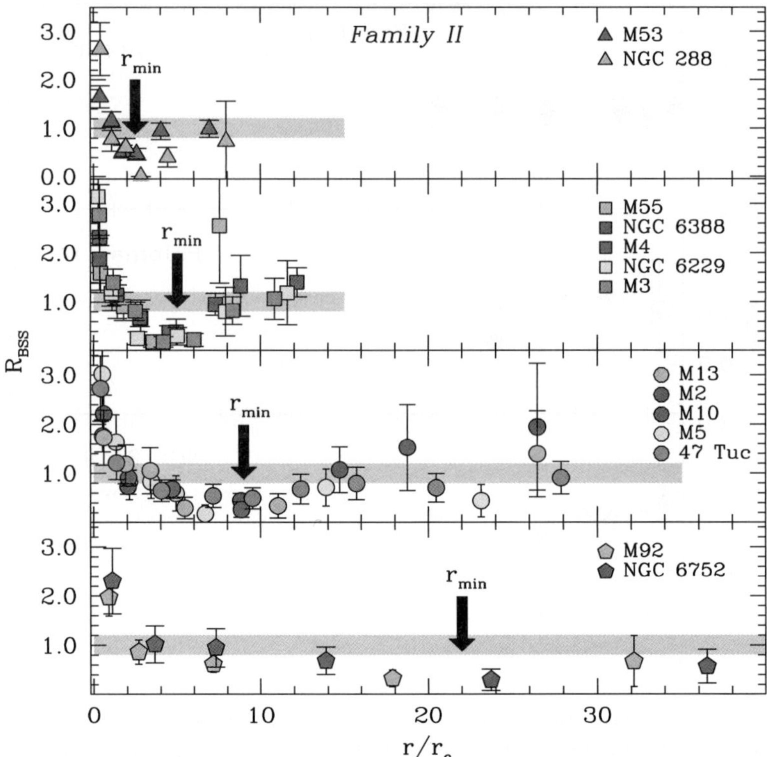

Figure 3. BSS radial distribution observed in clusters of intermediate dynamical age (*Family II*). The distribution is clearly bimodal and the radial position of the minimum (marked with the arrow and labelled as r_{min}) clearly moves outward from top to bottom, suggesting that the bottom clusters are more dynamically evolved than the upper ones. For the sake of clarity, the grey bands schematically mark the distribution of the reference populations. From Ferraro *et al.* (2012).

(hence possibly at different stages of their dynamical evolution), but with nearly the same chronological age (12-13 Gyr; Marín-Franch *et al.* 2009), with the only exception of Palomar 14 which formed $\sim$ 10.5 Gyr ago Dotter *et al.* (2008). Ferraro et al. (2012) showed that once the radial distance from the centre is expressed in units of the core radius (thus to allow a meaningful comparison among the clusters), GCs can be efficiently grouped on the basis of the shape of their BSS radial distribution, and at least three families can be defined:

• *Family I* – the radial distribution of the BSS double normalized ratio (R_{BSS}) is fully consistent with that of the reference population (R_{pop}) over the entire cluster extension (see Figure 2);

• *Family II* – the distribution of R_{BSS} is incompatible with that of R_{pop}, showing a significant bimodality, with a central peak and an external upturn. At intermediate radii a minimum is evident and its position (r_{min}) can be clearly defined for each sub-group (see Figure 3);

• *Family III* – the radial distribution of R_{BSS} is still incompatible with that of the reference population, showing a well defined central peak with no external upturn (see Figure 4).

Previous preliminary analysis (Mapelli *et al.* 2004, Mapelli *et al.* 2006), Lanzoni *et al.* 2007a) of a few clusters indicated that BSSs generated by stellar collisions mainly/only

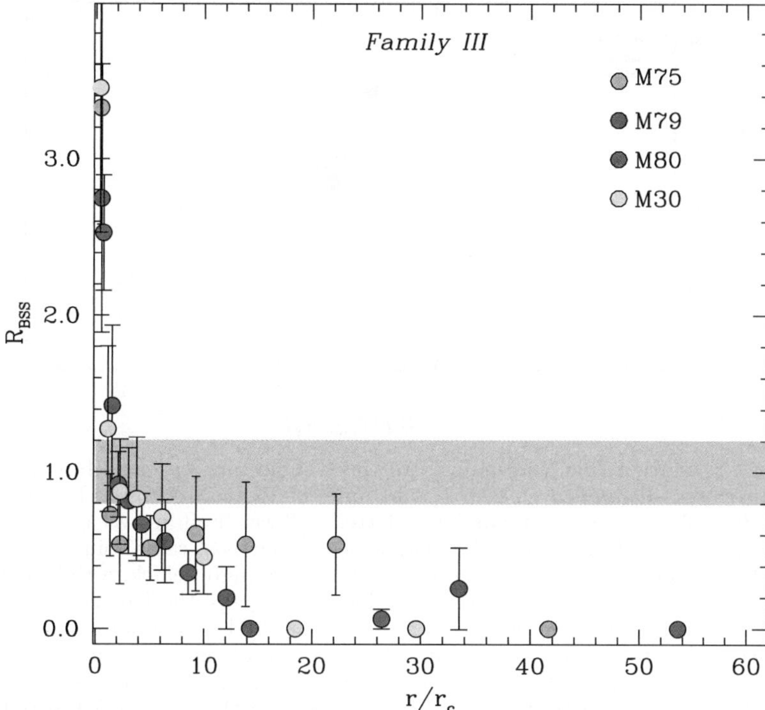

Figure 4. BSS radial distribution for dynamically old clusters (*Family III*): only a central peak is visible, while the external upturn is no more present because of the dynamical friction action out to the cluster outskirts. From Ferraro *et al.* (2012)

contribute to the central peak of the distribution, while the portion beyond the observed minimum is populated by MT-BSSs which are evolving in isolation in the cluster outskirts and have not yet suffered the effects of dynamical friction. Overall, the BSS radial distribution is primarily modelled by the long-term effect of dynamical friction acting on the cluster binary population (and its progeny) since the early stages of cluster evolution. In fact, what we call MT-BSS today is the by-product of the evolution of a $\sim 1.2M_\odot$ binary that has been orbiting the cluster and suffering the effects of dynamical friction for a significant fraction of the cluster lifetime. The efficiency of dynamical friction decreases for increasing radial distance from the centre, as a function of the local velocity dispersion and mass density. Hence, dynamical friction first segregates (heavy) objects orbiting close to the centre and produces a central peak in their radial distribution. As the time goes on, the effect extends to larger and larger distances, thus yielding to a region devoid of these stars (i.e., a dip in their radial distribution) that progressively propagates outward. Simple analytical estimate of the radial position of this dip turned out to be in excellent agreement with the position of the minimum in the *observed* BSS radial distributions ($r_{\min}$), despite a number of crude approximation (see, e.g. Mapelli *et al.* 2006). Moreover, a progressive outward drift of $r_{\min}$ as a function of time is confirmed by the results of direct N-body simulations that follow the evolution of $\sim 1.2M_\odot$ objects within a reference cluster over a significant fraction of its lifetime (Miocchi *et al.* 2015).

In light of these considerations, the three families defined in Figs. 2–4 correspond to GCs of increasing dynamical ages. Hence, the shape of the BSS radial distribution turns out to be a powerful dynamical-age indicator. A flat BSS radial distribution (consistent

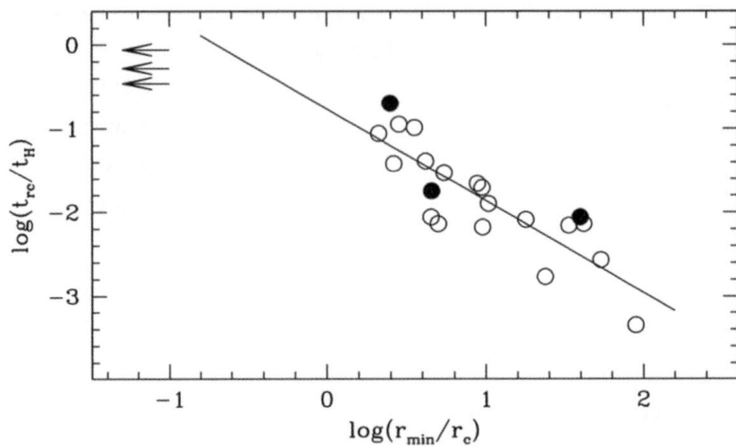

Figure 5. Core relaxation time (normalized to the Hubble time t_H) as a function of the time hand of the proposed *dynamical clock* (r_{min}, in units of the core radius). Dynamically young systems (*Family I*) show no minimum and are plotted as lower-limit arrows at $r_{\mathrm{min}}/r_c = 0.1$. For dynamically old clusters (*Family III*), the distance of the farthest radial bin where no BSSs are observed has been adopted as r_{min}. As expected for a meaningful clock, a tight anticorrelation is found: clusters with relaxation times of the order of the age of the Universe show no signs of BSS segregation (hence their BSS radial distribution is flat and r_{min} is not definable; see Fig. 2), whereas for decreasing relaxation times the radial position of the minimum increases progressively. The solid line correspond to the best-fit relations. Empty circles are data from Ferraro et al. (2012), filled circles are three clusters (NGC362, NGC5824 and NGC5466) published after Ferraro et al. (2012).

with that of the reference population; see *Family I* in Fig. 2) indicates that dynamical friction has not played a major role yet even in the innermost regions, and the cluster is still dynamically young. This interpretation is confirmed by the absence of statistically significant dips in the BSS distributions observed in dwarf spheroidal galaxies (Mapelli *et al.* 2009; Monelli *et al.* 2012): these are, in fact, collisionless systems where dynamical friction is expected to be highly inefficient. In more evolved clusters (*Family II*), dynamical friction starts to be effective and to segregate BSSs that are orbiting at distances still relatively close to the centre: as a consequence, a peak in the centre and a minimum at small radii appear in the distribution, while the most remote BSSs are not yet affected by the action of dynamical friction (this generates the rising branch of the observed bimodal BSS distributions; see upper panel in Fig. 3). Since the action of dynamical friction progressively extends to larger and larger distances from the centre, the dip of the distribution progressively moves outward (as seen in the different groups of *Family II* clusters; Fig. 3, panels from top to bottom). In highly evolved systems dynamical friction already affected even the most remote BSSs, which started to gradually drift toward the centre: as a consequence, the external rising branch of the radial distribution disappears (as observed for *Family III* clusters in Fig. 4). All GCs with a single-peak BSS distribution can therefore be classified as "dynamically old".

Interestingly, this latter class includes M30, a system that already suffered core collapse which is considered as a typical symptom of extreme dynamical evolution (Meylan & Heggie 1997). The proposed classification is also able to shed light on a number of controversial cases debated into the literature. In fact, M4 turns out to have an intermediate dynamical age, at odds with previous studies suggesting that it might be in a PCC state (Heggie & Giersz 2008). On the other hand, NGC 6752 turns out to be in a quite advanced state of dynamical evolution, in agreement with its observed double

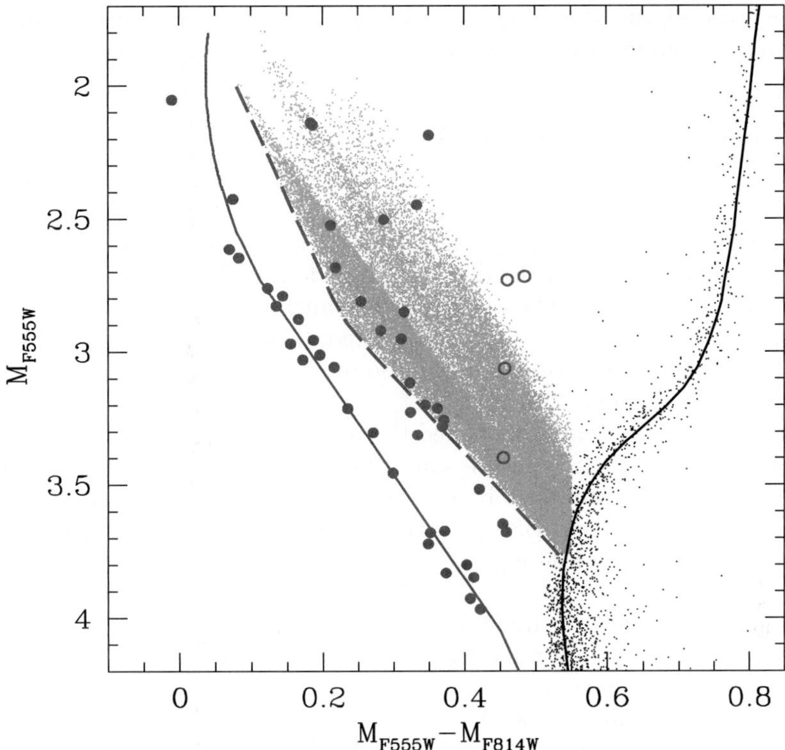

Figure 6. Distribution of the synthetic MT-binaries (grey dots) bluer than the cluster MS turnoff, in the absolute CMD of M30. The dashed line marks the "low-luminosity boundary" of the distribution. The BSSs observed by F09 along the red and the blue sequences are shown as solid circles. The positions of 4 additional red-BSSs, not considered in F09, are also shown as empty circles. Clearly, the distribution of synthetic BSSs well samples the location of the observed red-BSSs, and its low-luminosity boundary nicely follows the red-BSS sequence. The solid line is the collisional 2 Gyr-isochrone by Sills *et al.* (2009), which has been found to nicely fit the observed blue-BSS sequence.

King profile indicating that the cluster core is detaching from the rest of the cluster structure (Ferraro *et al.*2003b). Finally this approach might provide the key to discriminate between a central density cusp due to core collapse (as for M30) and that due to the presence of an exceptional concentration of dark massive objects (neutron stars and/or the still elusive intermediate-mass black holes; (see the case of NGC 6388, Lanzoni *et al.* 2007c).

The quantization in distinct age-families is of course an over-simplification, while the position of $r_{\min}$ is found to vary with continuity as a sort of clock time-hand. This allowed Ferraro et al. (2012) to define the first empirical clock able to measure the dynamical age of a stellar system from pure observational quantities (the *"dynamical clock"*): as the engine of a chronometer advances the clock hand to measure the time flow, in a similar way the progressive sedimentation of BSSs towards the cluster centre moves $r_{\min}$ outward, thus marking its dynamical age. This is indeed confirmed by the tight correlations found between the clock-hand ($r_{\min}$) and the central and half-mass relaxation times ($t_{\rm rc}$ and $t_{\rm rh}$, respectively), which are commonly used to measure the cluster dynamical evolution time-scales. The correlation found with $t_{\rm rc}$ is shown in Figure 5. Note that, while $t_{\rm rc}$ and $t_{\rm rh}$ provide an indication of the relaxation timescales at specific radial distances from the cluster centre ($r_{\rm c}$ and $r_{\rm h}$, respectively), the dynamical clock here defined provides instead a measure of the global dynamical evolution of the systems, because the BSS radial distribution simultaneously probes all distances from the cluster centre.

3. The discovery of the double BSS sequence

While the proposed BSS formation mechanisms could be separately at work in clusters with different densities (Ferraro *et al.* 1995, Ferraro *et al.* 2006b, Lovisi *et al.* 2013) a few pieces of evidence are now emerging suggesting that they could also act simultaneously within the same cluster. Indeed the discovery of a double BSS sequence in M30 (Ferraro *et al.* 2009b, hereafter F09) indicates that this can be the case: two well distinct BSS sequences, almost parallel and similarly populated, have been found in the CMD of M30 (F09). Similar features have been detected also in NGC 362 by Dalessandro *et al.* (2013) and NGC 1261 by Simunovic *et al.* (2014).

Such a discovery has potentially opened the possibility to photometrically distinguish collisional BSSs from MT BSSs in the same cluster. In fact, the blue-BSS sequence observed in M30 is nicely reproduced by collisional isochrones (Sills *et al.* 2009) with ages of 1-2 Gyr. In contrast, the red-BSS population is far too red to be consistent with collisional isochrones of any age. F09 suggested that the location of the red-BSS population in the CMD correspond to the extrapolation of the the the "low-luminosity boundary" outlined by the MT binary populations simulated by Tian *et al.* (2006) for the open cluster M67. Indeed, by using binary evolution models specifically computed for M30, Xin *et al.* (2015) demonstrated that the distribution of synthetic MT-BSSs in the CMD nicely matches the observed red BSS sequence (see Figure 6), thus providing strong support to the MT origin for these stars. In addition the CMD distribution of synthetic MT-BSSs never attains the observed location of the blue BSS sequence, thus reinforcing the hypothesis that the latter formed through a different channel (likely collisions).

Acknowledgements

Most of the results discussed in this talk have been obtained within the project Cosmic-Lab (PI: Ferraro, see http:://www.cosmic-lab.eu), a 5-year project funded by the European European Research Council under the 2010 Advanced Grant call (contract ERC-2010-AdG-267675). I warmly thank the other team members involved in this research: Barbara Lanzoni, Emanuele Dalessandro, Alessio Mucciarelli, Yu Xin, Licai Deng, Giacomo Beccari, Paolo Miocchi, Mario Pasquato.

References

Alessandrini, E., Lanzoni, B., Miocchi, P., Ciotti, L., & Ferraro, F. R,. 2014, *ApJ*, 795, 169
Bailyn, C. 1992, *ApJ*, 392, 519
Bailyn, C. D., 1995, *ARAA*, 33, 133
Beccari, G., Sollima, A., Ferraro, F. R., *et al.*, 2011, *ApJ*, 737, L3
Beccari, G., Dalessandro, E., Lanzoni, B., *et al.*, 2013, *ApJ*, 776, 60
Bellazzini, M., Pasquali, A., Federici, L., Ferraro, F. R., & Pecci, F. F. 1995, *ApJ*, 439, 687
Benz, W. & Hills, J. G., 1987, *ApJ*, 323, 614
Buonanno, R., Corsi, C. E., Buzzoni, A., *et al.*, 1994, *A&A*, 290, 69
Chen, X. F., & Han, Z. W., 2009, *MNRAS*, 395, 1822
Contreras Ramos, R., Ferraro, F. R., Dalessandro, E., Lanzoni, B., & Rood, R. T., 2012, *ApJ*, 748, 91
Dalessandro, E., Lanzoni, B., Ferraro, F. R., Vespe, F., Bellazzini, M., & Rood, R. T., 2008a, *ApJ*, 681, 311

Dalessandro, E., Lanzoni, B., Ferraro, F. R., Rood, R. T., Milone, A., Piotto, G., & Valenti, E., 2008b, *ApJ*, 677, 1069

Dalessandro, E., Ferraro, F. R., Massari, D., Lanzoni, B., Miocchi, P., Beccari, G., Bellini, A., Sills, A., Sigurdsson, S., Mucciarelli, A., & Lovisi, L., 2013, *ApJ*, 778, 135

De Marco, O., Lanz, T., Ouellette, J. A., Zurek, D., & Shara, M. M., 2004, *ApJ*, 606, L151

Davies, M. B., Piotto, G., & de Angeli, F., 2004, *MNRAS*, 349, 129

Dotter, A., Sarajedini, A., & Yang, S.-C., 2008, *AJ*, 136, 1407

Ferraro F. R., Pecci F. F.,Cacciari C., Corsi C., Buonanno R., Fahlman G. G., & Richer H. B., 1993, *AJ*, 106, 2324

Ferraro, F. R., Fusi Pecci, F., & Bellazzini, M., 1995, *A&A*, 294, 80

Ferraro, F. R., *et al.*,Paltrinieri, B., Fusi Pecci, F., Cacciari, C., Dorman, B., Rood, R. T., Buonanno, R., Corsi, C. E., Burgarella, D., & Laget, M., 1997, *A&A*, 324, 915

Ferraro, F. R., Paltrinieri, B., Rood, R. T., & Dorman, B., 1999a, *ApJ*, 522, 983

Ferraro, F. R., Messineo, M., Fusi Pecci, F., *et al.*, 1999b, *AJ*, 118, 1738

Ferraro, F. R., D'Amico, N., Possenti, A., Mignani, R. P., & Paltrinieri, B., 2001, *ApJ*, 561,337

Ferraro, F. R., Sills, A., Rood, R. T., Paltrinieri, B., & Buonanno, R., 2003a, *ApJ*, 588, 464

Ferraro, F. R., Possenti, A., Sabbi, E., *et al.*, 2003b, *ApJ*, 595, 179

Ferraro, F. R., Beccari, G., Rood, R. T., Bellazzini, M., Sills, A., & Sabbi, E. , 2004, *ApJ*, 603, 127

Ferraro, F. R., Sollima, A., Rood, R. T., Origlia, L., Pancino, E., & Bellazzini, M., 2006a, *ApJ*, 638, 433

Ferraro, F. R., Sabbi, E., Gratton, R., *et al.*, 2006b, *ApJ*, 647, L53

Ferraro, F. R., Dalessandro, E., Mucciarelli, A., *et al.*, 2009a, *Nature*, 462, 483

Ferraro, F. R., Beccari, G., Dalessandro, E., *et al.*, 2009b, *Nature*, 462, 1028 (F09)

Ferraro, F. R., Lanzoni, B., Dalessandro, E., *et al.*, 2012, *Nature*, 492, 393, (Ferraro et al. (2012))

Fiorentino, G., Lanzoni, B., Dalessandro, E., *et al.*, 2014, *ApJ*, 783, 34

Fusi Pecci, F., Ferraro, F. R., Corsi, C. E., Cacciari, C., & Buonanno, R., 1992, *ApJ*, 104, 1831

Gilliland, R. L., Bono,G., Edmonds, P. D., *et al.*, 1998, *ApJ*, 507, 818

Glebbeek, E., Pols, O. R., & Hurley, J. R., 2008, *A&A*, 488, 1007

Gosnell, N. M., Mathieu, R. D., Geller, A. M., Sills, A., Leigh, N., & Knigge, C., 2014, *ApJ*, 783, L8

Heggie, D. C., & Giersz, M., 2008, *MNRAS*, 389, 1858

Hut, P., McMillan, S., & Romani, R. W., 1992, *ApJ*, 389, 527

Hills, J., & Day, C., 1976, Astron. Lett., 17, 87

Hurley, J. R., Tout, C. A., Aarseth, S. J., & Pols, O. R., 2001, *MNRAS*, 323, 630

Knigge, C., Leigh, N., & Sills, A., 2009, *Nature*, 457, 288

Lanzoni, B., Dalessandro, E., Perina, S., Ferraro, F. R., Rood, R. T., & Sollima, A., 2007a, *ApJ*, 670, 1065

Lanzoni, B., Sanna, N., Ferraro, F. R., *et al.*, 2007b, *ApJ*, 663, 1040

Lanzoni, B., Dalessandro, E., Ferraro, F. R., *et al.*, 2007c, *ApJ*, 668, L139

Lanzoni, B., Ferraro, F. R., Dalessandro, E., *et al.*, 2010, *ApJ*, 717, 653

Leigh, N., Sills, A., & Knigge, C., 2007, *ApJ*, 661, 210

Leigh, N., Knigge, C., Sills, A., Perets, H. B., Sarajedini, A., & Glebbeek, E., 2013, *MNRAS*, 428, 897

Lenoard, P. J. T., 1989, *AJ*, 98, 217

Leonard, P. J. T., & Linnell, A. P., 1992, *AJ*, 103, 1928

Lombardi, Jr., J. C., Rasio, F. A., & Shapiro, S. L., 1995, *ApJ*, 445, L117

Lombardi, Jr., J. C., Warren, J. S., Rasio, F. A., Sills, A., & Warren, A. R., 2002, *ApJ*, 568, 939

Lovisi, L., Mucciarelli, A., Lanzoni, B., Ferraro, F. R., Dalessandro, E., & Monaco, L., 2013, *ApJ*, 772, 148

Mapelli, M., Sigurdsson, S., Colpi, M., Ferraro, F. R., Possenti, A., Rood, R. T., Sills,A., & Beccari, G., 2004, *ApJ*, 605, L29

Mapelli, M., Sigurdsson,S., Ferraro, F. R., Colpi, M., Possenti, A., & Lanzoni, B., 2006, *MNRAS*, 373, 361

Mapelli, M., Ripamonti, E., Battaglia, G., *et al.*, 2009, *MNRAS*, 396, 1771

Miocchi, P., Pasquato, M., Lanzoni, B., *et al.* 2015, *ApJ*, 799, 44
Marín-Franch, A., Aparicio, A., Piotto, G., *et al.*, 2009, *ApJ*, 694, 1498
Mathieu, R. D., & Geller, A. M., 2009, *Nature*, 462, 1032
McCrea, W. H., 1964, *MNRAS*, 128, 147
Meylan, G. & Heggie, D. C. 1997, *ARAA*, 8, 1
Monelli, M., Cassisi, S., Mapelli, M., *et al.*, 2012, *ApJ*, 744, 157
Ouellette, J. A. & Pritchet, C. J., 1998, *AJ*, 115, 2539
Paresce, F., de Marchi, G.,& Ferraro, F. R., 1992, *Nature*, 360, 46
Pooley, D., & Hut, P., 2006, *ApJ*, 646, L143
Piotto, G., *et al.*, 2004, *ApJ*, 604, L109
Preston, G. W., & Sneden, C., 2000, *AJ*, 120, 1014
Ransom, S. M., Hessels, J. W. T., Stairs, I. H., *et al.*, 2005, Science, 307, 892
Renzini, A., & Fusi Pecci, F., 1988, *ARAA*, 26, 199
Shara, M. M., Saffer, R. A., & Livio, M., 1997, *ApJ*, 489, L59
Sills, A., Karakas, A., & Lattanzio, J., 2009, *ApJ*, 692, 1411
Simunovic, M., Puzia, T. H., & Sills, A., 2014, *ApJ*, 795, L10
Sollima, A., Lanzoni, B., Beccari, G., Ferraro, F. R., & Fusi Pecci, F., 2008, *A&A*, 481, 701
Tian, B., Deng, L.-C., Han, Z.-W., & Zhang, X.-B., 2006, *A&A*, 455, 247
Xin, Y., Ferraro, F. R., Lu, P., *et al.* 2015, *ApJ*, 801, 67

Star clusters and black holes in galaxies across cosmic time
Proceedings IAU Symposium No. 312, 2014
Y. Meiron, S. Li, F.-K. Liu & R. Spurzem, eds.
© International Astronomical Union 2016
doi:10.1017/S1743921315007784

Intermediate-mass black holes in globular clusters: observations and simulations

Nora Lützgendorf[1], Markus Kissler-Patig[2], Karl Gebhardt[3], Holger Baumgardt[4], Diederik Kruijssen[5], Eva Noyola[3], Nadine Neumayer[6], Tim de Zeeuw[7,8], Anja Feldmeier[7], Edwin van der Helm[8], Inti Pelupessy[8] and Simon Portegies Zwart[8]

[1]ESA, Space Science Department, Keplerlaan 1, NL-2200 AG Noordwijk, The Netherlands
email: `nluetzge@cosmos.esa.int`

[2]Gemini Observatory, Northern Operations Center, 670 N. A'ohoku Place, Hilo, Hawaii, 96720, USA

[3]Department of Astronomy, University of Texas at Austin, Austin, TX 78712, USA

[4]School of Mathematics and Physics, University of Queensland, Brisbane, QLD 4072, Australia

[5]Max-Planck Institut für Astrophysik, Karl-Schwarzschild-Straße 1, D-85748, Garching, Germany

[6]Max-Planck-Institute for Astronomy, Königstuhl 17, 69117, Heidelberg, Germany

[7]European Southern Observatory, Karl-Schwarzschild-Straße 2, D-85748 Garching, Germany

[8]Leiden Observatory, Leiden University, PO Box 9513, NL-2300 RA, Leiden, The Netherlands

Abstract. The study of intermediate-mass black holes (IMBHs) is a young and promising field of research. If IMBHs exist, they could explain the rapid growth of supermassive black holes by acting as seeds in the early stage of galaxy formation. Formed by runaway collisions of massive stars in young and dense stellar clusters, intermediate-mass black holes could still be present in the centers of globular clusters, today. Our group investigated the presence of intermediate-mass black holes for a sample of 10 Galactic globular clusters. We measured the inner kinematic profiles with integral-field spectroscopy and determined masses or upper limits of central black holes in each cluster. In combination with literature data we further studied the positions of our results on known black-hole scaling relations (such as $M_\bullet - \sigma$) and found a similar but flatter correlation for IMBHs. Applying cluster evolution codes, the change in the slope could be explained with the stellar mass loss occurring in clusters in a tidal field over its life time. Furthermore, we present results from several numerical simulations on the topic of IMBHs and integral field units (IFUs). We ran N-body simulations of globular clusters containing IMBHs in a tidal field and studied their effects on mass-loss rates and remnant fractions and showed that an IMBH in the center prevents core collapse and ejects massive objects more rapidly. These simulations were further used to simulate IFU data cubes. For the specific case of NGC 6388 we simulated two different IFU techniques and found that velocity dispersion measurements from individual velocities are strongly biased towards lower values due to blends of neighboring stars and background light. In addition, we use the Astrophysical Multipurpose Software Environment (AMUSE) to combine gravitational physics, stellar evolution and hydrodynamics to simulate the accretion of stellar winds onto a black hole.

Keywords. stars: kinematics and dynamics, methods: numerical, black hole physics

---

1. Introduction

Intermediate-mass black holes (IMBHs) provide the missing link between supermassive black holes and stellar-mass black holes. Their masses range from a few hundred solar masses up to $10^5 M_\odot$. Possible formation scenarios are remnants of population III stars

(Madau & Rees 2001) or runaway merging in young dense star clusters (e.g. Portegies Zwart *et al.* 2004). For supermassive black holes, correlations between the black-hole mass ($M_\bullet$) and several properties of their host system, such as velocity dispersion (σ, e.g. Ferrarese & Merritt 2000) or total mass (M, e.g. Häring & Rix 2004) are observed. Extrapolating these correlations to low mass systems ($\sigma \sim 10 - 30$ km s^{-1}) suggests that IMBHs could reside in today's massive globular clusters. Detecting and measuring these black holes would provide important data points to the $M_\bullet - \sigma$ relation at the lower mass end. Furthermore, IMBHs could explain the rapid growth of supermassive black holes at high redshift by acting as seeds in the early universe.

Many attempts have been made to detect IMBHs in globular clusters. These range from spectroscopic and photometric velocity measurements (e.g. Noyola *et al.* 2008) to X-ray and radio observations in order to find signatures of accretion (e.g. Strader *et al.* 2012). The results are disputed and the question whether IMBHs exist in globular clusters is not yet resolved. All methods bring their caveats. The low gas content in globular clusters makes the detection of accretion signatures rather difficult. The X-ray and radio signals from low and irregular accretion rates are not yet well understood. On the other hand, kinematic signatures suffer from shot noise caused by a few bright stars or contamination from background light. It is therefore crucial to advance the field of IMBHs in all directions. Simulations on globular clusters and observing techniques provide a valuable tool to verify observing results and to understand the internal processes of globular clusters.

Previous numerical work on IMBHs in globular clusters has been performed by Baumgardt *et al.* (2005) and Noyola & Baumgardt (2011) who found that the surface brightness profiles of clusters hosting an IMBH exhibit weak central cusps, in contrast to core-collapsed clusters with very steep profiles and pre-core collapsed systems with no cusp at all. However, Trenti *et al.* (2010) and Vesperini & Trenti (2010) showed that the shallow cusp may also form as a transition state of globular clusters undergoing core collapse and therefore cannot be a sufficient criterion for a globular clusters hosting an IMBH. Furthermore, it has been shown that IMBHs prevent a cluster from undergoing core collapse and reduce the degree of mass segregation compared with non-IMBH clusters (Baumgardt *et al.* 2004; Gill *et al.* 2008). Trenti *et al.* (2007) predicted that clusters with a high ratio of core radius to half-mass radius are good candidates for hosting an IMBH. This was challenged by Hurley (2007), who showed that the ratios observed for Galactic globular clusters can be explained without the need for an IMBH when treating model data as if they were observational data.

2. Observations

We used integral field spectroscopy provided by the FLAMES (Fiber Large Array Multi Element Spectrograph, Pasquini *et al.* 2002) instrument mounted on UT2 at the Very Large Telescope (VLT) to obtain integrated light spectra from 10 Galactic globular clusters. These clusters have been selected by their mass and the shape of their density profiles to be good candidates for hosting IMBHs. *N*-body simulations have shown that an IMBH in the core of a globular cluster prevents the cluster to undergo core collapse and produces a large core with a shallow cusp in its density profile (Baumgardt *et al.* 2005).

In addition to the spectroscopic data we used HST images from the archive for each cluster. From these images we obtained the color magnitude diagram, the photometric center and the surface brightness profile. The center determination is a crucial step as the shape of the surface brightness profile and the kinematic profile depend on its position.

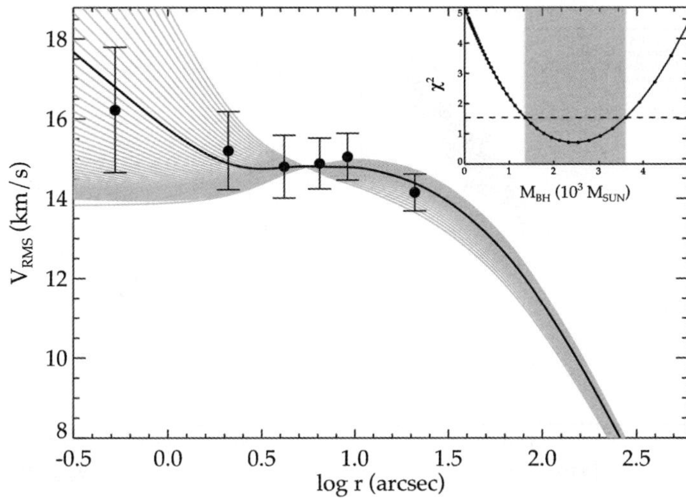

Figure 1. Velocity-dispersion profile of NGC 6266 overplotted by Jeans models with different black-hole masses. The χ^2 values is shown in the upper right and the best fit model indicated by the black sold line.

We therefore used several methods to determine the center and to confirm our results (e.g. isodensity contours, pie wedges, Lützgendorf *et al.* 2011). Another important part of the photometry is the surface-brightness or density profile of the globular cluster. This is used as an input for the dynamical modeling. The profile is obtained by applying radial bins around the photometric center and computing the density/brightness in each bin with a combination of star counts and pixel statistics.

For the spectroscopic data, the large integral field unit ARGUS was pointed at the center of each globular cluster and if necessary a mosaic was produced to cover a substantial area of the core radius. The three-dimensional data cube of the observations was used to construct a velocity map to check for rotation or spurious kinematic features. To compare with dynamical models, however, a velocity dispersion profile is needed. To obtain this, radial bins were applied around the photometric center and the spectra falling in these bins were combined. From the combined spectrum we measured the line broadening by cross correlating with a template spectrum which gives the direct value of the velocity dispersion from the integrated light. The uncertainties for each velocity dispersion measurement were acquired by running Monte Carlo simulations on the simulated IFU using the information from the high-resolution HST image.

The resulting velocity dispersion profile was then used to determine the possible existence and mass of an intermediate-mass black hole in the center. For this, the derived density profile of each cluster was used as an input for Jeans models. These models, derived from the collisionless Boltzmann equation, use the density profile of a stellar system to predict the second moment of the velocity distribution, i.e., the velocity dispersion. A central mass, such as a central black hole can be added to the model. By applying χ^2 statistics we found the model that fits the data best. Figure 1 shows the fit of the globular cluster NGC 6266 with an IMBH signature. The profile clearly rises and requires a model with a black-hole mass of $\sim 3000 \ M_\odot$. This method was applied to all clusters in the sample and for each one we reported an upper limit or a value on the black hole mass (Lützgendorf *et al.* 2011, 2012, 2013b; Feldmeier*et al.* 2013).

Using the data points and upper limits of our sample together with measurements from the literature, we compared the correlation between black-hole mass and host system

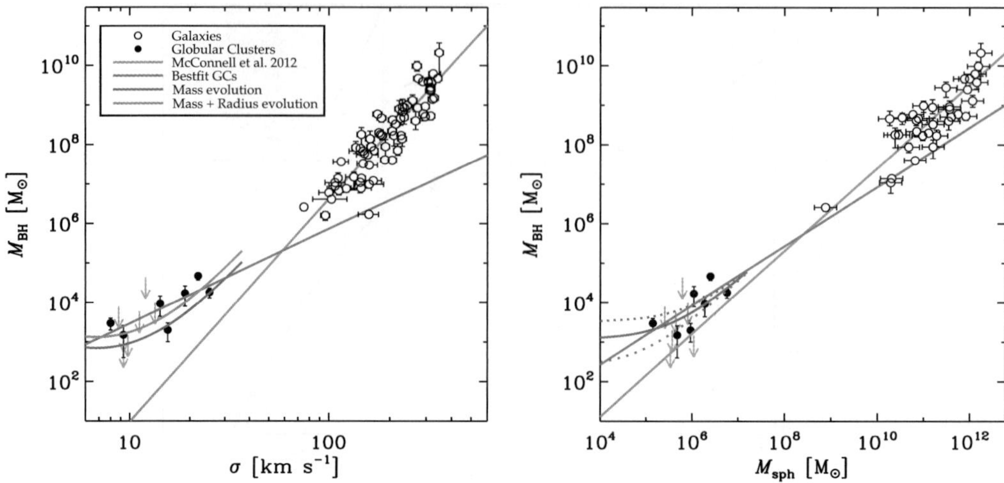

Figure 2. $M_\bullet - \sigma$ and $M_\bullet - M$ relation for supermassive black holes and IMBHs. The clear offset and shallower slope is well reproduced by cluster evolution models considering mass evolution only (magenta line) as well as models that additionally take the radius evolution (i.e. expansion) into account (orange line Kruijssen & Lützgendorf 2013).

properties such as velocity dispersion and total mass with those measured for supermassive black holes in galaxies. In Lützgendorf *et al.* (2013c) we analyzed this small and challenging data set by applying survival analysis in combination with Markov Monte Carlo Chains in order to account for uncertainties and upper limits. We computed correlation coefficients and linear regressions for the major correlations such as $M_\bullet - \sigma$, $M_\bullet - L$ and $M_\bullet - M$ as well as possible dependencies on half-mass radius, metallicity and galactocentric distance. We found that the major correlations are prominent but shallower than those for supermassive black holes (Figure 2). In a follow up paper (Kruijssen & Lützgendorf 2013) we showed that a possible explanation for the offset more shallow correlation is the severe mass loss and expansion of globular clusters during their life time in a tidal field which leads to a reduction of the velocity dispersion and total mass of the system (see Figure 2).

3. Simulations

Because of large uncertainties in the measurements and contradicting results from different methods in the field of IMBHs, it is crucial to support the observing strategies, analysis and physics of the observations with sophisticated simulations. The next sections describe the different simulations that were performed to compare observations and verify our observing techniques.

3.1. *N-body simulations*

N-body simulations are a valuable tool to understand the physics of self gravitating systems such as globular clusters and to identify possible observables of IMBHs in their centers. In Lützgendorf *et al.* (2013a) we ran N-body simulations based on the GPU (Graphic Processing Unit)-enabled version of the collisional N-body code NBODY6 (Aarseth 1999; Nitadori & Aarseth 2012) on GPU graphic cards at the Headquarters of the European Southern Observatory (ESO) in Garching and the University of Queensland in Brisbane. This code uses a Hermite integration scheme with variable time steps. Furthermore, it treats close encounters between stars by applying KS (Kustaanheimo & Stiefel 1965)

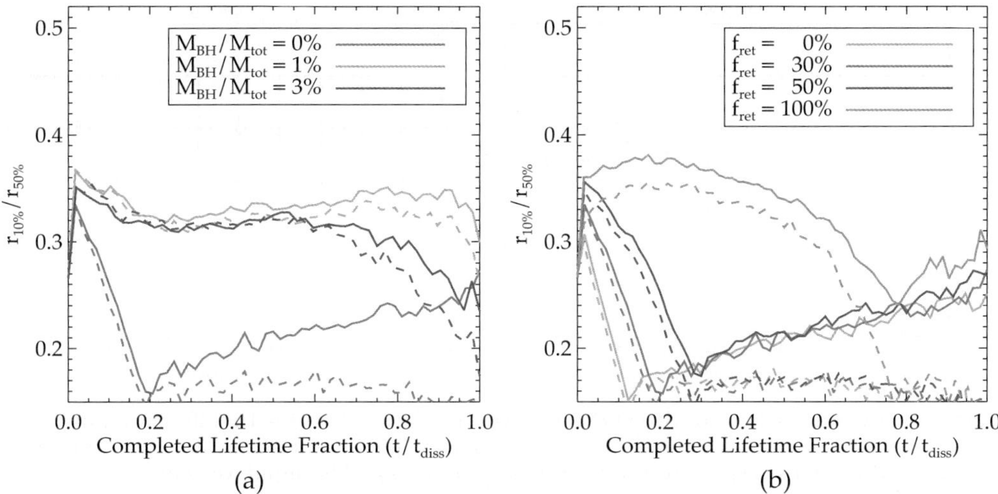

Figure 3. Characteristic radii ratios ($r_{10\%}/r_{50\%}$, i.e., deprojected radii containing 10% and 50% of the stars) as a function of the completed lifetime fraction for simulations with different a) IMBH masses, b) black-hole retention fractions. The dashed lines indicate the $r_{L,10\%}/r_{L,50\%}$ (deprojected radii containing 10% and 50% of the mass) evolution when taking the Lagrangian radii.

and chain regularizations and accounts for stellar evolution (Hurley *et al.* 2000). The regularization procedures are crucial for following orbits of tightly bound binaries over a cluster lifetime accurately and treat strong binary-single and binary-binary interactions properly. The simulations were carried out with particle numbers of $N = 32\,768$ (32k), $65\,536$ (64k), and $131\,072$ (128k) stars.

We have investigated the effect of intermediate-mass black holes, stellar-mass black hole retention fractions, and primordial binary fractions on the properties of globular clusters evolving in a tidal field. We studied the effect of the different initial conditions on the cluster lifetime, remnant fraction, mass function, and structural parameters. In addition, we compared the results of the simulations with observational data from the literature and found good agreement. Owing to the specific shape of the King profile, we found the concentration parameter c to be a poor representation of the cluster's internal properties. Especially after core collapse, a King model is not able to reproduce the central cusp in any of our models and the concentration is systematically underestimated. For that reason we also computed the ratio of the radius containing 10% and 50% of the stars in the cluster $r_{10\%}$ and $r_{50\%}$, which is a more accurate quantity than the parametric King fit. Figure 3 shows the evolution of the ratio of these two non-parametric radii with time. The core collapse is shown as a prominent dip and is prevented (or delayed) by the presence of an IMBH in the center or a large number of stellar-mass black holes. The simulations showed further that the remnant fraction decreases faster when an IMBH is present in the center. This can be explained with the lower degree of mass segregation which increases the chance of finding high-mass objects in the outskirts of the cluster where they can be easily removed.

3.2. *IFU simulations*

Another important part in the field of research on IMBHs is provided by observations of the integrated cluster light. Integral field spectroscopy has become a popular tool in astronomy over the last years. The ability of simultaneously retrieving spatial and spectral data of an object has significant advantages for dynamical and stellar

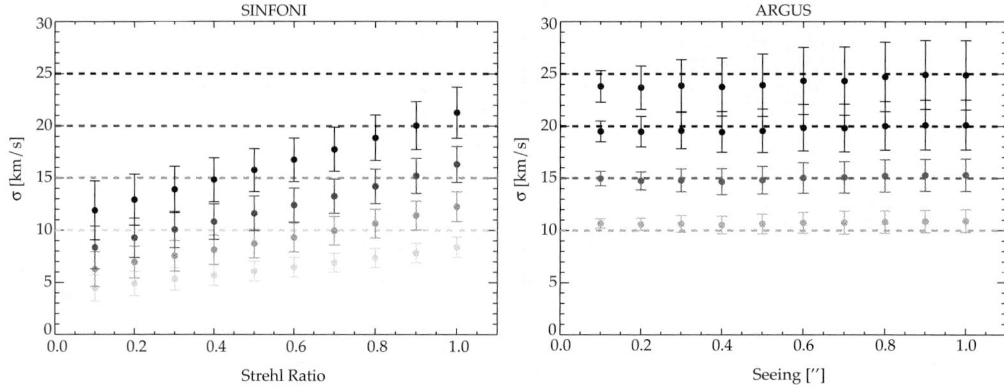

Figure 4. IFU simulations for SINFONI and ARGUS (central bin) with different input velocity dispersions as a function of Strehl ratio (the amount of light contained in the diffraction-limited core of the PSF, with respect to the total flux) and seeing. The input velocity dispersions are shown as dashed lines, the measured velocity dispersions as dots in the corresponding color.

population studies. However, it is not yet clear what is the effect of this observing method on semi-resolved systems such as Galactic globular clusters. For this reason it is crucial to simulate the observations in order to understand possible biases. We have developed a tool that allows to simulate IFU observations using star catalogs for different distances and observing conditions (i.e. seeing, Strehl ratio). The program uses a Moffat PSF that is applied to each star and integrated in the grid of the IFU. Using N-body simulations and N-body realizations we have tested the outcome of IFU observations for clusters with different properties, distances, and seeing. The results will be presented in Lützgendorf *et al.* (2015b, in prep). In addition we investigated the specific case of NGC 6388 where two different methods using IFUs brought very different results. While Lützgendorf *et al.* (2011) found a steeply rising velocity dispersion profile and therefore a strong signature for an IMBH in its center, Lanzoni *et al.* (2013) measured a central velocity dispersion by taking individual velocities from adaptive optics supported SINFONI observations that is 40% lower than the value measured with integrated light. We reproduced both observations using the IFU simulation code and find the SINFONI observations being biased towards lower velocity dispersions due to blends of neighboring stars and background light (Lützgendorf *et al.* 2015a, submitted, see Figure 4).

3.3. *AMUSE simulations*

Recently, many IMBH detections were challenged by contradicting measurements from different observing techniques. Especially the discrepancy between kinematic black-hole measurements and the absence of strong X-ray and radio emission from the centers of globular clusters remain an unsolved mystery (Strader *et al.* 2012). The main uncertainty when translating X-ray and radio flux measurements to black-hole masses is the amount of gas that is accessible for the black hole to accrete. We set out to investigate the effect of stellar winds on the accretion flow of the black hole. By using the Astrophysical Multipurpose Software Environment (AMUSE, Portegies Zwart *et al.* 2009, 2013; Pelupessy *et al.* 2013) we combined gravitational physics, stellar evolution and hydrodynamics into a single simulation of stars interacting with a black hole in the center of a globular cluster. The first application of this code will be presented in Lützgendorf 2015c (in prep.) where the accretion rate of the supermassive black hole in the Milky Way from stellar winds of the surrounding S-Stars is studied (Figure 5). The S-Star system in the Galactic center is well observed and constrained and provides the ideal laboratory to verify our

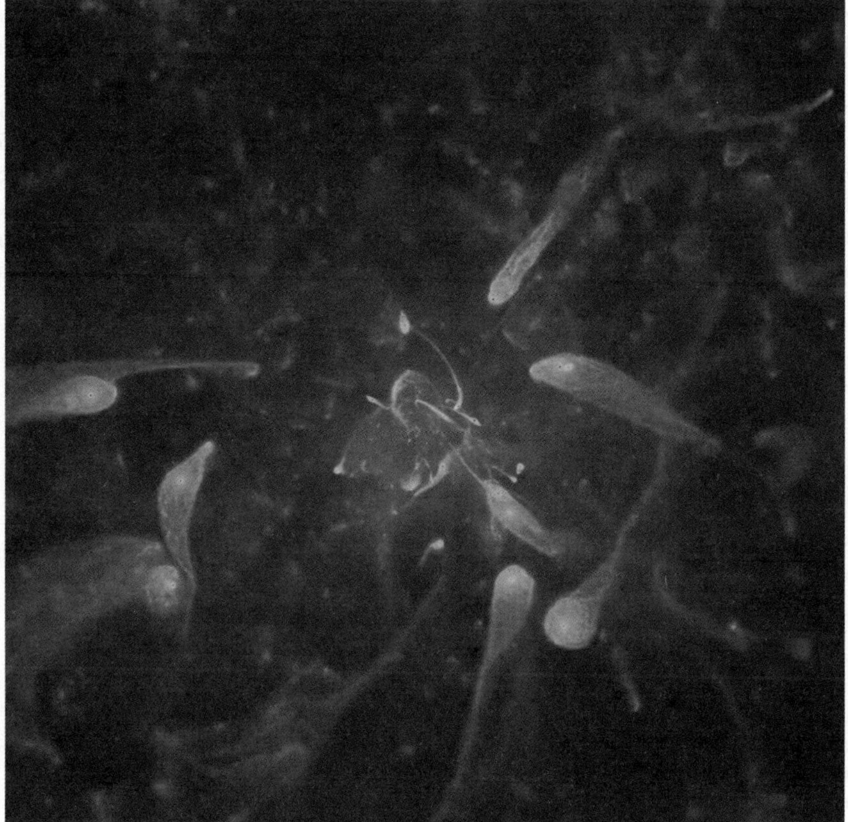

Figure 5. Snapshot of the AMUSE simulation showing all 27 S-Stars and their stellar winds while orbiting the supermassive black hole in the center (Lützgendorf, 2015c, in prep).

code. In this work we study the influence of individual stars on the accretion rate as well as temperature and density evolution in the system. The final outcome of the simulations will be compared to accretion rates computed from X-ray observations of the galactic center. From this, we will be able to extend the simulations to actual star cluster sizes and to pin down accretion rates expected from such a system.

4. Summary

We present the work done by our group on the field of IMBHs in globular clusters over the past years. Using integral field spectroscopy in combination of high-resolution HST imaging we have investigated the presence of IMBHs in a sample of 10 galactic globular clusters. We found about a third of them show signatures of an IMBH in the center. Using our data points in combination with literature data we put the IMBH mass estimates in context with properties of their host systems and compared this to existing scaling relations. We found an offset, shallower correlation for the main relations ($M_\bullet - \sigma$, $M_\bullet - L$ and $M_\bullet - M$) which can be explained by the mass loss of globular clusters located in a tidal field over its life time. To support our observations we performed N-body simulations on globular clusters with and without a central IMBH. We found that clusters with IMBHs do not undergo core collapse and lose massive stars such as stellar remnants more rapidly. Further simulations aimed on reproducing IFU simulations by creating fake data sets from realistic N-body simulations and realizations with different distances and observing conditions. The simulations show that internal properties are in general well reproduced with IFU observations but that large uncertainties due to shot noise are unavoidable. In the specific case of NGC 6388 we found a strong bias in the velocity dispersion when computing it from individual velocities due to blend effects and background contamination. Finally, we used the Astrophysical Multipurpose Software Environment (AMUSE) to simulate the accretion of stellar winds onto a supermassive black hole to reproduce the S-Star system in the Galactic center. We plan to apply the code developed here to globular clusters and intermediate mass black holes in future work.

References

Aarseth, S. J. 1999, *PASP*, 111, 1333
Baumgardt, H., Makino, J., & Ebisuzaki, T. 2004, *ApJ*, 613, 1143
Baumgardt, H., Makino, J., & Hut, P. 2005, *ApJ*, 620, 238
Feldmeier, A., Lützgendorf, N., Neumayer, N., *et al.* 2013, *A&A*, 554, AA63
Ferrarese, L. & Merritt, D. 2000, *ApJ*, 539, L9
Gill, M., Trenti, M., Miller, M. C., *et al.* 2008, *ApJ*, 686, 303
Häring, N. & Rix, H.-W. 2004, *ApJ*, 604, L89
Hurley, J. R. 2007, *MNRAS*, 379, 93
Hurley, J. R., Pols, O. R., & Tout, C. A. 2000, *MNRAS*, 315, 543
Kruijssen, J. M. D. & Lützgendorf, N. 2013, *MNRAS*, 434, L41
Kustaanheimo, P. & Stiefel, E. 1965, J. Reine Angew. Math., 218, 204
Lanzoni, B., Mucciarelli, A., Origlia, L., *et al.* 2013, *ApJ*, 769, 107
Lützgendorf, N., Baumgardt, H., & Kruijssen, J. M. D. 2013a, *A&A*, 558, A117
Lützgendorf, N., Kissler-Patig, M., Gebhardt, K., *et al.* 2013b, *A&A*, 552, A49
Lützgendorf, N., Kissler-Patig, M., Gebhardt, K., *et al.* 2012, *A&A*, 542, A129
Lützgendorf, N., Kissler-Patig, M., Neumayer, N., *et al.* 2013c, *A&A*, 555, A26
Lützgendorf, N., Kissler-Patig, M., Noyola, E., *et al.* 2011, *A&A*, 533, A36
Madau, P. & Rees, M. J. 2001, *ApJ*, 551, L27
Nitadori, K. & Aarseth, S. J. 2012, *MNRAS*, 424, 545
Noyola, E. & Baumgardt, H. 2011, *ApJ*, 743, 52
Noyola, E., Gebhardt, K., & Bergmann, M. 2008, *ApJ*, 676, 1008
Pasquini, L., Avila, G., Blecha, A., *et al.* 2002, The Messenger, 110, 1
Pelupessy, F. I., van Elteren, A., de Vries, N., *et al.* 2013, *A&A*, 557, A84
Portegies Zwart, S., McMillan, S., Harfst, S., *et al.* 2009, New Astronomy, 14, 369
Portegies Zwart, S., McMillan, S. L. W., van Elteren *et al.* 2013, Comp. Phys. Comm., 183, 456
Portegies Zwart, S. F., Baumgardt, H., Hut, P. *et al.* 2004, *Nature*, 428, 724
Strader, J., Chomiuk, L., Maccarone, T. J., *et al.* 2012, *ApJ*, 750, L27
Trenti, M., Ardi, E., Mineshige, S., & Hut, P. 2007, *MNRAS*, 374, 857
Trenti, M., Vesperini, E., & Pasquato, M. 2010, *ApJ*, 708, 1598
Vesperini, E. & Trenti, M. 2010, *ApJ*, 720, L179

Star clusters and black holes in galaxies across cosmic time
Proceedings IAU Symposium No. 312, 2014
Y. Meiron, S. Li, F.-K. Liu & R. Spurzem, eds.

© International Astronomical Union 2016
doi:10.1017/S1743921315007796

Searching for IMBHs in Galactic globular clusters through radial velocities of individual stars

Barbara Lanzoni

Department of Physics and Astronomy, University of Bologna, Viale Berti Pichat 6/2, 40127
Bologna, Italy
email: barbara.lanzoni3@unibo.it

Abstract. I present an overview of our ongoing project aimed at building a new generation of velocity dispersion profiles ad rotation curves for a representative sample of Galactic globular clusters, from the the radial velocity of hundreds of individual stars distributed at different distances from the cluster center. The innermost portion of the profiles will be used to constrain the possible presence of intermediate-mass black holes. The adopted methodology consists of combining spectroscopic observations acquired with three different instruments at the ESO-VLT: the adaptive-optics assisted, integral field unit (IFU) spectrograph SINFONI for the innermost and highly crowded cluster cores, the multi-IFU spectrograph KMOS for the intermediate regions, and the multi-fiber instrument FLAMES/GIRAFFE-MEDUSA for the outskirts. The case of NGC 6388, representing the pilot project that motivated the entire program, is described in some details.

Keywords. globular clusters: general, globular clusters: individual (NGC 6388), stars: kinematics, techniques: spectroscopic, instrumentation: adaptive optics, black hole physics

1. Introduction

Confirming the existence of intermediate mass ($10^3 - 10^4 M_\odot$) black holes (IMBHs) would have a dramatic impact on a number of open astrophysical problems, ranging from the formation of supermassive BHs and their co-evolution with galaxies, to the origin of ultraluminous X-ray sources in nearby galaxies, up to the detection of gravitational waves (e.g. Gebhardt *et al.* 2005). However, the evidence gathered so far in support of the existence of IMBHs are inconclusive and controversial (see, e.g. Noyola *et al.* 2010, Anderson & van der Marel 2010). Globular clusters (GCs) are thought to be the best places where to search for these elusive objects. In fact, the extrapolation of the "Magorrian relation" (Magorrian *et al.* 1998) down to the IMBH masses naturally leads to the GC mass regime. Moreover, numerical simulations have shown that the cores of dense star clusters are the ideal habitat for the formation of IMBHs (e.g. Portegies Zwart *et al.* 2004). For these reasons, the recent years have seen an increasing number of works dedicated to the search for IMBHs in GCs, exploiting all observational channels, ranging from the detection of X-ray and radio emission (see Strader *et al.* 2012 and Kirsten & Vlemmings 2012, and references therein), to the detailed study of the shape of GC density and velocity dispersion profiles (e.g., Gebhardt *et al.* 2000; Gerssen *et al.* 2002, Lanzoni *et al.* 2007, Noyola *et al.* 2010, Lützgendorf *et al.* 2012).

In spite of such an effort, however, no firm conclusions could be drawn to date. This is mainly because of the great difficulties encountered from both the theoretical and the observational points of view.

2. Methodology

To search for IMBHs in Galactic GCs (GGCs) we follow the "dynamical approach", i.e., we perform detailed studies of the cluster structure and dynamics with the goal of identifying the central density and velocity dispersion (VD) cusps theoretically predicted for stellar systems hosting an IMBH (e.g. Baumgardt, Makino, & Hut 2005; Miocchi 2007).

Despite its importance, very little is empirically known to date about GGC internal dynamics, especially in the most crowded central regions. Recent results suggest that some insight can be obtained from the observations of exotic stellar populations, like blue straggler stars and millisecond pulsars (e.g., Ferraro *et al.* 2003, Ferraro *et al.* 2009a, Ferraro *et al.* 2009b, Ferraro *et al.* 2012). However, a detailed knowledge of the VD profile and rotation curve is still missing in the vast majority of GGCs because of observational difficulties. The velocity components on the plane of the sky can be obtained from internal proper motion measurements. However, this requires high-precision photometry and astrometry on quite long temporal baselines, together with precise estimates of GC distances. While this is starting to be feasible (thanks to the combination of multi-epoch HST observations and the improved techniques of data analysis; Anderson & van der Marel 2010; Massari *et al.* 2013, Bellini *et al.* 2014), it is still very challenging in the high density central regions, where the crucial dynamical information constraining the presence of an IMBH is expected. The line of sight (*los*) velocity is in principle easier to obtain (through spectroscopy) and measurable in any cluster region and in GCs at any distance within the Galaxy. In practice, however, the standard approach commonly used in extra-Galactic astronomy (i.e. measuring the line broadening of integrated light spectra) is prone to a severe "shot noise bias": if a few bright stars are present, the acquired spectrum is completely dominated by their light and the line broadening is therefore not produced by the true VD of the underlying stellar population (e.g. Dubath, Meylan & Mayor 1997). The alternative approach is to measure the dispersion about the mean of the radial velocities of statistically significant samples of individual stars. While this is safe from obvious biases, it is observationally challenging, especially in the highly crowded centers of GCs, requiring multi-object and spatially resolved spectroscopy.

Motivated by the astrophysical importance of determining the kinematical properties of GGCs, and because of the known shot noise bias affecting the integrated light spectroscopy method, we designed an ambitious project aimed at measuring the radial velocity of resolved stars along the entire radial extension of a representative sample of GGCs, by means of a multi-instrument approach: adaptive-optics (AO) assisted spectroscopy with SINFONI in the innermost, highly crowded regions, KMOS and FLAMES observations in the intermediate and external regions, respectively. Two large programmes are currently running at the Very Large Telescope of the European Southern Observatory (Prop. ID: 193.D-0232 and 195.D-0750, PI: F. R. Ferraro), collecting data for a representative sample of ~ 30 GGCs. Here we present the results obtained for the pilot project on NGC 6388 (see Lanzoni *et al.* 2013 and Lapenna *et al.* 2015).

3. The pilot project: NGC 6388

By using high spatial resolution ($0.025'' \times 0.027''$/pix) data acquired with the HST/ACS High Resolution Channel (HRC), we discovered a shallow central density cusp in the inner $1''$ of this cluster and, from the comparison with King models including a central IMBH (Miocchi 2007), we concluded that a $\sim 6 \times 10^3 M_\odot$ BH could be hidden in NGC 6388 (Lanzoni *et al.* 2007).

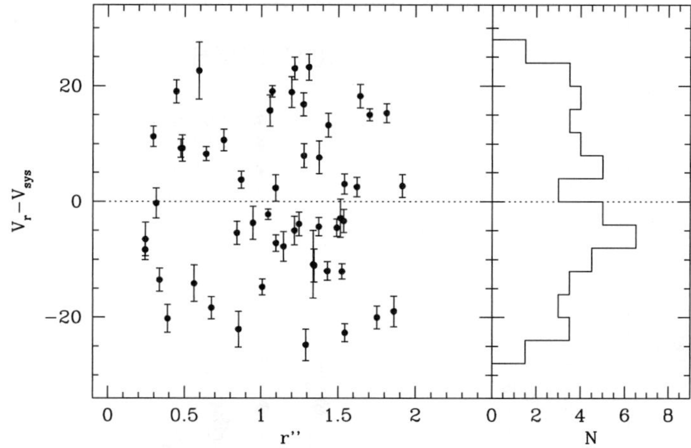

Figure 1. *Left panel:* distribution of the radial velocities measured for the 52 SINFONI targets in NGC 6388, referred to the cluster systemic velocity Vsys and plotted as function of the distance from the centre. *Right panel:* histogram of the radial velocity distribution. The apparent bimodality indicates the presence of systemic rotation.

To further investigate this possibility we then adopted the multi-instrument approach described above for determining the cluster VD profile and search for the central cusp expected in the presence of an IMBH.

SINFONI The innermost cluster regions have been properly investigated by exploiting the high spatial resolution capabilities of SINFONI (Eisenhauer *et al.* 2003), a near-IR (1.1-2.45 μm) integral field spectrograph fed by an AO module and mounted on the YEPUN (VLT-UT4) telescope at the ESO Paranal Observatory. By using the 100 mas plate-scale and the K-band grating (sampling the 1.95-2.45 μm wavelength range), we acquired spectra at a resolution $R = 4000$ for ~ 60 stars located in a $3.2'' \times 3.2''$ region centered on the cluster gravity center (as quoted in Lanzoni *et al.* 2007). From the $^{12}C^{16}O$ band-heads (Origlia *et al.* 1997) and a few atomic lines we then measured the radial velocity (V_r) of 52 giant stars located within $\sim 2''$ from the center of NGC 6388.

KMOS KMOS is a second generation spectrograph equipped with 24 IFUs that can be allocated within a $7.2'$ diameter field of view. Each IFU covers a projected area on the sky of about $2.8'' \times 2.8''$, and it is sampled by an array of 14×14 spatial pixels (hereafter spaxels) with an angular size of $0.2''$ each. KMOS is equipped with four gratings providing a maximum spectral resolution R between ~ 3200 and 4200 over the 0.8-2.5 μm wavelength range. We have used the YJ grating and observed in the 1.00-1.35 μm spectral range at a resolution $R \approx 3400$, corresponding to a sampling of about 1.75 Å pixel^{-1}, i.e. ~ 46 km s^{-1} pixel^{-1} at 1.15 μm. By measuring several atomic lines for each star, we finally attained an average uncertainty of ~ 3 km s^{-1} in the radial velocity estimates. With KMOS we thus measured V_r for a total of 82 giant stars located within $\sim 70''$ from the center of NGC 6388.

FLAMES The spectra of individual stars out to $\sim 600''$ have been acquired by using the ESO VLT multi-object spectrograph FLAMES, in the MEDUSA UVES+GIRAFFE combined mode. This provides 132 fibers to observe an equivalent number of targets in one single exposure, over a field of view of $25'$ in diameter. Four pointings of 2320 s each have been obtained with the GIRAFFE grating HR21 (which samples the Ca II triplet spectral range at a resolution $R = 16200$) and the UVES setup Red Arm 580, covering the wavelength range $4800\text{Å} < \lambda < 6800\text{Å}$ at a resolution $R = 47000$. The targets have been selected from the combined HST/ACS-WFC and ESO/WFI photometric catalog

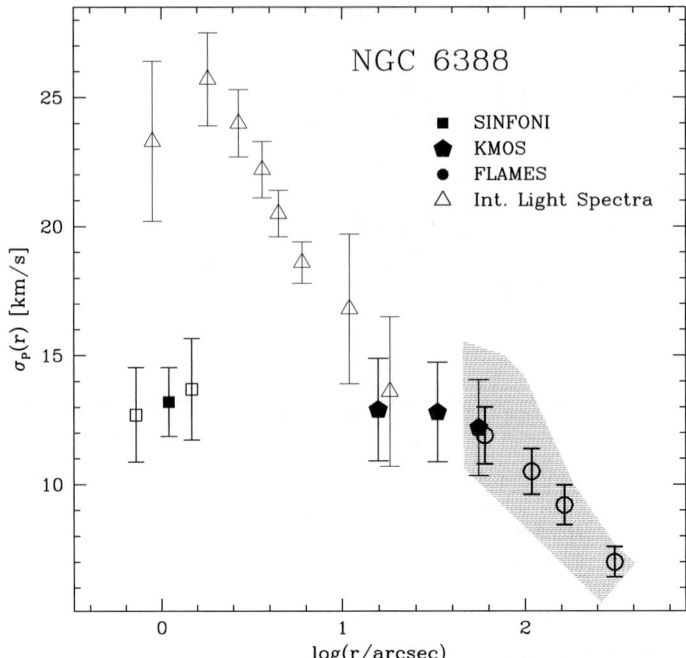

Figure 2. Line-of-sight velocity dispersion profile of NGC 6388 computed from the radial velocities of ~ 400 individual stars as measured with with SINFONI (squares), KMOS (pentagons), and FLAMES (empty circles). The the gray region indicates the dispersion obtained for different choices of the radial bins. The velocity dispersion profile obtained from integrated-light spectra (Lützgendorf *et al.* 2011) is also shown for comparison (empty triangles).

discussed in Lanzoni *et al.* (2007) and Dalessandro *et al.* (2008), considering only isolated objects (with no brighter stars within a circle of $1''$ radius), having $V < 17$ and being located along the canonical evolutionary sequences of the color-magnitude diagram (CMD). By also including two additional data-sets retrieved from the ESO Archive, we finally obtained the radial velocities of about 270 member stars of the cluster.

In all cases, radial velocities have been measured by adopting the Fourier cross-correlation method (Tonry *et al.* 1979) as implemented in the *fxcor* IRAF task. The observed spectrum is cross-correlated with a template of known radial velocity and a cross-correlation function (i.e., the probability of correlation as a function of the pixel shift) is computed. Then, this is fitted by using a Gaussian profile and its peak value is derived. Once the spectra are wavelength calibrated, the pixel shift obtained by the Gaussian fit is converted into radial velocity. We employed different reference template spectra according to the adopted spectral configuration and stellar type.

The systemic velocity of NGC 6388 has been computed by conservatively using all stars with radial velocities between 60 and 105 km s^{-1}, deriving a value of $V_r = 82.0 \pm 0.5$ km s^{-1}, in agreement with the literature (Harris 1996). Hereafter, $\widetilde{V_r}$ will indicate radial velocities referred to the cluster systemic velocity: $\widetilde{V_r} \equiv V_r - V_{\rm sys}$.

Systemic rotation. The radial velocity distribution for the 52 SINFONI targets is plotted in Fig. 1. The histogram in the right-hand panel shows a clear bimodality, indicating the presence of systemic rotation. To more quantitatively investigate this possibility, we used the method fully described in Bellazzini *et al.* 2012. We considered a line passing through the cluster center with position angle PA varying between $0°$ (North direction) and $90°$ (East direction), by steps of $15°$. For each value of PA, such a line splits the

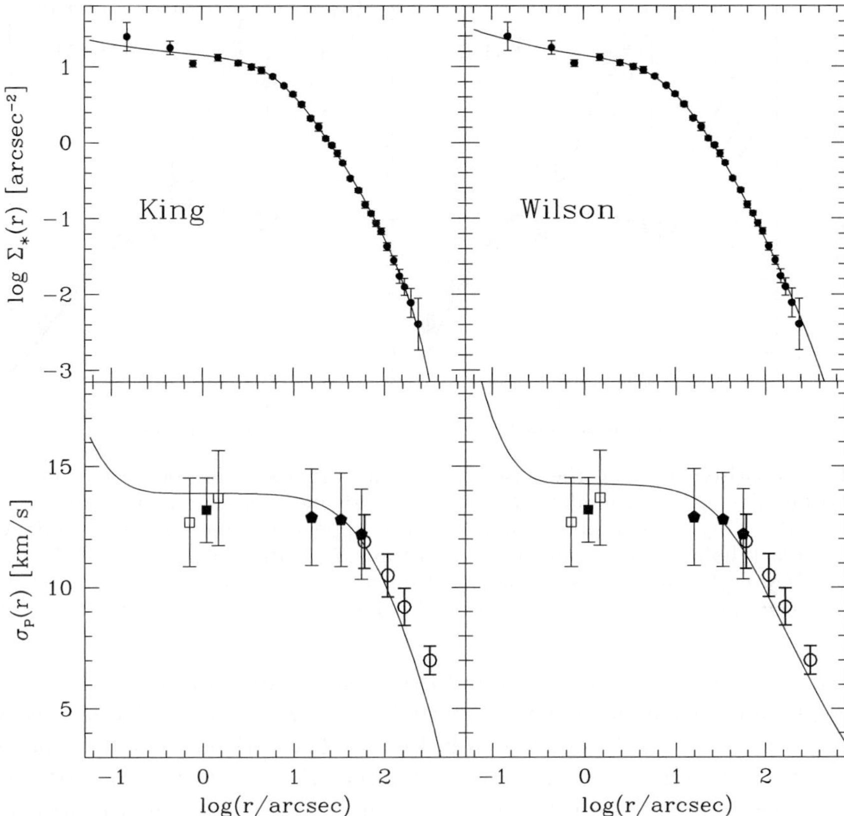

Figure 3. Comparison between the observations and the best-fit self-consistent King and Wilson models with central IMBH (left- and right-hand panels, respectively). The observed density profile (solid circles in the upper panels) is from Lanzoni *et al.* (2007). The observed velocity dispersion profiles are the same as in Fig. 2. An IMBH mass of $\sim 2000 M_\odot$ is assumed in these models.

SINFONI sample in two. The difference $\Delta\langle\widetilde{V_r}\rangle$ between the mean radial velocity of the two sub-samples was computed and has been found to show a coherent sinusoidal behavior, which is a signature of rotation. The best-fit position angle of the rotation axis is $\mathrm{PA_0} = 11°$ and the amplitude of the rotation curve $A_{\mathrm{rot}} = 8.5$ km s^{-1}. The Kolmogorov-Smirnov probability, however, indicates a not very high statistical significance, possibly due to the limited sample. Some hints of rotation are found also from the KMOS and FLAMES samples, but again the significance is weak. Additional data symmetrically sampling the cluster are needed before drawing any firm conclusion about ordered rotation in NGC 6388.

Velocity dispersion. To compute the projected VD profile, the surveyed area has been divided in a set of concentric annuli, chosen as a compromise between a good radial sampling and a statistically significant number (> 50) of stars. In each radial bin, σ_p has been computed from the dispersion of the values of $\widetilde{V_r}$ measured for all the stars in the annulus, by following the Maximum Likelihood method described in Walker *et al.* (2006; see also Martin *et al.* 2007; Sollima *et al.* 2009). An iterative 3σ clipping algorithm was applied in each bin. The error estimate is performed by following Pryor & Meylan (1993). The resulting VD profile is shown in Figure 2. Given the number of stars in the SINFONI data set (52 objects in total) we computed $\sigma_p(r)$ by considering both one single

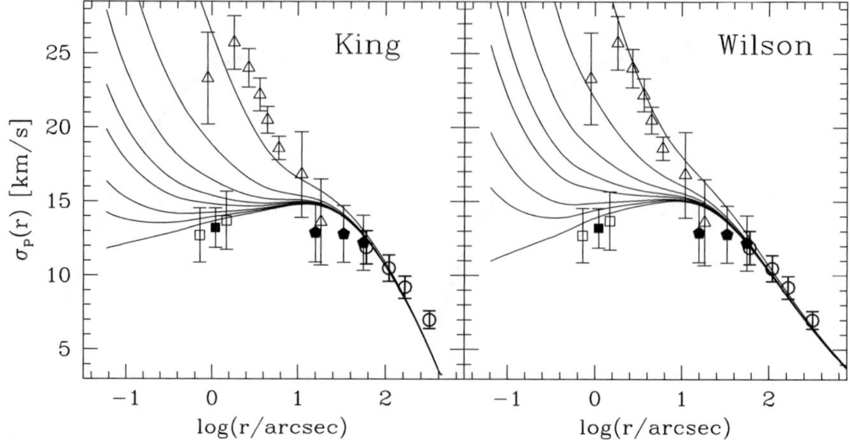

Figure 4. Comparison between the observed velocity dispersion profiles (the same as in Fig. 2) and two families of Jeans models. *Left panel:* solid lines correspond to Jeans models calculated from the King density profile shown in the upper-left panel of Fig. 3 and by assuming different black hole masses: from bottom to top, $M_{\rm BH}^{\rm J}/M_{\rm BH}^{\rm sc} = 0, 0.5, 1, 2, 3, 5, 10, 30$, with $M_{\rm BH}^{\rm sc} = 2147 M_\odot$. *Right panel:* the same as in the left panel, but for Jeans models calculated from the Wilson density profile shown in the upper-right panel of Fig. 3. In this case, $M_{\rm BH}^{\rm sc} = 2125 M_\odot$.

central bin (solid square in the figure) and two separate annuli (26 stars at $r < 1.2''$, 26 stars beyond; empty squares). For the external data set we tried different sets of radial bins, finding consistent results (grey area in Fig. 2). The derived values are shown in the figure as black diamonds and empty circles for the KMOS and the FLAMES samples, respectively. We find that the velocity dispersion of NGC 6388 has a central ($r \sim 1''$) value of ~ 13 km s^{-1}, stays approximately flat out to $r \sim 60''$, and then decreases to ~ 7 km s^{-1} at $200'' < r < 600''$. If no σ clipping algorithm is applied, the VD profile remains almost unchanged, with the only exception of the outermost data-point that rises to $\sim 8.9 \pm 0.8$ km s^{-1}.

The inner part ($r < 10''$) of the VD profile is clearly incompatible with the one obtained by Lützgendorf *et al.* (2011, hereafter L11) from the line broadening of integrated-light spectra, which shows a steep rise toward the cluster centre, up to values of $23 - 25$ km s^{-1} (empty triangles in Fig. 2), suggesting the presence of an IMBH of $(1.7 \pm 0.9) \times 10^4 M_\odot$. While the exact reason for such a disagreement is not completely clear, a detailed comparison between our radial velocity measurements and L11 radial velocity map suggests that their measure is affected by the shot noise bias mentioned above (see Sect. 4.1 and Fig. 12 in Lanzoni *et al.* 2013).

Comparison with models. In order to constraint the presence and the mass of a central IMBH in NGC 6388, we followed two different and complementary approaches. First, starting from a family of self-consistent models admitting a central IMBH, we selected the one that best reproduces the observed density and VD profiles. This provided us with the corresponding structural and kinematical parameters in physical units, including the IMBH mass. Second, we used the observed density profile and included a variable central point mass, to solve the spherical Jeans equation (this is what is done also in L11). We therefore obtained the corresponding family of VD profiles, which were then compared to the observations to constrain the BH mass. The first approach is more realistic in terms of the stellar mass-to-light ratio (M/L), since it takes into account various populations of stars with different masses and different radial distributions (as it is indeed expected and observed in mass segregated GCs). Moreover it has the advantage that the models stem from a distribution function which is known a priori. Hence, the model consistency (i.e., the non-negativity of the distribution function in the phase-space) is under control and guaranteed by construction. However, the validity of the results is limited to the case in which the assumed models are the correct representation of the true structure and dynamics of the cluster. The second approach is more general, but we assume a constant M/L ratio and the resulting models could be non-physical (e.g. Binney & Tremaine 1987).

In all cases we assumed spherical symmetry and velocity isotropy. This seems to be reasonable for NGC 6388, which does not show significant ellipticity (Harris 1996) and is thought to be quite dynamically evolved (Ferraro *et al.* 2012; see also L11). In both approaches we considered King (1966) and Wilson (1975) distribution functions, which are known to well reproduce the observed density profiles of GGCs (e.g., McLaughlin & van der Marel 2005; Miocchi *et al.* 2013).

As shown in Figures 3 and 4, our observed velocity dispersion profile is consistent with both the absence of a central IMBH, and the presence of a BH with a mass up to $\sim 2000 M_\odot$ (the latter possibility is mainly supported by the cusp in the density profile). The absence of a central IMBH in NGC 6388 is consistent with the results obtained from both X-ray and radio observation, which put an upper limit of $\sim 600 M_\odot$ to a possible compact dark mass in this cluster (Nucita *et al.* 2008; Cseh *et al.* 2010; Bozzo *et al.* 2011).

4. Summary

We are using a multi-instrument approach (SINFONI+KMOS+FLAMES at the ESO-VLT) to measure the radial velocities of hundred individual stars distributed along the entire extension of ~ 30 Galactic GCs. The shape of the innermost portion of the velocity dispersion profile (and density distribution) will be used to constrain the presence of IMBHs in these systems. The pilot project on NGC 6388 demonstrates that this is the right route to properly determine GGC internal dynamics and pursue the "dynamical approach" to the search for IMBHs. Our study shows that no IMBH (or a $2000 M_\odot$ BH, at most) is required to account for the observations in this cluster.

Most of the results discussed in this talk have been obtained within the project Cosmic-Lab (PI: Ferraro, see http://www.cosmic-lab.eu), a 5-year project funded by the European European Research Council under the 2010 Advanced Grant call (contract ERC-2010-AdG-267675). I warmly thank the other team members involved in this research: Francesco Ferraro, Elena Valenti, Alessio Mucciarelli, Livia Origlia, Emilio Lapenna, Emanuele Dalessandro, Michele Bellazzini, Paolo Miocchi, Davide Massari and Cristina Pallanca.

References

Anderson, J. & van der Marel, R. P. 2010, *ApJ*, 710, 1032
Baumgardt H., Makino J., & Hut P., 2005, *ApJ*, 620, 238
Bellini, A., Anderson, J., van der Marel, R. P., *et al.* 2014, *ApJ*, 797, 115
Binney, J. & Tremaine, S. 1987, Princeton, NJ, Princeton University Press
Bellazzini, M., Bragaglia, A., Carretta, E., *et al.* 2012, *A&A*, 538, A18
Bozzo, E., Ferrigno, C., Stevens, J., *et al.* 2011, *A&A*, 535, L1
Cseh, D., Kaaret, P., Corbel, S., *et al.* 2010, *MNRAS*, 406, 1049
Dubath, P., Meylan, G., & Mayor, M. 1997, *A&A*, 324, 505
Eisenhauer, F. *et al.*, 2003, SPIE, 4841, 1548
Ferraro, F. R., Possenti, A., Sabbi, E., *et al.* 2003, *ApJ*, 595, 179

Ferraro, F. R., Beccari, G., Dalessandro, E., *et al.* 2009a, *Nature*, 462, 1028

Ferraro, F. R., Dalessandro, E., Mucciarelli, A., *et al.* 2009b, *Nature*, 462, 483

Ferraro, F. R. *et al.*, 2012, *Nature*, 492, 393

Dalessandro, E., Lanzoni, B., Ferraro, F. R., *et al.* 2008, *ApJ*, 677, 1069

Gebhardt, K., Pryor, C., O'Connell, R. D., Williams, T. B., & Hesser, J. E. 2000, *AJ*, 119, 1268

Gebhardt, K., Rich, R. M., & Ho, L. C. 2005, *ApJ*, 634, 1093

Gerssen, J., van der Marel, R. P., Gebhardt, K., *et al.* 2002, *AJ*, 124, 3270

Harris, W. E. 1996, *AJ*, 112, 1487

King I. R., 1966, *AJ*, 71, 64

Kirsten, F. & Vlemmings, W. H. T. 2012, *A&A*, 542, A44

Lanzoni, B., Dalessandro, E., Ferraro, F. R., *et al.* 2007, *ApJ*, 668, L139

Lanzoni, B., Mucciarelli, A., Origlia, L., *et al.* 2013, *ApJ*, 769, 107

Lapenna, E., Origlia, L., Mucciarelli, A., *et al.* 2015, *ApJ*, 798, 23

Lützgendorf, N., Kissler-Patig, M., Noyola, E., *et al.* 2011, *A&A*, 533, A36 (L11)

Lützgendorf, N., Kissler-Patig, M., Gebhardt, K., *et al.* 2012, *A&A*, 542, A129

Massari, D., Bellini, A., Ferraro, F. R., *et al.* 2013, *ApJ*, 779, 81

Magorrian, J., Tremaine, S., Richstone, D., *et al.* 1998, *AJ*, 115, 2285

Martin, N. F., Ibata, R. A., Chapman, S. C., Irwin, M., & Lewis, G. F. 2007, *MNRAS*, 380, 281

McLaughlin, D. E. & van der Marel, R. P. 2005, *ApJS*, 161, 304

Miocchi, P. 2007, *MNRAS*, 381, 103

Miocchi, P. *et al.*, 2013, *ApJ*, 774, 151

Noyola, E., Gebhardt, K., Kissler-Patig, M., *et al.* 2010, *ApJ*, 719, L60

Nucita, A. A., de Paolis, F., Ingrosso, G., Carpano, S., & Guainazzi, M. 2008, *A&A*, 478, 763

Origlia, L., Ferraro, F. R., Fusi Pecci, F., & Oliva, E. 1997, *A&A*, 321, 859

Portegies Zwart, S. F., Baumgardt, H., Hut, P., Makino, J., & McMillan, S. L. W. 2004, *Nature*, 428, 724

Pryor, C. & Meylan, G. 1993, *Structure and Dynamics of Globular Clusters*, 50, 357

Sollima, A., Bellazzini, M., Smart, R. L., *et al.* 2009, *MNRAS*, 396, 2183

Strader, J., Chomiuk, L., Maccarone, T. J., *et al.* 2012, *ApJ*, 750, L27

Tonry, J. & Davis, M., 1979, *AJ*, 84, 1511

Walker, M. G., Mateo, M., Olszewski, E. W., *et al.* 2006, *AJ*, 131, 2114

Wilson, C. P. 1975, *AJ*, 80, 175

Star clusters and black holes in galaxies across cosmic time
Proceedings IAU Symposium No. 312, 2014
Y. Meiron, S. Li, F.-K. Liu & R. Spurzem, eds.

© International Astronomical Union 2016
doi:10.1017/S1743921315007802

On the uniqueness of kinematical signatures of intermediate-mass black holes in globular clusters

Alice Zocchi, Mark Gieles and Vincent Hénault-Brunet

Department of Physics, University of Surrey, Guildford, Surrey, GU2 7XH, United Kingdom
email: a.zocchi@surrey.ac.uk

Abstract. Finding an intermediate-mass black hole (IMBH) in a globular cluster (GC), or proving its absence, is a crucial ingredient in our understanding of galaxy formation and evolution. The challenge is to identify a unique signature of an IMBH that cannot be accounted for by other processes. Observational claims of IMBH detection are often based on analyses of the kinematics of stars, such as a rise in the velocity dispersion profile towards the centre. In this contribution we discuss the degeneracy between this IMBH signal and pressure anisotropy in the GC. We show that that by considering anisotropic models it is possible to partially explain the innermost shape of the projected velocity dispersion profile, even though models that do not account for an IMBH do not exhibit a cusp in the centre.

Keywords. Globular clusters: general, globular clusters: individual:NGC 5139, stars: kinematics, black hole physics

---

1. Introduction

The task of finding an intermediate-mass black hole (IMBH) in a globular cluster is very challenging, because the predicted signatures of an IMBH are degenerate with alternative scenarios. IMBH detections in globular clusters are mostly claimed on the basis of the discovery of a shallow cusp in the surface brightness profile and a rise in the velocity dispersion profile towards the centre (e.g., see Noyola *et al.* 2008, van der Marel & Anderson 2010, Noyola *et al.* 2010). However, similar features can also be produced by different processes, and conclusive evidence for the existence of IMBHs in globular clusters is still lacking. For example, mass segregation, core collapse, or the presence of binary stars in the centre can also generate a shallow cusp in the surface brightness profile, as shown by means of dedicated *N*-body simulations (Vesperini & Trenti 2010). Some kinematical properties can be explained by the presence of pressure anisotropy, without a central intermediate-mass black hole (van der Marel & Anderson 2010; Zocchi *et al.* 2012). In order to find conclusive evidence of the presence of IMBH in globular clusters, it is important to fully understand the internal dynamics of these systems: in this contribution we focus on the role played by radially-biased pressure anisotropy.

2. Why is pressure anisotropy relevant?

Galactic globular clusters are characterised by different relaxation conditions. For many of them, the relevant relaxation times are shorter than their age, so that they are commonly considered to be close to thermodynamical equilibrium, with an isotropic velocity distribution that is close to Maxwellian. However, some large globular clusters have very long relaxation times, and their structure might be more similar to that of elliptical galaxies, for which pressure anisotropy is thought to play an important role.

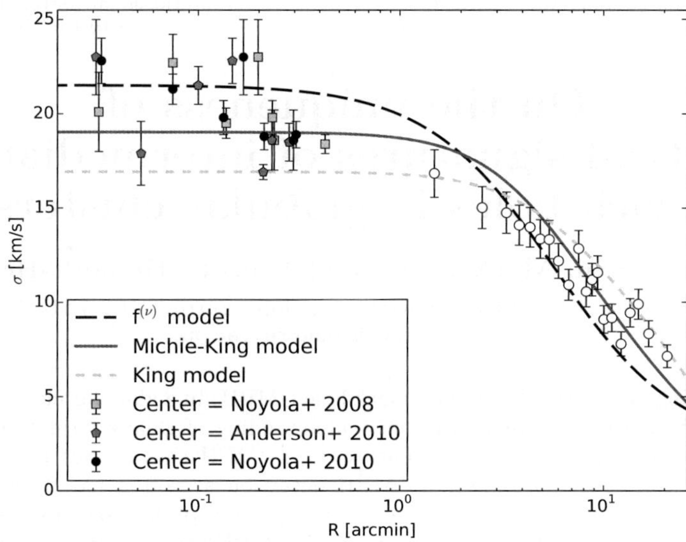

Figure 1. Velocity dispersion profile of ω Cen. Best-fit $f^{(\nu)}$ (long-dashed line), Michie-King (solid line), and King (short-dashed line) models profiles are shown along with line-of-sight velocity dispersion data. Empty circles represent the velocity dispersion profile calculated from velocities of single stars as in Bianchini *et al.* (2013). Filled symbols represent the velocity dispersion calculated from integrated spectra observations carried out by Noyola *et al.* (2010), and binned with respect to the position of the centre proposed by Noyola *et al.* 2008 (squares), van der Marel & Anderson 2010 (pentagons), and Noyola *et al.* 2010 (circles); the shape of these profiles is highly dependent on the position of the centre, and only in one case a cusp is visible.

Several studies have suggested the presence of pressure anisotropy in globular clusters (Ibata *et al.* 2011, Zocchi *et al.* 2012, Bellazzini *et al.* 2015), by analysing sets of line-of-sight kinematical data. The only data that would enable us to directly measure the presence of anisotropy in these stellar systems are proper motions (see for example Bianchini *et al.* 2013), but unfortunately these are available only for a few GCs†.

Confirmations of the importance of this physical ingredient arrive also from numerical simulations. Lützgendorf *et al.* (2011) showed that primordial pressure anisotropy can last for long in the outer parts of globular clusters, while in the central region it is not stable, and is washed away very quickly. We recently found that anisotropy can originate in the early phases of the life of GCs, even for systems that are originally isotropic, as the result of two-body relaxation that scatters stars out of the core on radial orbits; for clusters located in an external tidal field, it is erased in the final stages, just before their complete dissolution (Zocchi *et al.*, in prep).

Even if pressure anisotropy might not concern the very central part of GCs, where we look for the presence of an IMBH, we need to determine its role in shaping the *projected* kinematical quantities that we measure, especially when using only line-of-sight kinematical data. Indeed, the presence of radially-biased anisotropy has the effect of increasing the velocity dispersion that is measured when looking towards the centre of the system, where the line-of-sight is aligned with the radial direction; similarly, in the outermost parts, where the line-of-sight is parallel to the tangential direction, the measured velocity dispersion would be smaller (the opposite is true for

† Measurement of proper motions for several Galactic globular clusters are becoming available via the HSTPROMO collaboration (Bellini *et al.* 2014), and even more will be available in the near future thanks to the observations carried out by the Gaia satellite.

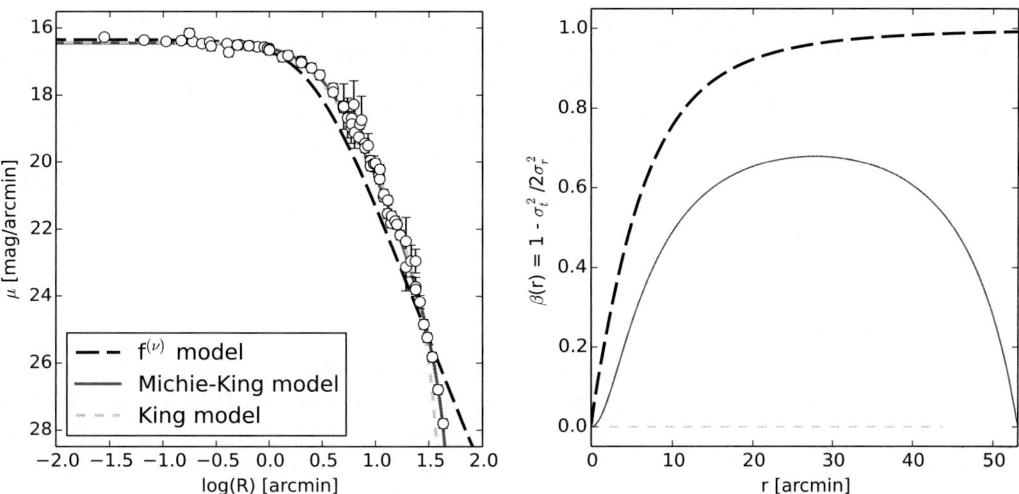

Figure 2. Left panel: best-fit surface brightness profiles and data from Trager *et al.* (1995) and Noyola *et al.* (2008). Right panel: best-fit models anisotropy profiles. Lines are as in Fig. 1.

tangentially-biased anisotropic systems). It is therefore crucial to take anisotropy into account when looking for the presence of an IMBH in the centre of globular clusters.

3. The case of ω Cen

To illustrate the effect that anisotropy has on projected velocity dispersion profiles, we analyse the data available for the massive globular cluster ω Cen (NGC 5139), for which several studies have been carried out, looking for the presence of an IMBH at its centre (Noyola *et al.* 2008, van der Marel & Anderson 2010, Noyola *et al.* 2010).

In this contribution we consider three different families of dynamical models; for simplicity, and to show more clearly the contribution of anisotropy, we chose to consider only single-mass, spherical, non-rotating models. The nontruncated radially-biased anisotropic $\mathbf{f}^{(\nu)}$ **models** (see Bertin & Trenti 2003, and references therein) were constructed to describe the products of (incompletely) violently relaxed elliptical galaxies, and describe systems that are isotropic in the centre, and anisotropic in the outer parts; the anisotropy profile has roughly the same shape for all the models in the family. Truncated radially-biased anisotropic **Michie-King models** (Michie 1963) describe systems that are isotropic in the centre and near the tidal radius, and anisotropic in the intermediate radial range. Depending on the value of a model parameter, namely the anisotropy radius r_a, it is possible to obtain different shapes for the anisotropy profile, with different values for the maximum of the anisotropy β: when r_a is very small, the maximum of β tends to 1, while for very large values of r_a the profile becomes the same as for an isotropic model. For comparison, we also consider isotropic **King models** (King 1966).

We performed fits of these models to the surface brightness and velocity dispersion profiles of ω Cen. Figure 1 shows the best-fit velocity dispersion profiles along with line-of-sight data (see caption for a description). The effect of anisotropy is clearly visible: the central velocity dispersion is larger for the anisotropic $f^{(\nu)}$ models, and smaller for King isotropic models. By considering anisotropic models it is therefore possible to partially explain the innermost shape of the velocity dispersion profile, even though models that do not account for an IMBH exhibit a flat profile and no cusp in the centre. Anyway, it is clear that, by taking anisotropy into account, a smaller IMBH mass would be needed to match a central cusp. The left panel of Fig. 2 shows the best-fit surface brightness profiles: the models represent reasonably well the data; only the $f^{(\nu)}$ model shows some significant deviations from the observed profile. The right panel of Fig. 2 shows the anisotropy profiles of the best-fit models; we checked that the anisotropic models are stable against radial orbit instability, by computing the value of the parameter κ (1.75 and 1.36 for $f^{(\nu)}$ and Michie-King models, respectively) introduced by Polyachenko & Shukhman (1981). This amount of energy in radial orbits is consistent with the build-up of radial orbits that we find in our N-body models (Zocchi et al. in prep).

4. Conclusion

Pressure anisotropy plays an important role in the dynamics of globular clusters, and it should be taken into account to properly describe these systems. Here we showed that models with isotropic velocity distributions in the core and radial anisotropy in the outer parts can describe reasonably well the surface brightness and velocity dispersion profiles of ω Cen, without the need of the presence of an IMBH. The models we used do not take into account some of the known complexities of globular clusters: by considering rotation and by including a range of stellar masses, it will be possible to give a more accurate representation of the dynamics of this system.

The uncertainty in the position of the centre of the cluster prevents an accurate determination of kinematic and photometric profiles. In the future, we plan to adopt a discrete fitting approach, and we will determine the position of the centre as a fitting parameter.

References

Bellazzini, M., Mucciarelli, A., Sollima, A., Catelan, M., Dalessandro, E., Correnti, M., D'Orazi, V., Cortés, C., & Amigo, P. 2015, *MNRAS*, 446, 3130

Bellini, A., Anderson, J., van der Marel, R. P., Watkins, L. L., King, I. R., Bianchini, P., Chanamé, J., Chandar, R., Cool, A. M., Ferraro, F. R., Ford, H., & Massari, D. 2014, *ApJ*, 797, 115

Bertin, G. & Trenti, M. 2003, *ApJ*, 584, 729

Bianchini, P., Varri, A. L., Bertin, G., & Zocchi, A. 2013, *ApJ*, 772, 67

Ibata, R., Sollima, A., Nipoti, C., Bellazzini, M., Chapman, S. C., & Dalessandro, E. 2011, *ApJ*, 738, 186

King, I. R. 1966, *AJ*, 71, 64

Lützgendorf, N., Kissler-Patig, M., Noyola, E., Jalali, B., de Zeeuw, P. T., Gebhardt, K., & Baumgardt, H. 2011, *A&A*, 533, 36

Michie, R. W. 1963, *MNRAS*, 125, 127

Noyola, E., Gebhardt, K., & Bergmann, M. 2008, *ApJ*, 676, 1008

Noyola, E., Gebhardt, K., Kissler-Patig, M., Lützgendorf, N., Jalali, B., de Zeeuw, P. T., & Baumgardt, H. 2010, *ApJ*, 719, L60

Polyachenko, V. L. & Shukhman, I. G. 1981, *Soviet Astron.*, 25, 533

Trager, S. C., King, I. R., & Djorgovski, S. 1995, *AJ*, 109, 218

van der Marel, R. P. & Anderson, J. 2010, *ApJ*, 710, 1063

Vesperini, E. & Trenti, M. 2010, *ApJ*, 720, L179

Zocchi, A., Bertin, G., & Varri, A. L. 2012, *A&A*, 539, 65

Star clusters and black holes in galaxies across cosmic time
Proceedings IAU Symposium No. 312, 2014
Y. Meiron, S. Li, F.-K. Liu & R. Spurzem, eds.

© International Astronomical Union 2016
doi:10.1017/S1743921315007814

Spectroscopic study of formation, evolution and interaction of M31 and M33 with star clusters

Zhou Fan[1] and Yanbin Yang[2]

[1]Key Laboratory of Optical Astronomy, National Astronomical Observatories, Chinese Academy of Sciences, 20A Datun Road, Chaoyang Dist., Beijing, China
email: zfan@bao.ac.cn

[2]GEPI, Observatoire de Paris, CNRS, 5 Place Jules Janssen, Meudon F92195, France

Abstract. The recent studies show that the formation and evolution process of the nearby galaxies are still unclear. By using the Canada France Hawaii Telescope (CFHT) 3.6m telescope, the PanDAS shows complicated substructures (dwarf satellite galaxies, halo globular clusters, extended clusters, star streams, etc.) in the halo of M31 to $\sim$ 150 kpc from the center of galaxy and M31-M33 interaction has been studied. In our work, we would like to investigate formation, evolution and interaction of M31 and M33, which are the nearest two spiral galaxies in Local Group. The star cluster systems of the two galaxies are good tracers to study the dynamics of the substructures and the interaction. Since 2010, the Xinglong 2.16m, Lijiang 2.4m and MMT 6.5m telescopes have been used for our spectroscopic observations. The radial velocities and Lick absorption-line indices can thus be measured with the spectroscopy and then ages, metallicities and masses of the star clusters can be fitted with the simple stellar population models. These parameters could be used as the input physical parameters for numerical simulations of M31-M33 interaction.

Keywords. galaxies: individual (M31) – galaxies: individual (M33) – galaxies: star clusters – globular clusters: general – star clusters: general

1. Introduction

From 2008 to 2011, the Pan-Andromeda Archaeological Survey (PanDAS), which covers > 300 $\deg^2$ M31-M33 field with CFHT 3.6m Telescope, was conducted by McConnachie *et al.* (2009). The survey discovered several new stellar stream features surrounding M31 and M33, respectively. It was found that dozens of dwarf galaxies and halo star clusters are distributed to the outer halo $\sim$ 150 kpc from the center of M31. Observational evidences indicated that the structures may be due to M33 orbits around M31. The M31-M33 interaction was suggested by HI observations previously. Since the star clusters are much brighter than stars, the halo star clusters are good tracers to study the dynamics of the interactions.

2. Observations and analysis

Since 2010, we have started to observe the spectroscopy of the star clusters in the halo of M31 with the red channel spectrograph of 6.5m MMT and BFOSC/OMR spectrograph of Xinglong 2.16m telescope. The star clusters of M31 were selected from RBC v.4 (http://www.bo.astro.it/M31; Galleti *et al.* 2004). The Lick absorption-line indices (see Worthey *et al.* 1994) of the spectroscopy were measured and fit with the stellar population model of Thomas *et al.* (2011), and then the ages, [Fe/H] and [α/Fe] can be derived, which provides the clues of the formation and evolution of the two galaxies. Meanwhile,

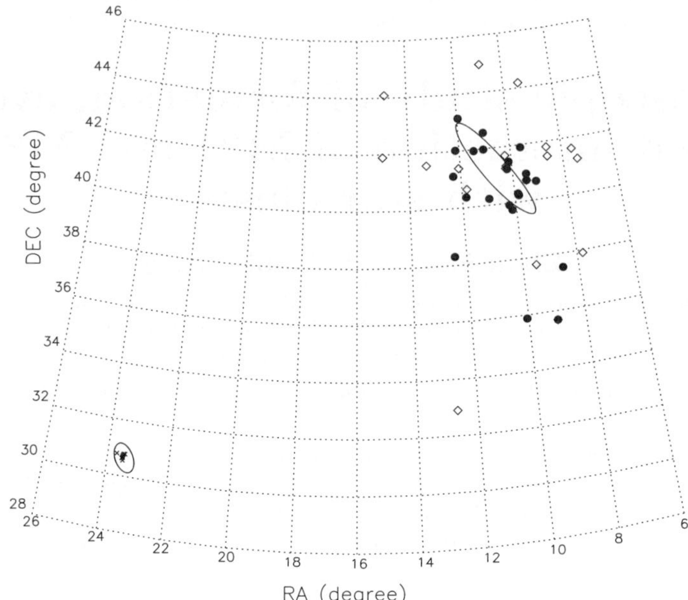

Figure 1. The distribution of star clusters that we have observed in M31 (larger ellipse) and M33 (smaller ellipse) field. The filled circles represent M31 star clusters observed with MMT; the open squares and crosses are star clusters of M31 and M33 observed with Xinglong 2.16m telescope, respectively.

the radial velocity V_r can be estimated by fitting the absorption lines, which can be used to study the dynamics of the star cluster system as well as for tracing the dynamics of M31-M33 interaction.

Figure 1 shows the distribution of star clusters that we have observed so far with MMT and Xinglong 2.16m telescope. The open squares represent the M31 clusters observed with Xinglong 2.16m telescope and analyzed in Fan *et al.* (2011) and Fan *et al.* (2012). The filled circles are M31 clusters observed with MMT and the crosses show M33 star clusters observed with BFOSC spectrograph of Xinglong 2.16m telescope in Nov. 2014.

In the future, more spectroscopic observations are planed to be conducted with Lijiang 2.4m, and Guoshoujing (LAMOST) telescopes. The sample will be enlarged and more informations will be derived for the numerical simulations of the interaction of M31 and M33 galaxies.

Acknowledgements

This research is supported by the National Natural Science Foundation of China Grant No. 11373003 and National Key Basic Research Program of China (973 Program) No. 2015CB857002.

References

Fan, Z., Huang, Y. F., Li, J. Z., Zhou, X., Ma, J., *et al.* 2011, *RAA*, 11, 1298
Fan, Z., Huang, Y. F., Li, J. Z., Zhou, X., Ma, J., & Zhao, Y. H. 2012, *RAA*, 12, 829
Galleti, S., Federici, L., Bellazzini, M., Fusi Pecci, F., & Macrina, S. 2004, *A&A*, 416, 917
McConnachie, A. W., *et al.* 2009, *Nature*, 461, 66
Perina, S., Cohen, J. G., Barmby, P., *et al.* 2010, *A&A*, 511, A23
Thomas, D., Maraston, C., & Johansson, J. 2011, *MNRAS*, 412, 2183
Worthey, G., Faber, S. M., Gonzalez, J. J., & Burstein, D. 1994, *ApJS*, 94, 687

Star clusters and black holes in galaxies across cosmic time
Proceedings IAU Symposium No. 312, 2014
Y. Meiron, S. Li, F.-K. Liu & R. Spurzem, eds.

© International Astronomical Union 2016
doi:10.1017/S1743921315007826

Evolution of binaries with compact objects in globular clusters

Natalia Ivanova

Dept. of Physics, University of Alberta, 11322-89 Ave, Edmonton, AB, T6G 2E7, Canada
email: `nata.ivanova@ualberta.ca`

Abstract. Dynamical interactions that take place between objects in dense stellar systems lead to frequent formation of exotic stellar objects, unusual binaries, and systems of higher multiplicity. They are most important for the formation of binaries with neutron stars and black holes, which are usually observationally revealed in mass-transferring binaries. Here we review the current understanding of compact object's retention, of the metallicity dependence on the formation of low-mass X-ray binaries with neutron stars, and how mass-transferring binaries with a black hole and a white dwarf can be formed. We discuss as well one old unsolved puzzle and two new puzzles posed by recent observations: what descendants do ultra-compact X-ray binaries produce, how are very compact triples formed, and how can black hole low-mass X-ray binaries acquire non-degenerate companions?

Keywords. binaries : close, globular clusters: general

1. Introduction

A low-mass X-ray binary (LMXB) is a binary star consisting of a neutron star (NS) or a black hole (BH) accretor, with a less-massive donor that fills its Roche lobe (RL). A donor can be a main-sequence (MS) star, a white dwarf (WD) or a red giant (RG). An isolated binary (a field binary) can become an LMXB via several possible scenarios (e.g., Bhattacharya & van den Heuvel 1991; Verbunt 1993; Tauris & van den Heuvel 2006). A typical formation scenario starts with a common envelope event, where the initially most massive star overfills its RL. The expulsion of the formed common envelope leads to the binary orbit tightening. Later, the initially most massive star explodes in a core-collapse (CC) Ib or Ic supernova and the compact object is formed. Further orbit's shrinkage can take place due to tides, magnetic braking and gravitational waves in, likely, an eccentric binary; a second episode of a common envelope event can take place as well. When the initially less massive stars overfills its RL, the system is revealed as an X-ray source. It is essential that a typical formation scenario of a field LMXB includes several phases or processes which are not yet fully understood – tides, common envelope events, magnetic braking and natal supernova kicks (see, e.g., reviews Eggleton 2006; Tauris & van den Heuvel 2006; Zahn 2008; Knigge *et al.* 2011; Ivanova *et al.* 2013). We further distinguish LMXBs by how they appear – whether they have a stable disk accretion and therefore can be seen persistently, or if they raise drastically their X-ray luminosity in outbursts, and therefore are transient sources.

Globular clusters (GCs) contain about two orders of magnitude more LMXBs per stellar mass than the Milky Way (MW) population contains of "native" LMXBs (Grindlay 1984; Verbunt & van den Heuvel 1995; Bildsten & Deloye 2004). This overabundance has been attributed to the role that dynamical encounters play in the formation of LMXBs in GCs. Indeed, studies have shown strong positive correlation between the stellar encounter rate and the number of close X-ray binaries in GCs (Pooley *et al.* 2003; Bahramian *et al.*

2013). Additionally, the *metallicity dependence* was found – that a metal-rich GC in the Galaxy and in M31 is ~ 3 times more likely to contain a bright LMXB than a metal-poor GC (Grindlay 1993; Bellazzini *et al.* 1995). Observations of GCs around distant galaxies have also shown that the fraction of GCs hosting LMXBs is larger by a factor of 3 in metal-rich than in metal-poor GCs (Kundu *et al.* 2002; Sarazin *et al.* 2003; Kim *et al.* 2006; Sivakoff *et al.* 2007; Paolillo *et al.* 2011; Kim *et al.* 2013).

It is interesting that most of the very uncertain physics involved in the formation of "native" field LMXBs – e.g., common envelope phase, binary modification due to natal supernova kicks, tides – is either not involved in the formation scenario of a dynamically formed LMXB, or affects the formation scenario with a different, often lesser, impact. For example, supernova kicks affect GCs LMXBs mainly by reducing the retention of compact objects in GCs after their formation (Verbunt & Hut 1987; Drukier 1996). The formation path of an LMXB in a GC, on the other hand, can take a short-cut by pairing a NS or a BH via a dynamical encounter. To understand the puzzles that observations bring us, we need to analyze how the dynamical formation channels – binary exchanges, tidal captures, physical collisions and triple formations – perform with different type of donors.

2. Formation and retention of compact objects

The mean three-dimensional pulsar birth velocity is 400 km/s, and a pulsar two-dimensional speed can be as large as 1600 km/s (Hobbs *et al.* 2005). The escape velocity from a GC is an order of magnitude less than pulsars' mean kick velocities. If all NSs are born with kicks we derived for pulsars, only a tiny fraction of them could remain in a GC (Pfahl *et al.* 2002). Indeed, a typical GC with current mass of $2 \times 10^5 M_\odot$ and Kroupa's initial mass function (Kroupa 2002), forms about 3000 NSs. If the escape velocity is 40km/s, only one NS can be retained if all stars were single at birth (Ivanova *et al.* 2008b). Being born in a binary changes the post-kick system's space velocity (e.g., Brandt & Podsiadlowski 1995; Pfahl *et al.* 2002), and if all stars would be initially in binaries, a $2 \times 10^5 M_\odot$ GC could retain 15 NSs (Ivanova *et al.* 2008b). This is still too little to explain the observations. Indeed, Ter 5, which is about $2 \times 10^6 M_\odot$, is estimated to contain 150 recycled pulsars (Bagchi *et al.* 2011).

The problem is considered now to be resolved by assuming that the observed high pulsar velocities are only related to the case of a CC supernova. Indeed, there is another formation channel for NSs, electron capture (EC) supernova, that takes places when a degenerate ONeMg core reaches $1.38M_\odot$ (Miyaji *et al.* 1980; Nomoto 1984; Timmes & Woosley 1992). EC supernova takes place in a narrow range of masses in single stars, where this range depends slightly on metallicity and the used stellar code (for example, this range is $7.7 - 8.3M_\odot$ at $Z = 0.02$ and $6.2 - 6.8M_\odot$ at Z=0.0005 using the STARS/ev code, Pols *et al.* 1995). In binary stars, the range of initial stellar masses that can produce EC supernova is larger, due to mass transfer history (Podsiadlowski *et al.* 2004). In addition, for stars in binaries, EC NSs can be formed due to accretion on a companion WDs, or due to a merger of two WDs. As a result, in field, 10-15% of all NSs could be formed via various EC channels (Ivanova *et al.* 2008b). Probably the most famous example is Crab Supernova (Kitaura *et al.* 2006; Tominaga *et al.* 2013). The most important feature of EC supernovae for GCs is that the explosion is much weaker (e.g., Dessart *et al.* 2006), and hence their kicks are believed to be weaker as well. Under assumption of lower kicks, most of retained NSs in a GC come from different EC supernovae channels. The ratio of CC NSs to EC NSs in a GC is about 1 to 30-200 instead of about 10 to 1 in the field (Ivanova *et al.* 2008b). Since post-EC NS masses are low, $1.22 - 1.27 M_\odot$, a low-mass dominated NSs mass function can be expected in GCs.

The understanding of BHs retention has recently drastically changed. It can be estimated that each $150 - 200\ M\odot$ of the current, "aged", stellar mass, have produced a BH in the past. This is assuming Kroupa's initial mass function and no mass loss from a GC except that due to stellar evolution. Due to lower kicks, the retention fraction immediately after formation is much higher than for NSs, 30-40%, if the escape velocity from a GC is 50 km/s (Belczynski *et al.* 2006). It was considered previously that equipartition between the stars in a GC would lead to the formation of a BH sub-cluster in a center of a GC (Spitzer instability), followed by rapid BH population depletion via an ejection of nearly all BHs (Spitzer 1969; Kulkarni *et al.* 1993). However, recent detailed numerical simulations of BH subclusters have shown that up to 20% of the BHs may remain in massive BH subclusters, and these subclusters do not reach equipartition (O'Leary *et al.* 2006). Monte Carlo simulations of a whole GC have shown that up to 25% of initial BHs can remain and participate in interactions with other stars (Downing *et al.* 2010), or that even more than a half of formed BHs can remain, with no evidence for the Spitzer instability (Morscher *et al.* 2013). It can be now assumed that a large number of initially retained BHs can remain in GCs, 10-60%, depending on cluster mass and its initial virial radius (Morscher *et al.* 2014).

3. Formation channels and metallicity dependence

The observed metallicity dependence for bright LMXBs can be interpreted as a combination of the metallicity dependence due to the encounter rates, and the metallicity dependence due to the mass transfer – whether the formed LMXBs appear as persistent or transient sources, how bright they are when 'on', and how long they can remain as bright sources.

Typical dynamical formation channels of a binary with a compact object are: (i) via binary exchanges (BEs), (ii) physical collisions (PCs) with a giant or (iii) via tidal captures (TCs) (e.g., see Verbunt & Lewin 2006, and references therein). A dynamical event provides a *direct* LMXB formation if a newly formed binary can start the mass transfer within the Hubble time while evolved in isolation. Not all formed binaries with a NS and a BH will become LMXBs directly, but the formed binaries can serve as seed binaries for further dynamical encounters (hardening, triples formation) and eventually can also start the mass transfer.

For MW GCs, bright LMXBs are those that have their X-ray luminosity $L_X > 10^{36}$ erg s^{-1}. 18 bright transient or persistent LMXBs have been observed in MW GCs. For the 13 known by 2004, a compilation is provided in Verbunt & Lewin 2006. Two more transients were found in Ter 5 (Bordas *et al.* 2010; Bahramian *et al.* 2014), one more in NGC 6440 (Heinke *et al.* 2010), and the first transient was found in each of two GCs, NGC 6388 and M 28 (Bozzo *et al.* 2011; Eckert *et al.* 2013; Papitto *et al.* 2013).

The separation between metal-rich (also known as red) and metal-poor (also known as blue) MW GCs is not strictly defined, and is usually taken at [Fe/H] $= -1$ (a gap in the distribution of metallicities is observed between [Fe/H] $= -1.2$ and [Fe/H] $= -0.8$, and this gap disappears if the sample is limited to low-reddening clusters, Vanderbeke *et al.* 2014a,b). Here, we will refer to the GCs that have [Fe/H] between -1.5 and -0.5 as "intermediate" metallicity GCs, if [Fe/H] < -1.5 as "distinctly metal-poor" GCs, and if [Fe/H] > -0.5 as "distinctly metal-rich" GCs.

From 18 known bright MW GCs LMXBs only 2 are in "distinctly metal-poor" GCs (both are in M 15), 8 are in the "intermediate" metallicity GCs (LMXBs in M 28,

NGC 1851, NGC 6388, NGC 6652, NGC 6712, Ter 1, Ter 2, Ter 6) and another 8 are in "distinctly metal-rich" GCs (LMXBs in Ter 5, NGC 6440, NGC 6441, NGC 6624, Lil 1), where two GCs contain more than one bright LMXB (Ter 5 and NGC 6440).

If we look at "classes" of bright MW LMXBs by their donor-type (note that this is also strongly linked to their orbital periods), we can see that 6 likely have a WD donor (Verbunt & Lewin 2006; Zurek *et al.* 2009; Altamirano *et al.* 2010). These binaries are called ultra-compact X-ray binaries (UCXBs). 3 of these UCXBs are in "intermediate" GCs (NGC 1851, NGC 6712, Ter 2), one in a "distinctly metal-poor" GC (M 15) and two in "distinctly metal-rich" GCs (NGC 6624, NGC 6440). 2 of bright MW LMXBs likely have MS donors, both are persistent and are located in GCs with [Fe/H] > -1 (NGC 6441, Bałucińska-Church *et al.* 2004, NGC 6652, Engel *et al.* 2012). 5 of bright LMXBs have most likely either a subgiant or a giant companion (Verbunt & Lewin 2006; Testa *et al.* 2012; Papitto *et al.* 2013), one in a "distinctly metal-poor" GC (M 15), two in "distinctly metal-rich" GCs (Ter 5, NGC 6440) and two are in the "intermediate" GCs (M 28, Ter 6). In addition, 5 bright LMXBs are transients where neither the companion nor orbital period has yet been determined, all of which are located in metal-rich GCs with [Fe/H] $\gtrsim -1$. Among LMXBs where we have some knowledge of the donor, most are persistent (8 persistent versus 5 transient LMXBs). Transient sources tend to have a longer orbital period, both for non-degenerate donors (4 out of 5 LMXBs with giant donors are transient, while all LMXBs with MS donors are persistent) and degenerate donors (the UCXB with the largest orbital period, in NGC 6440, is transient).

As we can see, for MW GCs, the 3 times greater probability of hosting an LMXB by a red GC than by a blue GC is *cumulative*. The ratio would change depending on what is defined as a metal-poor and a metal-rich GC, and is not satisfied for any class of donors separately. The most striking difference is for LMXBs with MS donors, but this is based on two LMXBs only.

Studies of extragalactic GCs have shown a similar preference for LMXBs to reside in metal-rich GCs. The ratio was found to hold across the range of X-ray luminosities, from 2×10^{37} erg s^{-1} to 5×10^{38} erg s^{-1} (Kim *et al.* 2013). For even higher X-ray luminosity, where the accretors are likely BHs, this ratio is less certain, but still is above one, $2.5^{+0.9}_{-1.1}$. There, it also was shown that the metallicity effect is not affected by other factors such as stellar age, GC mass, stellar encounter rate, and galacto-centric distance. The caveat in making a link between MW GCs LMXBs and extragalactic ones is that the the boundary between blue and red GCs is not the same, and for extragalactic GCs is done using colors (e.g., Brodie & Strader 2006), not [Fe/H] as for MW GCs. The relation between colors and [Fe/H] is however non-linear (Vanderbeke *et al.* 2014b).

It is important also that the population of detected extragalactic LMXBs is not similar to the population of bright GC LMXBs in our Milky Way. Indeed, in the Galaxy, the brightest observed GC UCXB 4U1820-303, in NGC 6624, has $L_x \approx 4 - 7 \times 10^{37}$ erg s^{-1} (Giacconi *et al.* 1974; van der Klis *et al.* 1993) (we note that this LMXBs could be special in other respects too, see §4). In extragalactic GC LMXBs, statistically significant studies were done for LMXBs above $L_x \approx 2 \times 10^{37}$ erg s^{-1} (Kim *et al.* 2013). Different ranges of X-ray luminosity are dominated by different donors (e.g., Fragos *et al.* 2008). For example, mass transfer in binaries with MS donors is driven by magnetic braking, and rarely produce X-ray luminosity $\gtrsim 2 \times 10^{37}$ erg s^{-1} (Fragos *et al.* 2008, 2009). Hence, MS donors most likely are not present among the X-ray sources that form X-ray luminosity functions of extragalactic GCs.

Among LMXBs, the evolution of UCXBs is best understood; predominantly they will be seen as bright persistent sources (e.g., Deloye & Bildsten 2003; Bildsten & Deloye 2004), although there are some discrepancies between theoretically expected and observationally derived mass-transfer rates for UCXBs with longer orbital periods (van Haaften *et al.* 2012b; Cartwright *et al.* 2013; Heinke *et al.* 2013).

The strongest uncertainty for extragalactic GC LMXBs is coming from our limited knowledge of the duty cycles in LMXBs with giant donors, and their luminosity during outburst (see overview of the problem in Fragos *et al.* 2008). Further, their outburst duration can be longer than the history of X-ray astronomy, and hence observationally they would be known as persistent. There are several X-ray binaries in the galaxy which confirm this suspicion as their outburst duration is at least 20 years – for example, GRS 1915+105 which turned on 1992 and has been active ever since (Castro-Tirado *et al.* 1992; Deegan *et al.* 2009), or the X-ray binary in Ter 1, which was active for about 20 years and then turned off between 1996 and 1999 (Guainazzi *et al.* 1999).

LMXBs with RG donors are most likely to be formed via BE encounters; the evolutionary expansion of a giant brings the system to the start of a mass transfer. The BEs formation channel does not have a metallicity dependence for the encounter rates (for GCs which otherwise are "dynamical twins"). The mass transfer slightly favors the appearance of metal-rich LMXBs – those can have the short duration of a persistent mass transfer, while metal-poor LMXBs with RG donors are likely always in quiescence (Ivanova *et al.* 2012). It is interesting that the observations of MW GCs show the opposite situation to theoretical expectations: the only persistent RG-LMXB is located in a metal-poor cluster M 15, although this can be related to the uncertainty in our understanding of the duty cycles.

For LMXBs with MS donors, the formation mechanisms include BEs and TCs. TCs, unlike BEs, can provide direct LMXB formation in most of the formed binaries, though cumulatively are less efficient for LMXBs formation than BEs (Ivanova *et al.* 2008b). In close NS-MS binaries, the most effective mechanism of orbital shrinkage is the magnetic braking, which operates only in stars with an outer convective zone. TCs are also more effective with stars that are convective (Portegies Zwart & Meinen 1993). However, most-massive MS stars in metal-poor GCs lack surface convective zone (Ivanova 2006). This slight difference between stars which are convective at the surface, and those which are radiative, for a small mass-range, affects the following: (i) the rate of the orbital shrinkage, where MS-NS binaries with a larger post-BE or post-TC orbital period can start the MT if their MS star is metal-rich; (ii) the formation rate via TCs, which are more efficient in metal-rich GCs; and (iii) metal-poor MS-LMXBs are expected to be in quiescence, where persistent MS-LMXBs are expected to be observed only in metal-rich globular clusters. The dynamically formed NS-MS binaries that would have started the mass transfer if the donor is metal-rich and have failed to start the mass-transfer with a metal-poor donor, will start the MT when the donor leaves the MS.

The best-understood formation mechanism of GC LMXBs is for UCXBs – via PC of a NS with a RG (as was initially proposed in Verbunt 1987, and see most recent details in Ivanova *et al.* 2005; Lombardi *et al.* 2006). If PC is between a NS and a subgiant, it often leads to a direct LMXB formation (Ivanova *et al.* 2005). In UCXBs, post-PC binary orbital evolution is solely due to angular momentum loss via gravitational wave radiation, same mechanism operates during the mass transfer, and there is no known effects through which the metallicity could affect the appearance of the X-ray binary during the mass transfer. The theoretically predicted number of UCXBs is consistent with the observed bright LMXBs in MW GCs, in case if at least several per cent of ever formed NSs is retained Ivanova *et al.* (2005).

In extragalactic GC LMXBs, which are generally more luminous than "bright" LMXBs in MW GCs, the potential donors are likely RGs and WDs. For neither of the donors, as discussed above, metallicity affects the formation rate *per giant*. The dynamical properties of blue and red extragalactic GCs in the studied sample were found to be statistically identical (Kim *et al.* 2013). The total formation rate depends however also on the number of encounter participants, which are NSs and RGs. Theoretically, it is expected that slightly more of NSs can be formed in a metal-poor GC than in a similar metal-rich GC (Ivanova *et al.* 2008b) – this trend is opposite to the observed metallicity dependence.

From stellar evolution, we know that metal-poor MS stars evolve faster than metal-rich MS stars of the same mass. Therefore, metal-poor GCs of the same age have stars of a lower mass at their MS turn-off. Lifetime of the RG stage is shorter for metal-poor stars as well. Since RGs live shorter, a metal-poor GCs should contain RGs that have a smaller mass range than RGs in a metal-rich GCs. If one assumes that GCs of different metallicities had the same initial mass function, detailed calculations confirm that the fraction of RG stars that a metal-poor GC can have is about twice less than a metal-rich GC has (Ivanova *et al.* 2012). This directly affects the encounter rates, and increases it in a metal-rich GC, compared to a metal-poor GC, by a factor of two.

The mass of the RG also plays a role: first, the expulsion of a RG envelope, which is more massive in a metal-rich RG, results in a smaller post-PC separation, and in a direct LMXB formation in a larger fraction of encounters; and second, the effective cross-section – how likely each RG is to participate in a dynamical encounter – is higher in metal-rich GCs. In total, this results in about 3 times more frequent formation of UCXBs in metal-rich clusters than in metal-poor, and can explain the observed metallicity dependence (Ivanova *et al.* 2012). The RG lifetime affects RG-LMXBS as well, as the duration of a RG-NS mass transfer is shorter in metal-poor GCs.

Between all GC LMXB classes, the best theoretical understanding, as well as the best agreement on numbers with the observations of MW GC LMXBs, we have is for the formation and evolution of UCXBs. It is thought that millisecond pulsars (MSPs) are related to LMXBs, and transitional systems between LMXBs and binary MSPs (bMSPs) are now known (e.g., Wijnands & van der Klis 1998; Archibald *et al.* 2009; Papitto *et al.* 2013). Simulations of GCs have shown good agreement between the number and period distribution for all theoretically produced radio bMSPs, but with the striking exception for radio bMSPs presumably produced by UCXBs (Ivanova *et al.* 2008a). The lifetime of an UCXB at $L_x > 10^{36}$ erg s^{-1} is $\tau \sim 10^8$ yr. This implies that for each currently observed bright UCXB, 100 time more bMSPs should have been created, or about 600 bMSPs with a small orbital period and a low-mass companion should have been detected in MW GCs. However, we do not observe such a radio bMSP population in GCs, nor in the overall number of MSPs, nor for expected companion-mass and orbital period parameter space for bMPSs. This problem is by no means confined only to GCs, there is a lack of post-UCXBs radio bMSPs in the galactic "field" population as well (Deloye 2008).

The interesting case that can provide in future a clue on the fate of post-UCXB systems is the discovery of PSR J1719-1438 – a "field" MSP which has the orbital period of 2.2 hour, its companion mass is about that of Jupiter, but the minimum companion density is 23 g cm^{-3}, suggesting that it may be an ultralow-mass carbon white dwarf (Bailes *et al.* 2011). This system is very puzzling, as it takes more than a Hubble time to get to this period under conventional UCXB evolution (van Haaften *et al.* 2012a), while this evolution can be deemed possible if additional donor wind, irradiation and donor's evaporation are taken into account (Benvenuto *et al.* 2012). While this may explain why we do not see post-UCXB system as bMSPs, it still does not explain the mismatch in the total number of MSPs. On the other hand, it is plausible that NSs that were spun up in an UCXB binary are less likely to be detected in radio, and instead can be detected in γ-rays, as, for example, in case of PSR J1311-3430 that has a 93 min orbital period and very low-mass helium companion (Pletsch *et al.* 2012; Romani *et al.* 2012).

4. Role of triples

In a hierarchical triple, a distant third body exerts tidal forces on the inner binary. As a result, there is a cyclic exchange of the angular momentum between an inner binary and a third body, causing variations in the eccentricity and inclination of the stars orbits (Kozai 1962; Ford *et al.* 2000; Blaes *et al.* 2002). If Kozai mechanism is coupled with dissipative tidal friction (Kozai Cycle with Tidal Friction, KZTF), the inner binary can be driven to start the RL overflow of one of the companions, with the subsequent either a merger or a start of the stable mass transfer (Eggleton & Kisseleva-Eggleton 2006; Fabrycky & Tremaine 2007). If the timescale of KZTF is shorter than the timescale of dynamical encounters, KZTF can provide a formation channel for LMXBs in GCs.

The interesting case is LMXB 4U 1820-303 in NGC 6624. Binary orbital period of this system is ~ 685s (Stella *et al.* 1987; Anderson *et al.* 1997). Theoretically predicted period increase for this system, if it is a binary, is $\dot{P}/P > 8 \times 10^{-8}$ yr^{-1} (Rappaport *et al.* 1987). However, from observations, the orbital period is decreasing as $(\dot{P}/P)_{\text{obs}} = -(5.3\pm0.3)\times10^{-8}$yr^{-1} (van der Klis *et al.* 1993; Peuten *et al.* 2014). In addition, 4U 1820-303 has the luminosity variation by a factor of ~ 2 at a super-orbital period $P \sim 170$d (Chou & Grindlay 2001). Chou & Grindlay (2001) have suggested that this system could be a hierarchical triple with a 1.1 day outer orbital period.

Prodan & Murray (2012) have found that they can explain this system to be a triple, where the inner binary is composed of $1.4M_\odot$ NS and $0.067M_\odot$ WD, and the third companion has mass of $0.55M_\odot$. However, the outer semi-major axis is small, $1.5R_\odot$, and the outer orbital period is only 0.15d. The re-analysis of simulations performed in Ivanova *et al.* (2008b) have shown that none of the dynamically formed triples with a NS was anywhere close to the observed compactness, with the tightest dynamically formed triple having its outer orbital period about a day, and most of dynamically formed triples were even much wider. This should come as no surprise, as an inner binary in a dynamically formed hierarchical triple would be rarely substantially tighter than the tightest of the two binaries that have participated in the encounter. Too tight binaries both have a small encounter rate and can be easily destroyed via collisions during an encounter (Fregeau *et al.* 2004). A possible explanation is that this hierarchical triple, formed initially dynamically, was substantially wider than now, and then was tightened through either *external* common envelope event, initiated by an outer companion, or a *double* common envelope event. The latter is a bit less likely, also in that case the seed inner WD-NS binary could not have been formed via a PC.

Prodan & Murray (2014) had also suggested that 3 more GC UCXBs could be triple systems. This, if confirmed, would bring the fraction of triple UCXBs to 2/3 of all MW GC UCXBs. It also may suggest that yet unknown formation channel is as efficient as the current favorite formation channel of UCXBs – PC, or that most of NS-WD binary, formed initially via a PC, form subsequently a triple system. How plausible is that? Theoretically expected formation rates suggests that each NS binary has a 5% chance to form a hierarchically stable triple in a dense GC like 47 Tuc per Gyr, and as much as 15% chance per Gyr in a cluster like Ter 5. So, in a dense cluster, most of seed NS binaries could become a member of a triple system at some point (Ivanova 2008, also reanalysis of simulations presented in Ivanova *et al.* 2008b). The formation of the triples with very short outer orbital periods is however as difficult dynamically as for the described above 4U 1820-303.

5. LMXBs with BH accretors

Observational frequency of very bright extragalactic GC LMXBs, with $L_x \gtrsim 10^{39}$ erg s^{-1} indicating possible BH accretors, is about $0.7 \div 2 \times 10^{-9}$ per $M_\odot$ in massive GCs with an average $M_V \approx -9$ (Kim $et\ al.$ 2006; Sivakoff $et\ al.$ 2007; Humphrey & Buote 2008; Kim $et\ al.$ 2013). BH LMXBs formed via binary exchanges are likely to have low duty cycles and not be detected as bright LMXBs (Kalogera $et\ al.$ 2004; Fragos $et\ al.$ 2008, 2009). TCs by a BH likely destroy the capturing star (Kalogera $et\ al.$ 2004). This leaves as most likely donors WDs. A BH-WD LMXB spends an order of 5×10^5 yr at $L_x \gtrsim 10^{39}$ erg s^{-1}. Assuming that about 10% of all formed BHs is retained, the observationally inferred formation rate of LMXBs is $\sim 4 \times 10^{-3}$ per Gyr per each retained BH.

It has been shown that PCs with giants, the most effective mechanism to form NS-WD LMXBs, can not provide direct formation of BH-WD LMXBs at the observationally inferred rate. A dynamically formed binary with a BH is several times more massive than any non-BH binary in a GC. If a dynamically formed BH-WD binary participates in a binary-binary encounter, the chance to form a hierarchically stable triple is substantial. If Kozai timescale is short compared to the characteristic time this triple has before its next dynamical encounter, triple induced mass transfer can transform the inner BH-WD binary into LMXBs (Ivanova $et\ al.$ 2010). Interestingly, Maccarone $et\ al.$ (2010) have found a large X-ray flux variation in RZ 2109, such that this source is consistent with being in a triple. RZ 2109 was the first discovered extragalactic GC X-ray source which is believed to contain a stellar mass BH (Zepf $et\ al.$ 2008).

The observations of bright BH LMXBs in extragalactic GCs can be extrapolated to our MW GCs. It is important that after the "bright" stage at $L_x \gtrsim 10^{39}$ erg s^{-1}, an ultra-compact X-ray BH-WD binary will remain a mass-transferring system for many gigayears. With uniform formation rate, and average GC age of 10 Gyr, an average BH-WD LMXB will exist $\sim 10^4$ times longer than a bright BH-WD LMXB, and will spend most of its life at extremely low mass transfer rates. The expected frequency per dense and massive MW GCs is then of order 10^{-5} per $M_\odot$. This implies that almost every massive GC in the Galaxy might contain a very faint BH-WD LMXB! Note that only 7 MW GCs are above $M_V = -9$, to have BHs retention comparable to massive extragalactic GCs.

The recent radio observations of MW GCs clusters have indeed indicated that some of them contain faint potential BH-LMXBs. Five candidates are under investigation – two LMXBs in M22 (Strader $et\ al.$ 2012), one in M62 (Chomiuk $et\ al.$ 2013), in 47 Tuc (Miller-Jones $et\ al.$ in prep.) and in one more GC (Shishkovsky $et\ al.$ in prep). Interesting that only in two cases WD companions are plausible, in M22 and in 47 Tuc. The rest of the companions are likely non-degenerate – MS stars, or giants, or "red stragglers". A BH LMXB with a non-degenerate donor can continue mass transfer at a very low rate for several gigayears, comparable to that of BH-WD binaries, hinting that the dynamical formation rate of BH binaries with non-degenerate donors is comparable to that of BH UCXBs. The formation booster with help of triples has not yet been shown to work effectively for non-degenerate companions. Therefore, the puzzle of the formation of BH LMXBs with non-degenerate companions in GCs remains to be explained.

References

Altamirano D. $et\ al.$, 2010, $ApJL$, 712, L58
Anderson S. F. $et\ al.$, 1997, $ApJL$, 482, L69
Archibald A. M. $et\ al.$, 2009, Science, 324, 1411
Bagchi M., Lorimer D. R., & Chennamangalam J., 2011, $MNRAS$, 418, 477
Bahramian A. $et\ al.$, 2014, ApJ, 780, 127

Bahramian A., Heinke C. O., Sivakoff G. R., & Gladstone J. C., 2013, *ApJ*, 766, 136
Bailes M. *et al.*, 2011, Science, 333, 1717
Bałucińska-Church M., Church M. J., & Smale A. P., 2004, *MNRAS*, 347, 334
Belczynski K., Sadowski A., Rasio F. A., & Bulik T., 2006, *ApJ*, 650, 303
Bellazzini M. *et al.*, 1995, *ApJ*, 439, 687
Benvenuto O. G., De Vito M. A., & Horvath J. E., 2012, *ApJL*, 753, L33
Bhattacharya D. & van den Heuvel E. P. J., 1991, *Phys.Rep.*, 203, 1
Bildsten L. & Deloye C. J., 2004, *ApJL*, 607, L119
Blaes O., Lee M. H., & Socrates A., 2002, *ApJ*, 578, 775
Bordas P. *et al.*, 2010, *The Astronomer's Telegram*, 2919, 1
Bozzo E. *et al.*, 2011, *A&A*, 535, L1
Brandt N. & Podsiadlowski P., 1995, *MNRAS*, 274, 461
Brodie J. P. & Strader J., 2006, *ARAA*, 44, 193
Cartwright T. F. *ct al.*, 2013, *ApJ*, 768, 183
Castro-Tirado A. J., Brandt S., & Lund N., 1992, *IAU circular*, 5590, 2
Chomiuk L. *et al.*, 2013, *ApJ*, 777, 69
Chou Y. & Grindlay J. E., 2001, *ApJ*, 563, 934
Deegan P., Combet C., & Wynn G. A., 2009, *MNRAS*, 400, 1337
Deloye C. J., 2008, in *AIP Conference Series*, Vol. 983, 40 Years of Pulsars, Bassa C., Wang Z., Cumming A., Kaspi V. M., eds., pp. 501–509
Deloye C. J. & Bildsten L., 2003, *ApJ*, 598, 1217
Dessart L. *et al.*, 2006, *ApJ*, 644, 1063
Downing J. M. B., Benacquista M. J., Giersz M., & Spurzem R., 2010, *MNRAS*, 407, 1946
Drukier G. A., 1996, *MNRAS*, 280, 498
Eckert D. *et al.*, 2013, *The Astronomer's Telegram*, 4925, 1
Eggleton P., 2006, *Evolutionary Processes in Binary and Multiple Stars*
Eggleton P. P. & Kisseleva-Eggleton L., 2006, *Ap&SS*, 304, 75
Engel M. C. *et al.*, 2012, *ApJ*, 747, 119
Fabrycky D. & Tremaine S., 2007, *ApJ*, 669, 1298
Ford E. B., Kozinsky B., & Rasio F. A., 2000, *ApJ*, 535, 385
Fragos T. *et al.*, 2008, *ApJ*, 683, 346
Fragos T. *et al.*, 2009, *ApJL*, 702, L143
Fregeau J. M., Cheung P., Portegies Zwart S. F., & Rasio F. A., 2004, *MNRAS*, 352, 1
Giacconi R. *et al.*, 1974, *ApJ Supp*, 27, 37
Grindlay J. E., 1984, *Advances in Space Research*, 3, 19
Grindlay J. E., 1993, in *ASP Conference Series*, Vol. 48, The Globular Cluster-Galaxy Connection, Smith G. H., Brodie J. P., eds., p. 156
Guainazzi M., Parmar A. N., & Oosterbroek T., 1999, *A&A*, 349, 819
Heinke C. O. *et al.*, 2010, *ApJ*, 714, 894
Heinke C. O. *et al.*, 2013, *ApJ*, 768, 184
Hobbs G., Lorimer D. R., Lyne A. G., & Kramer M., 2005, *MNRAS*, 360, 974
Humphrey P. J. & Buote D. A., 2008, *ApJ*, 689, 983
Ivanova N., 2006, *ApJ*, 636, 979
Ivanova N., 2008, in *Multiple Stars Across the H-R Diagram*, Hubrig S., Petr-Gotzens M., Tokovinin A., eds., p. 101
Ivanova N. *et al.*, 2010, *ApJ*, 717, 948
Ivanova N. *et al.*, 2012, *ApJL*, 760, L24
Ivanova N., Heinke C. O., & Rasio F. A., 2008a, in *AIP Conference Series*, Vol. 983, 40 Years of Pulsars, Bassa C., Wang Z., Cumming A., Kaspi V. M., eds., pp. 442–447
Ivanova N. *et al.*, 2008b, *MNRAS*, 386, 553
Ivanova N. *et al.*, 2013, *Astronomy and Astrophysics Review*, 21, 59
Ivanova N. *et al.*, 2005, *ApJL*, 621, L109
Kalogera V., King A. R., & Rasio F. A., 2004, *ApJL*, 601, L171
Kim D.-W. *et al.*, 2013, *ApJ*, 764, 98

Kim E. *et al.*, 2006, *ApJ*, 647, 276
Kitaura F. S., Janka H.-T., & Hillebrandt W., 2006, *A&A*, 450, 345
Knigge C., Baraffe I., & Patterson J., 2011, *ApJ Supp*, 194, 28
Kozai Y., 1962, *AJ*, 67, 591
Kroupa P., 2002, *Science*, 295, 82
Kulkarni S. R., Hut P., & McMillan S., 1993, *Nature*, 364, 421
Kundu A., Maccarone T. J., & Zepf S. E., 2002, *ApJL*, 574, L5
Lombardi, Jr. J. C. *et al.*, 2006, *ApJ*, 640, 441
Maccarone T. J., Kundu A., Zepf S. E., & Rhode K. L., 2010, *MNRAS*, 409, L84
Miyaji S., Nomoto K., Yokoi K., & Sugimoto D., 1980, *PASJ*, 32, 303
Morscher M. *et al.*, 2014, ArXiv e-prints
Morscher M., Umbreit S., Farr W. M., & Rasio F. A., 2013, *ApJL*, 763, L15
Nomoto K., 1984, *ApJ*, 277, 791
O'Leary R. M. *et al.*, 2006, *ApJ*, 637, 937
Paolillo M. *et al.*, 2011, *ApJ*, 736, 90
Papitto A. *et al.*, 2013, *Nature*, 501, 517
Peuten M., Brockamp M., Küpper A. H. W., & Kroupa P., 2014, *ApJ*, 795, 116
Pfahl E., Rappaport S., & Podsiadlowski P., 2002, *ApJ*, 573, 283
Pletsch H. J. *et al.*, 2012, Science, 338, 1314
Podsiadlowski P. *et al.*, 2004, *ApJ*, 612, 1044
Pols O. R., Tout C. A., Eggleton P. P., & Han Z., 1995, *MNRAS*, 274, 964
Pooley D. *et al.*, 2003, *ApJL*, 591, L131
Portegies Zwart S. F., & Meinen A. T., 1993, *A&A*, 280, 174
Prodan S. & Murray N., 2012, *ApJ*, 747, 4
Prodan S. & Murray N., 2014, ArXiv e-prints
Rappaport S., Ma C. P., Joss P. C., & Nelson L. A., 1987, *ApJ*, 322, 842
Romani R. W. *et al.*, 2012, *ApJL*, 760, L36
Sarazin C. L. *et al.*, 2003, *ApJ*, 595, 743
Sivakoff G. R. *et al.*, 2007, *ApJ*, 660, 1246
Spitzer, Jr. L., 1969, *ApJL*, 158, L139
Stella L., Priedhorsky W., & White N. E., 1987, *ApJL*, 312, L17
Strader J. *et al.*, 2012, *Nature*, 490, 71
Tauris T. M. & van den Heuvel E. P. J., 2006, *Formation and evolution of compact stellar X-ray sources*
Testa V. *et al.*, 2012, *A&A*, 547, A28
Timmes F. X. & Woosley S. E., 1992, *ApJ*, 396, 649
Tominaga N., Blinnikov S. I., & Nomoto K., 2013, *ApJL*, 771, L12
van der Klis M. *et al.*, 1993, *A&A*, 279, L21
van Haaften L. M., Nelemans G., Voss R., & Jonker P. G., 2012a, *A&A*, 541, A22
van Haaften L. M., Voss R., & Nelemans G., 2012b, *A&A*, 543, A121
Vanderbeke J. *et al.*, 2014a, *MNRAS*, 437, 1725
Vanderbeke J. *et al.*, 2014b, *MNRAS*, 437, 1734
Verbunt F., 1987, *ApJL*, 312, L23
Verbunt F., 1993, *ARAA*, 31, 93
Verbunt F. & Hut P., 1987, in *IAU Symposium, Vol. 125, The Origin and Evolution of Neutron Stars*, Helfand D. J., Huang J.-H., eds., p. 187
Verbunt F. & Lewin W. H. G., 2006, *Globular cluster X-ray sources*, Lewin W. H. G., van der Klis M., eds., pp. 341–379
Verbunt F. & van den Heuvel E. P. J., 1995, *X-ray Binaries*, 457
Wijnands R. & van der Klis M., 1998, *Nature*, 394, 344
Zahn J.-P., 2008, in *EAS Publications Series*, Vol. 29, EAS Publications Series, Goupil M.-J., Zahn J.-P., eds., pp. 67–90
Zepf S. E. *et al.*, 2008, *ApJL*, 683, L139
Zurek D. R. *et al.*, 2009, *ApJ*, 699, 1113

Star clusters and black holes in galaxies across cosmic time
Proceedings IAU Symposium No. 312, 2014
Y. Meiron, S. Li, F.-K. Liu & R. Spurzem, eds.
© International Astronomical Union 2016
doi:10.1017/S1743921315007838

Monte Carlo modeling of globular star clusters: many primordial binaries and IMBH formation

Mirek Giersz[1], Nathan Leigh[2], Michael Marks[3], Arkadiusz Hypki[1] and Abbas Askar[1]

[1]Nicolaus Copernicus Astronomical Centre, Polish Academy of Sciences, ul. Bartycka 18, 00-716 Warsaw, Poland
emails: `mig@camk.edu.pl`

[2]Department of Physics, University of Alberta, CCIS 4-183, Edmonton, AB T6G 2E1, Canada

[3]Helmholtz-Institut für Strahlen- und Kernphysik, Nussallee 14-16, D-53115, Bonn, Germany

Abstract. We will discuss the evolution of star clusters with a large initial binary fraction, up to 95%. The initial binary population is chosen to follow the invariant orbital-parameter distributions suggested by Kroupa (1995). The Monte Carlo MOCCA simulations of star cluster evolution are compared to the observations of Milone *et al.* (2012) for photometric binaries. It is demonstrated that the observed dependence on cluster mass of both the binary fraction and the ratio of the binary fractions inside and outside of the half mass radius are well recovered by the MOCCA simulations. This is due to a rapid decrease in the initial binary fraction due to the strong density-dependent destruction of wide binaries described by Marks, Kroupa & Oh (2011). We also discuss a new scenario for the formation of intermediate mass black holes in dense star clusters. In this scenario, intermediate mass black holes are formed as a result of dynamical interactions of hard binaries containing a stellar mass black hole, with other stars and binaries. We will discuss the necessary conditions to initiate the process of intermediate mass black hole formation and the dependence of its mass accretion rate on the global cluster properties.

Keywords. stellar dynamics – globular clusters: general – binaries: general – methods: numerical

1. Introduction

Recent high resolution observations of globular clusters (GC) provide a very detailed picture of their physical status and show complex phenomena connected with multiple stellar populations, binary evolution and the Galactic tidal field. Despite such great observational progress there are many theoretical uncertainties connected with the origins of GCs and the properties of their primordial binary populations. To bridge the gap between present-day observed binary properties and their properties at the time of cluster formation, we need to discriminate between different theories and models by means of numerical dynamical simulations of GCs. Based on then available observations of the late-type stellar binary population in the Galactic field, Kroupa (1995) suggested that, taking into account dynamical processing, the initial binary population in star clusters is largely invariant, in the sense that almost every star forms in a binary system with invariant formal distribution functions (due to energy and angular momentum conservation, the physics of molecular clouds, all of which are the same everywhere, except perhaps in very intense star bursts). These distribution functions are parent distribution functions, from which a particular case is discretized, or rendered, and their suggested invariance is tightly connected to the notion of an invariant IMF (Kroupa & Petr-Gotzens 2011).

These parent functions can have different properties in different mass ranges (brown dwarfs, late and early type stars). The Kroupa (1995) set-up of the primordial binary populations has been shown to work well for the late-type stellar population (solar-type stars and below) in young star forming regions (Marks & Kroupa 2012, Marks *et al.* 2014) and the Galactic field (Marks & Kroupa 2011). Such distribution functions are needed to initialize N-body models in order to study how young and old clusters evolve into the field and associations. The simulations of GCs described in this talk use the Kroupa (1995) initial binary population and were conducted by the Monte Carlo code MOCCA (Hypki & Giersz 2013, Giersz *et al.* 2013 and references therein). The results of the simulations were compared to observational data using the photometric binaries provided in Milone *et al.* (2012), and to the initial dissolution rate of primordial binaries found using N-body simulations in Marks, Kroupa & Oh (2011).

The presence of intermediate mass black holes (IMBH) in the cores of some GCs has been debated for a long time. There are many theoretical arguments in favor of the formation of IMBHs in the centers of GCs (e.g. Lützgendorf *et al.* 2013 and references therein), but there is yet no observational confirmation of the presence of any IMBH in any Galactic GCs. All proposed scenarios for the formation of IMBHs in GCs require special initial conditions: 1) the formation of very massive Population III stars (Madau & Rees 2001), 2) runaway merging of main sequence stars in young and very dense star clusters (Portegies Zwart *et al.* 2004, Gürkan *et al.* 2004), 3) accretion of residual gas on stellar mass black holes (BH) formed from the first generation stars (Leigh *et al.* 2013a) and 4) tidally stripped parent galaxy cores of neighboring dwarf galaxies (e.g. Baumgardt *et al.* 2003). A new scenario for IMBH formation is proposed in this talk and does not need very specific initial conditions. An IMBH is built-up only via binary dynamical interactions and mass transfer (also induced by dynamical interactions) in binaries.

2. Method

The MOCCA code used for the star cluster simulations presented here is the Monte Carlo code based on Hénon's implementation of the Monte Carlo method (Hénon 1971), which was further substantially developed by Stodółkiewicz in the early eighties (Stodolkiewicz 1986). This method can be regarded as a statistical way of solving the Fokker-Planck equation. A star cluster is treated as a set of spherical shells, each of which represents an individual object: star, binary or a group of the same objects. Relaxation of a given object with all other objects in the system is approximated via the interaction of two neighboring shells. There are two independently developed Monte Carlo codes: by Fred Rasio's group (Morscher *et al.* 2014 and reference therein) and by my group (Giersz *et al.* 2013 and reference therein). Actually, there are two more Monte Carlo codes, which were recently developed by Vasiliev (2015) for non spherical stellar systems and by Sollima & Mastrobuono Battisti (2014) for a realistic treatment of the tidal field.

The basic assumptions behind the Monte Carlo method are the following: 1) spherical symmetry, which makes it easy to quickly compute the gravitational potential and stellar orbits at any place in the system. That is a very severe assumption, e.g. the evolution of a system with rotation cannot be investigated. In specified spherically symmetric potentials every star is characterized by its mass, energy and angular momentum. Each star moves on a rosette orbit. There is no need for integration of an orbit in the Monte Carlo method, since it is easy to calculate the position of each star along its orbit. In this way, the Monte Carlo method is very fast; 2) cluster evolution is driven by two-body relaxation, and the time step at each position in the system is proportional to the local relaxation time. That is the second reason why the Monte Carlo method is so fast. Generally, we cannot follow any physical processes with characteristic time scales much shorter then the relaxation time.

The MOCCA code (Giersz *et al.* 2013) is a "kitchen sink" code, which is able to follow most physical processes that are important during star cluster dynamical evolution. For stellar and binary evolution, Jarrod Hurley's BSE code is used (Hurley *et al.* 2000, Hurley *et al.* 2002), for the scattering experiments John Fregeau's Fewbody code is used (Fregeau *et al.* 2004), and the realistic description of escape processes in tidally limited clusters is done on the basis of the Fukushige & Heggie (2000) theory. The MOCCA code provides as many details as N-body codes. It can follow the time evolution and movement of particular objects. The MOCCA code is extremely fast. It needs about a day to complete the evolution of a realistic globular cluster. So, instead of just one N-body model, hundreds or thousands of models can be computed with different initial conditions. The MOCCA code is ideal either for dynamical models of a particular cluster or for large surveys.

3. Results

The results presented in this talk came as a byproduct from two projects carried out with Nathan Leigh and other collaborators. The aim of the first project (Leigh *et al.* 2013b) was to explain an observed anti-correlation between cluster mass and binary fraction (Milone *et al.* 2012) and between the strength of low-mass star depletion in the present-day mass function (MF) and the cluster concentration (De Marchi *et al.* 2007). The aim of the second project (Leigh *et al.* 2014) was to constrain the initial properties of primordial binaries by comparison with observations (Milone *et al.* 2012) and to check if star cluster simulations with initial conditions drawn from the invariant Kroupa (1995) distributions are able to recover the observed spatial distributions of binaries. All together, for these projects about 400 models of star clusters (SC) were simulated by the MOCCA code.

The model parameters run for the mentioned projects are as follows. Generally, more massive models also have larger concentrations (measured as the ratio between the tidal and half-mass radii - $R_{plum} = R_t/R_h$), with some of them being extremely concentrated (R_{plum} as large as 125). Most of the initial models had binary fractions equal to 0.1, but for some of them it varied between 0.3 to 0.95. The binary period distribution was uniform in the logarithm of the semi-major axis up to 100AU. For a substantial number of models, instead of the flat semi-major axis distribution, the Kroupa (1995) period distribution was used, up to $\log(P) = 8.3$. Also, different IMFs were used: the Kroupa canonical - a two segmented IMF (Kroupa 2001), the Kroupa standard - a three segmented IMF (Kroupa, Tout & Gilmore 1993), and the two segmented modified Kroupa IMF (different power-law indexes). Supernovae (SN) natal kick velocities for neutron stars (NS) and BHs were modified for some models according to the mass fallback procedure described by Belczynski *et al.* (2002).

3.1. *Extremely Large Initial Binary Fraction*

The results of the MOCCA computations for GCs with the Kroupa (1995) primordial binary population and their comparison with observational data (Milone *et al.* 2012) are described in detail in Leigh *et al.* (2014). Here we concentrate on the evolution of the binary parameters in evolving clusters described by tidally filling models (with R_{plum} controlled by the $W_0 = 6$ King model), and strongly concentrated tidally under-filling $W_0 = 6$ King models with $R_{plum} = 50$. Also, we present a comparison between the MOCCA results and the results of the BiPoS code developed by Michael Marks, based

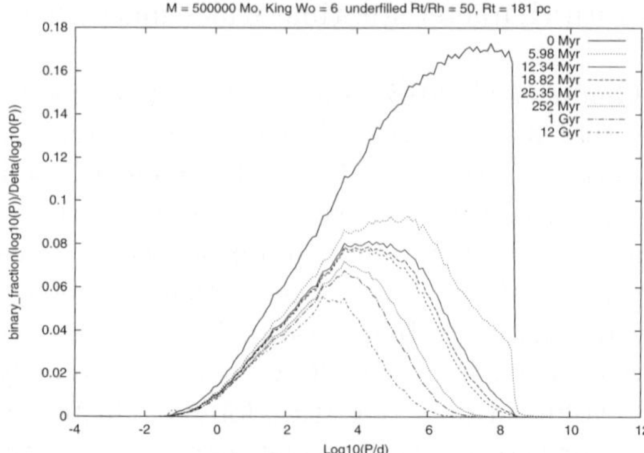

Figure 1. Time evolution of the binary period distribution in one of our tidally under-filling star cluster models.

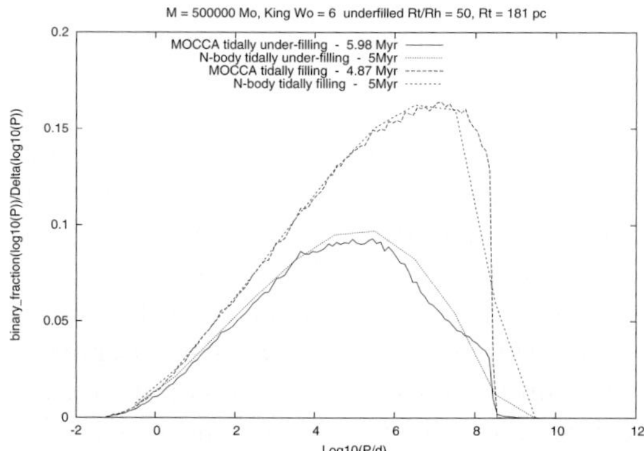

Figure 2. The binary period distribution for tidally filling and tidally under-filling models for MOCCA and N-body (using the semi-analytical BiPoS code) after 5 Myr computation time.

on Marks, Kroupa & Oh (2011) and Marks & Kroupa (2011). The BiPoS code offers an analytic description for the processing of the Kroupa (1995) initial binary population seen in N-body computations with initial masses up to $10^{3.5} M_\odot$. BiPoS thus evolves the initial binary population efficiently for cluster ages up to 5 Myr, the time for which their computations were run.

Fig. 1 shows the evolution of the primordial binary period distributions for one of our tidally under-filling models. As expected, the rate of evolution of the period distribution strongly depends on the initial cluster concentration. The larger the concentration, the larger the change in the period distribution. For more strongly concentrated clusters, the period distribution after about 10 Myr of cluster evolution is similar to the period distribution after a few Gyr of evolution in an initially tidally filling cluster, known as the density degeneracy (Marks & Kroupa 2012, Marks *et al.* 2014). The maximum of the period distribution moves towards smaller periods as the cluster ages and the distribution shape becomes quickly bell-shaped, as observed for Galactic field populations. The initial change in the distribution is rapid, occurring on a crossing-time scale (Marks, Kroupa & Oh 2011), after which the cluster enters a phase of slow, two-body relaxation driven binary processing. At 12 Gyr, the maximum of the period distributions for the tidally under-filling and the tidally filling models are about 10^3 days and 10^6 days, respectively.

Fig. 2 shows a comparison between the binary distributions at about 5 Myr for the tidally filling and tidally under-filling MOCCA models and the BiPoS code. The MOCCA and N-body period distributions agree remarkably well, both for the tidally filling and under-filling models, and so do the distributions for binary binding energy, mass ratio and eccentricity. This is somewhat unexpected, since Marks, Kroupa & Oh (2011) find that very fast destruction of binaries takes place on a dynamical time scale and strongly depends on the cluster concentration. The larger the cluster concentration, the larger the rate of binary destruction. Processes with characteristic time scales much shorter than the local relaxation time, and comparable to the dynamical time scale, are in principle not well followed by the Monte Carlo method. And yet, the good agreement between BiPoS and MOCCA suggests that the probability of binary dynamical interactions occurring in MOCCA is properly computed (taking into account the local cluster properties), and so the binary destruction rates are well reproduced. In the MOCCA simulations, some cooling of the system is observed (a small decrease in R_h) initially because of binary disruption (up to a time of 5 Myr), but then heating of the system takes over due to binary hardening.

The results of comparing the observational data from Milone *et al.* (2012) to the results of our MOCCA simulations are discussed in great detail in Leigh *et al.* (2014). Here we will only summarize the main conclusions. Only the tidally under-filling models can reproduce the observations. For these models, both the binary fractions and the ratio of the binary fractions inside and outside R_h reproduce the observations reasonably well. The tidally filling models, however, failed to reproduce the observations. For these models, a correlation is observed between the cluster mass and the binary fraction outside R_h, instead of an anti-correlation. If the initial binary populations are indeed described by the Kroupa (1995) distribution, then our comparison of the MOCCA computations with Milone *et al.* (2012) suggests that globular clusters must have formed strongly tidally under-filling. This is necessary to create sufficient dynamical processing to reproduce the observed anti-correlation between the binary fraction outside R_h and the total cluster mass. It is worth noting that those MOCCA simulations with high initial concentrations, initial binary fractions of about 10% and a flat distribution in the logarithm of the semi-major axis cannot recover the observed anti-correlation for the binary faction outside R_h (Leigh *et al.* 2013b). Of course, this does not rule out other types of distributions of the primordial binary parameters, which might also provide a good fit to the observed binary properties.

3.2. *A New IMBH Formation Scenario*

As was already mentioned in Section 3, the new scenario for IMBH formation came as a byproduct of projects carried out with Nathan Leigh and collaborators (Leigh *et al.* 2013b, Leigh *et al.* 2014). While working with the data produced during the simulations done for the first project, we noticed rather unexpectedly that for some models a slow buildup of BH mass is observed.

As is illustrated in Fig. 3, there are two different regimes of BH mass buildup. The first starts later on in the cluster evolution and has a rather small rate of mass increase (SLOW scenario), while the second starts very early on in the cluster evolution with a very high rate of mass increase (FAST scenario). The BH mass buildup is observed for runs with and without mass fallback (Belczynski *et al.* 2002) and, what is very unexpected, also for simulations done for massive and dense open clusters (project with Christoph Olczak - work in progress). Generally, only a small fraction of all models show a significant buildup of BH mass, and hence IMBH formation. The process of IMBH formation is highly stochastic. A quick inspection of the runs allows us to formulate the following rules: the larger the initial cluster mass, the larger the probability of IMBH formation;

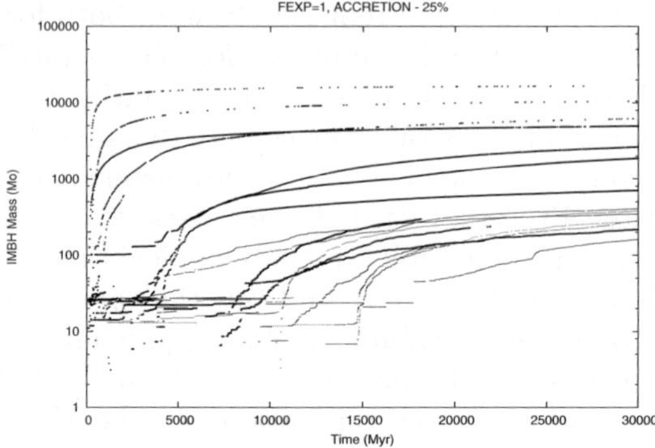

Figure 3. IMBH mass build up for models with reduced mass accretion onto the IMBH (25% of the standard BSE setup) and reduced standard star expansion after merger events (1/3 of the standard Fewbody setup). See description in the text.

the larger the initial concentration, the larger the probability of IMBH formation; the larger the initial concentration, the earlier and faster an IMBH is formed.

Most of our simulations were carried out with the assumption that 100% of the mass of a star colliding with a BH is accreted onto the BH, and the final size of an object formed in a collision is three times larger than the sum of the radii of the colliding stars ($F_{exp} = 3$). These are the standard assumptions applied in the BSE code (Hurley *et al.* 2000, 2002) and in the Fewbody code (Fregeau *et al.* 2004). As it was pointed out during the MODEST-14 conference, these assumptions might be too strong. The process of mass accretion onto the BH is very complicated, and recent simulations suggest that less than 50% of the incoming star mass is directly accreted onto the BH. So, in the next set of simulations, we weaken the standard assumptions - only 25% of the mass of the incoming star is accreted, and the size of the final merged object is just the sum of the radii of the colliding stars ($F_{exp} = 1$). As is clear from Fig. 3, an IMBH still forms. As expected, however, the IMBH formation is less efficient. Nevertheless, this strengthens the evidence in favor of our new scenario for IMBH formation.

Clear patterns in the subsequent binary evolution are apparent (not shown here) - e.g. shrinkage of the binary semi-major axis (binary hardening). This is due to dynamical interactions with other binaries and single stars, as well as binary evolution connected either with stellar evolution or gravitational wave radiation (GR). Binary mergers resulting purely from binary evolution are rare. More frequent are binary mergers connected with dynamical interactions. There are two kinds of dynamical mergers: mergers with one binary component (which preserve the binary), and total mergers in which all interacting stars merge into one object. The total binary merger events are crucial from the point of view of IMBH formation. Because of these interactions, an IMBH binary (when its mass is still low) cannot harden to the point of being able to escape from the cluster, which can occur if it receives a strong recoil during a subsequent dynamical interaction. So, the IMBH remains in the system and, consequently, is able to steadily grow in mass.

In the case of the SLOW scenario, the central cluster densities are not very high: only $10^5 \mathrm{M}_\odot \, \mathrm{pc}^{-3}$. Thus, a special set of initial conditions is not needed to form an IMBH. The situation is different for the case of the FAST scenario. The densities required for significant IMBH mass buildup to occur are very high, greater than $10^8 \mathrm{M}_\odot \, \mathrm{pc}^{-3}$. These extremely high densities are needed when BHs form a bound and very dense subsystem in the cluster center; mergers of binaries with BHs have to be more efficient than the removal of BHs from the cluster due to strong recoils in dynamical interactions. Such

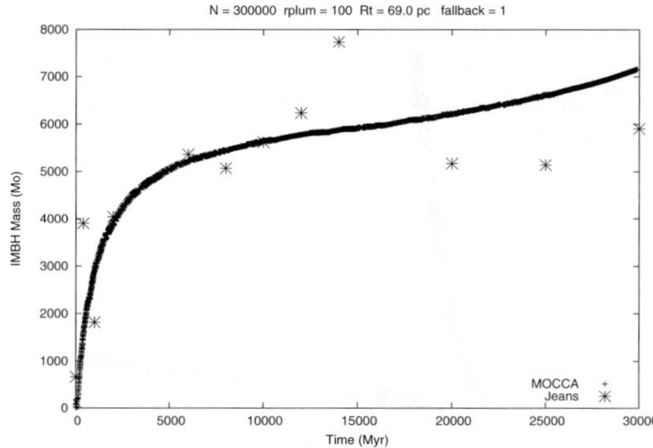

Figure 4. IMBH mass from our MOCCA simulations and from fitting Jeans models (Lützgendorf *et al.* 2013) to the SBP and VDP obtained from the MOCCA simulations.

high densities are not very probable in the GCs observed in the Milky Way, but they can occur in nuclear star clusters (NSC). Perhaps the FAST scenario for IMBH formation discussed here occurs commonly in the NSCs of low-mass galaxies.

The new scenario for IMBH formation can be summarized as follows:

- To initiate the process of BH mass growth, either at least one BH must be left in the cluster after the early phase of SN explosions, or a single BH must be formed via mergers during dynamical interactions. If several BHs remain in the system, the cluster density has to be extremely high for an IMBH to form, greater than $10^8 M_\odot\, pc^{-3}$;

- Next, the formation of a BH-any star binary forms via three-body interactions. The BH is the most massive object in the cluster, so there is a high probability that the BH will be exchanged into, or form a binary.

- Dynamical interactions with other binaries and stars:
    - orbit tightening leading to mass transfer from MS/RG/AGB companions;
    - exchanges and mergers, leaving the binary in tact;
    - total mergers in dynamical interactions or the emission of gravitational waves - in this case, the binary is destroyed and only the BH is left. The single BH is then formed a new binary via another three-body interaction, which is free to undergo subsequent dynamical interactions with other single and binary stars. In this way, the BH mass steadily increases.

The buildup of the IMBH mass for a model with N=300000, $R_{plum} = 100$ and mass fallback for SN natal kicks is shown in Fig 4. The mass of the IMBH increases up to 7000 $M_\odot$. The surface brightness profiles (SBP) and velocity dispersion profiles (VDP) constructed from this simulation were used by Nora Lützgendorf to fit the IMBH mass with her code based on Jeans' model (Lützgendorf *et al.* 2013). As seen in Fig. 4, the fit to the MOCCA data is rather good. MOCCA is able to reproduce more or less correctly the system structure around an IMBH. This is a rather unexpected result, given that MOCCA was not designed to model the relevant physics near a massive IMBH. In fairness though, the agreement is only good in some of our models. Definitely, more work is needed.

Fig. 5 shows the number of events in which mass is transferred onto an IMBH, either from its companion during binary evolution, or because of collisions with incoming stars for the SLOW scenario for standard and reduced accretion. Such mass accretion events are potentially observable because of associated electromagnetic or gravitational radiation. The number of events for collisions and GR are listed in Fig. 5. The numbers are substantial. Interestingly, there are more GR and binary evolution events for the case of

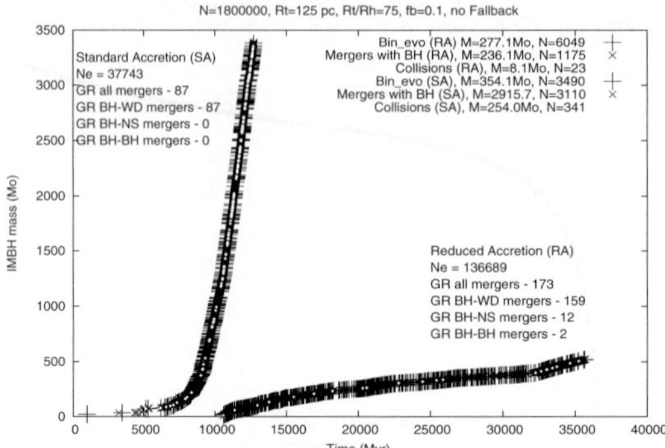

Figure 5. IMBH mass buildup for models with standard and reduced mass accretion onto the IMBH. Mass transfer events, mergers and collisions with the IMBH are depicted by different symbols (see insets). Also listed are the numbers of interactions, and the total mass gained by the IMBH in each dynamical interaction.

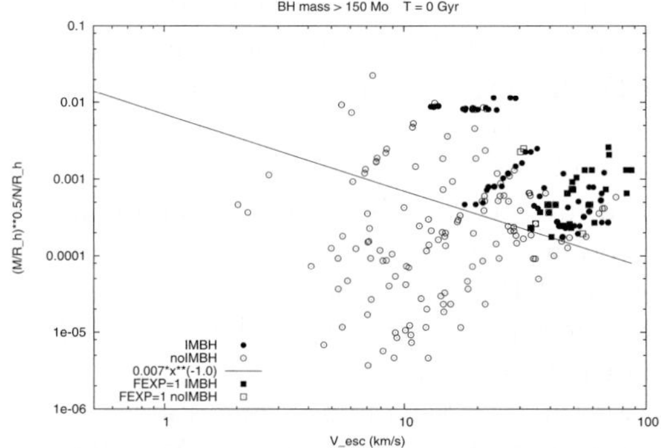

Figure 6. The interaction probability vs escape velocity at time $T = 0$ for all models (see the inset).

reduced mass accretion onto the IMBH. This is connected to the fact that mergers during binary dynamical interactions produce far fewer "total" mergers, due to the smaller size and hence cross-section of the merged object. Instead, there is a large number of GR events, mainly associated with BH-WD mergers. For the FAST scenario, physical collisions are the dominant type of interaction provided the IMBH is sufficiently massive (larger than 5000–7000 $M_\odot$).

Let's assume that the probability of IMBH formation depends mainly on the average binary interaction probability. Then we can compare models with and without IMBH formation to see if they occupy separate regions in Escape Velocity – Interaction Probability space. To calculate the interaction probability, we assume an interaction involving an average binary at the soft/hard boundary and an average single star, that occurs inside R_h. Fig. 6 shows the results of this exercise. Indeed, at time $T = 0$, there is a statistical boundary above which the probability for IMBH formation is substantial. This boundary line is drawn by eye. Models with substantially reduced mass accretion are only likely to form an IMBH for cluster escape velocities larger than about 30–40 km/s. The same exercise is repeated at time $T = 12$ Gyr (i.e. at the present-day), which shows that indeed some Galactic GCs occupy a region in Escape Velocity – Interaction Probability space for which the MOCCA models do produce an IMBH. These clusters include Omega Cen, 47Tuc, M22 and NGC6293.

Concluding, it is worth mentioning that models with reduced natal kicks (because of mass fallback) for BHs may still contain a substantial number of stellar mass BHs even after a Hubble time of cluster evolution. The number of retained BHs depends on the cluster mass and concentration, via the cluster half-mass relaxation time. The larger the half-mass relaxation time, the larger the number of retained BHs. This means that those clusters most likely to host stellar mass BHs are massive and have a low concentration, instead of massive and dense. Also, simulations of dense and massive GCs show that NSs and BHs can form in substantial numbers not only due to supernovae explosions, but also because of physical collisions and binary mergers later on in the cluster evolution. Interestingly, IMBH formation suppresses the production of any binaries with NS or BH companions, and even the formation of NSs and BHs themselves. This is because their progenitors (white dwarfs, NSs and BHs) are very quickly removed from the system via dynamical interactions with the IMBH. They are the most massive stars in the cluster, and therefore the probability for their involvement in dynamical interactions is high.

4. Conclusions

Here are our conclusions, which summarize the talk.

- The MOCCA code is able to follow the initial destruction of wide binaries. In support of this, the results are in good agreement with the semi-analytic code BiPoS, which is based on N-body simulations.

- If the initial binary population is described by the Kroupa (1995) distributions, then our comparison of the MOCCA computations with Milone *et al.* (2012) suggests that globular clusters need to have formed strongly tidally under-filling to produce sufficient dynamical processing, and hence to reproduce the observed anti-correlations between the binary fractions and the total cluster mass.

- NSs and BHs can form (in substantial numbers) in the course of star cluster evolution due to dynamical interactions (collisions and binary interactions);

- If the cluster density is large enough (about $10^5 M_\odot \, \text{pc}^{-3}$), an existing BH can experience substantial mass buildup due to collisions/mergers during dynamical interactions and mass transfer in binaries. The SLOW scenario is more probable;

- The process of BH mass buildup, and finally IMBH formation, is highly stochastic. The rate of IMBH mass buildup strongly depends on the cluster density. The larger the density, the higher the rate;

- There are frequent phases of mass transfer in binaries containing an IMBH and mergers of stars with an IMBH. Therefore, X-ray emissions and/or GR could be observable during these events.

Of course one should be aware about possible problems associated with the approximations used in the MOCCA code.

- The MOCCA code does not cope well with physical processes that have a characteristic time scale comparable to the dynamical time scale;

- The MOCCA code is not prepared to follow the dynamical evolution of extremely massive objects (larger than a few hundred $M_\odot$). Nevertheless, the initial IMBH mass buildup is modeled correctly;

- There are some doubts about the accuracy of the BSE code (Hurley *et al.* 2000, Hurley *et al.* 2002), and its ability to follow binary evolution and mass transfer involving extreme mass ratios and massive compact objects. The mass transferred onto an IMBH because of binary/stellar evolution is not the dominant source of mass accretion, so the process of IMBH mass buildup can occur even if binary evolution-induced mass transfer is completely switched off;

- Very large kicks associated with the mergers of BHs having misaligned spin vectors.

Acknowledgements

AH, MG and AA were partly supported by the Polish Ministry of Science and Higher Education and by the National Science Centre through the grants DEC-2011/01/N/ST9/06000 and DEC-2012/07/B/ST9/04412, respectively. NL is thankful for the generous support of an NSERC Postdoctoral Fellowship.

References

Baumgardt, H., Makino, J., Hut, P., McMillan, S., & Portegies Zwart, S. 2003 *ApJ* (Letters), 589, L25

Belczynski K., Kalogera V., & Bulik T. 2002, *ApJ*, 572, 407

De Marchi, G., Paresce, F., & Pulone, L. 2007, *ApJ* (Letters), 656, L65

Fregeau J. M., Cheung P., Portegies Zwart S. F. & Rasio F. A. 2004, *MNRAS*, 352, 1

Fukushige T. & Heggie D. C. 2000, *MNRAS*, 318, 753

Giersz, M., Heggie, D. C., Hurley, J. R., & Hypki, A. 2013, *MNRAS*, 431, 2184

Gürkan, M. A., Freitag, M., & Rasio, F. A. 2004, *ApJ*, 604, 632

Harris, W. E. 1996, *AJ*. 112, 1487 (2010 update)

Hénon, M. H. 1971, *Ap&SS*, 14, 151

Hurley J. R., Pols O. R. & Tout C. A. 2000, *MNRAS*, 315, 543

Hurley J. R., Tout C. A. & Pols O. R. 2002, *MNRAS*, 329, 897

Hypki, A. & Giersz, M. 2013, *MNRAS*, 429, 1221

Kroupa 1995, *MNRAS*, 277, 1507

Kroupa 2008, *MNRAS*, 322, 231

Kroupa P. & Petr-Gotzens M. G. 2011, *A&A*, 529, A92

Kroupa, P., Tout, C. A., & Gilmore, G. 1993, *MNRAS*, 262, 545

Kroupa, P., Weidner, C., Pflamm-Altenburg, J., Thies, I., Dabringhausen, J., Marks, M., & Maschberger, T. 2013, in: Oswalt, T. D. and Gilmore, G. (eds.), *Planets, Stars and Stellar Systems. Volume 5: Galactic Structure and Stellar Populations*, p. 17

Leigh, N. W. C., Böker, T., Maccarone, T. J., & Perets, H. B. 2013a, *MNRAS*, 429, 2997

Leigh, N. W. C., Giersz, Webb, J. J., Hypki, A., De Marchi, G., Kroupa, P., & Sills, A. 2013b, *MNRAS*, 436, 3399

Leigh, N. W. C., Giersz, M., Marks, M., Webb, J. J., Hypki, A., Heinke, C. O., Kroupa, P., & Sills, A. 2014, *arXiv1410.2248*

Lützgendorf, N., Kissler-Patig, M., Gebhardt, K., Baumgardt, H., Noyola, E., de Zeeuw, P. T., Neumayer, N., Jalali, B., & Feldmeier, A. 2013, *A&A*, 552, 49

Madau, P. & Rees, M. J. 2001, *ApJ* (Letters), 551, L27

Marks, M. & Kroupa, P. 2011, *A&A*, 417, 1702

Marks, M. & Kroupa, P. 2012, *A&A*, 543, A8

Marks, M., Kroupa, P., & Oh, S. 2011, *MNRAS*, 417, 1684

Marks, M., Leigh, N., Giersz, M., Pfalzner, S., Pflamm-Altenburg, J., & Oh, S. 2014, *MNRAS*, 431, 3503

&Milone, A. P., *et al.*, 2012, *A&A*, 540, 16

Morscher, M., Pattabiraman, B., Rodriguez, C. Rasio, F. A., & Umbreit, S. 2014, *arXiv1409.0866*

Portegies Zwart, S. F., Baumgardt, H., Hut, P., Makino, J., & McMillan, S. L. W. 2004, *NATURE*, 428, 724

Sollima, A. & Mastrobuono Battisti, A. 2014, *MNRAS*, 443, 3513

Stodółkiewicz, J. S. 1986, *AcA*, 36, 19

Vasiliev, E. 2015, *MNRAS*, 446, 3150

Star clusters and black holes in galaxies across cosmic time
Proceedings IAU Symposium No. 312, 2014
Y. Meiron, S. Li, F.-K. Liu & R. Spurzem, eds.

© International Astronomical Union 2016
doi:10.1017/S174392131500784X

Searching for intermediate mass black holes: understanding the data first

Paolo Bianchini, Mark Norris, Glenn van de Ven and Eva Schinnerer

Max-Planck Institute for Astronomy, Königstuhl 17, 69117 Heidelberg, Germany
email: bianchini@mpia.de

Abstract. The detection of intermediate mass black holes (IMBHs) in globular clusters has been hotly debated, with different observational methods delivering different outcomes for the same object. In order to understand these discrepancies, we construct detailed mock integral field spectroscopy (IFU) observations of globular clusters, starting from realistic Monte Carlo cluster simulations. The output is a data cube of spectra in a given field-of-view that can be analyzed in the same manner as real observations and compared to other (resolved) kinematic measurement methods. We show that the main discrepancies arise because the luminosity-weighted IFU observations can be strongly biased by the presence of a few bright stars that introduce a scatter in velocity dispersion measurements of several km s^{-1}. We show that this intrinsic scatter can prevent a sound assessment of the central kinematics, and therefore should be fully taken into account to correctly interpret the signature of an IMBH.

Keywords. black hole physics, globular clusters: general, stars: kinematics, instrumentation: spectrographs.

1. Introduction

The existence of intermediate mass black holes (IMBHs) with masses between those of stellar black holes ($M_\bullet < 100$ $M_\odot$) and those of supermassive black holes (SMBH, $M_\bullet > 10^5$ $M_\odot$) has been postulated. The extrapolation of the $M_\bullet - \sigma$ relation for galaxies, linking the mass of the central back hole to the velocity dispersion of the host stellar system (Ferrarese *et al.* 2000, Magorrian *et al.* 1998), suggests that globular clusters (GCs) represent the ideal environments to find central IMBHs with masses ranging from $10^3 - 10^4$ $M_\odot$. However, their detection has proven to be difficult, with contradictory results on their presence in local group GCs (e.g., Gebhardt *et al.* 2000, van den Bosch *et al.* 2006, Noyola *et al.* 2010, van der Marel *et al.* 2010, Lützgendorf *et al.* 2013, Lanzoni *et al.* 2013, den Brok *et al.* 2014).

The kinematic observations suggesting detections of IMBHs are primarily based on the search for a rise of the central velocity dispersion (e.g. Bahcall *et al.* 1976, Lützgendorf *et al.* 2013). This method is very challenging, since it requires both high spatial resolution, to resolve the very crowded central region of GCs (few central arcseconds), and very precise velocity measurements with accuracy ~ 1 km s^{-1}.

Two distinct observational strategies are employed for the kinematic detection of IMBHs: 1) measurements of velocities of resolved individual stars (line-of-sight velocities or proper motions), 2) unresolved kinematic measurements with integral field unit (IFU) spectroscopy, from line broadening of integrated spectra. These complementary methods can give significantly different observational outcomes when applied to the same object, making the detection of IMBHs highly ambiguous. In particular, integrated light spectroscopy seems to measure rising central velocity dispersions, favoring the presence of IMBHs (see for example, Noyola *et al.* 2010 for ω Cen, or Lützgendorf *et al.* 2011 for

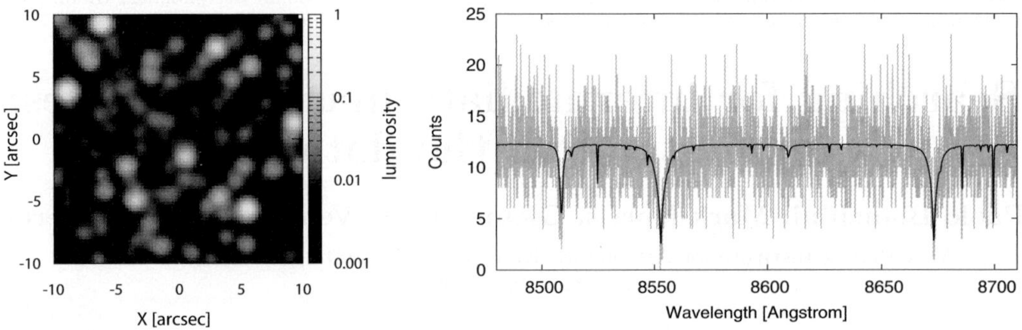

Figure 1. *Left panel:* Luminosity map of the central 20×20 arcsec^2 region of our simulated globular cluster placed at 10 kpc and observed with a seeing of 1 arcsec and an average signal–to-noise per Å of $S/N \approx 10$. *Right panel:* typical spectrum of a spaxel, obtained summing all the Doppler-shifted spectra falling in the spaxel, properly weighted by their PSF. The black line indicates the spectrum without noise, while the green line indicates the case of an observation with signal to noise per Å of $S/N \approx 10$.

NGC 6388), while resolved stellar kinematics do not confirm the presence of this signature (see van der Marel *et al.* 2010 for proper motion measurements of ω Cen, and Lanzoni *et al.* 2013 for discrete line-of-sight measurements in NGC 6388).

Our goal is to understand the systematic differences between the different observational methods, before undertaking any interpretation of the kinematic signatures connected to the presence of IMBHs. In particular, we wish to understand the biases that arise from applying IFU spectroscopy to systems with a (partially) resolved stellar population, like Galactic GCs. In order to do so, we develop a procedure to create detailed mock IFU observations of the center of GCs starting from realistic Monte Carlo cluster simulations. The output of our procedure is a data cube with spectra and luminosity information for every spaxel in a selected field-of-view, that will be analyzed in the same manner as real IFU observations to build mock velocity dispersion profiles in the central region of a GC.

2. Constructing a mock IFU observation

The starting point of our work are Monte Carlo cluster simulations, developed by Downing *et al.* (2010), providing a realistic description of a typical GC with initial number of particles of 2×10^6 drawn from a Plummer (1911) model, a Kroupa (2001) initial mass function, 10% primordial binary fraction, and metallicity [Fe/H]=-1.3. The simulations have no central IMBH and no internal rotation (note however that internal rotation is observed in several GCs, e.g. Bianchini *et al.* 2013, Fabricius *et al.* 2014, Kacharov *et al.* 2014). At 13 Gyr, the simulation is characterized by a total mass of $M \simeq 6.7 \times 10^5 \ M_\odot$ and a projected half light radius of $R_h \simeq 2.8$ pc. We place the simulated GC at 10 kpc from the observer with a global systemic line-of-sight velocity of 300 km s^{-1}, to match the typical properties of a Galactic GC.

We associate to each star (characterized by effective temperature T_{eff}, mass $M_\star$, luminosity $L_\star$, metallicity Z) a low-resolution broad-wavelength stellar spectrum, using GALEV evolutionary synthesis model (Kotulla *et al.* 2009). Then we associate a high-resolution spectrum in the wavelength range that will be used in our mock observations (calcium triplet, 8400-8800 Å) using the MARCS synthetic stellar library (Gustafsson *et al.* 2008) providing a resolving power of $R = \lambda/\Delta\lambda = 20\,000$, enough to measure the internal kinematics of GCs (typical velocity dispersions of 10 km s^{-1}). Finally, we Doppler-shift the spectra using the line-of-sight velocity from our cluster simulation.

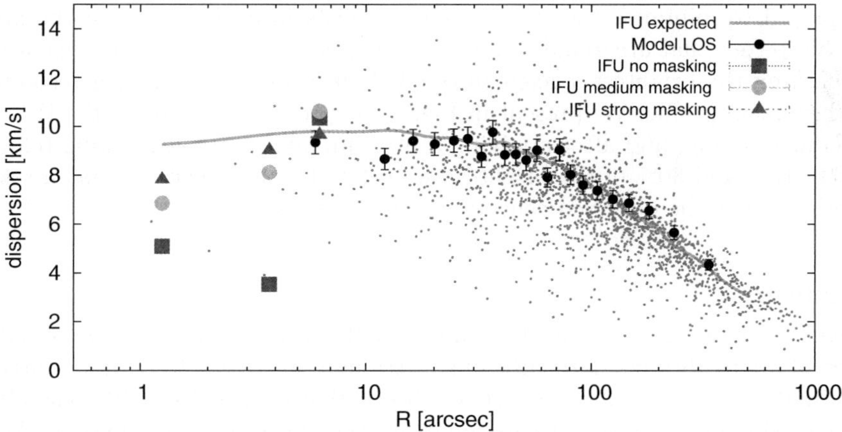

Figure 2. Velocity dispersion profiles for our mock IFU observation and our model GC. Both a discrete velocity dispersion profile based on giants stars (filled circles) and a luminosity-weighted profile (gray dots) constructed directly from our model are shown. The solid line is the profile expected for the typical kinematical tracer of an IFU observation (Bianchini *et al.* in prep.). The luminosity-weighted profile shows a high scatter due the luminosity differences between stars. The profiles in the central region are extracted from our mock IFU observation (blue squares), and recalculated after masking the brightest spaxels (green circles and red triangles).

We next define the observational setup of the simulated IFU instrument, selecting a 20×20 arcsec^2 field-of-view and a spaxel scale of 0.25 arcsec. After convolving each star with a Gaussian PSF, we sum the Doppler-shifted spectra of all the stars falling in each spaxel properly weighted by their PSF. In this way we have a spectrum and the corresponding luminosity information for each spaxel. Finally, we add Poisson noise to the final spectra in order to match the desired signal-to-noise ratio S/N. In Fig. 1 we show the final product of our mock IFU observation, consisting of a luminosity map, with seeing of 1 arcsec and average signal-to-noise per Å of $S/N = 10$ (left panel), and the typical spectrum of a spaxel (both with and without noise; right panel).

3. Analyzing the kinematics

After producing the data cube of our IFU observation, we can construct the one-dimensional velocity dispersion profiles. We divide our field of view in annular bins and sum all the spectra in each bin. With the penalized pixel-fitting (pPXF) program of Cappellari *et al.* (2004) we measure the velocity dispersion from the broadening of the lines of the summed spectra. The measured velocity dispersion profile of the central region can then be directly compared to the dispersion profiles expected from the model.

We construct three different velocity dispersion profiles from the model. The first is a typical line-of-sight velocity dispersion profile, constructed using the resolved kinematics of only the bright giant stars in our simulation. The second is a luminosity-weighted velocity dispersion profile, constructed taking into consideration all stars in our simulation and their corresponding luminosity. The third is the profile expected for the typical kinematical tracer of an IFU observation (Bianchini *et al.* in prep.). Fig. 2 shows that the luminosity-weighted profile is characterized by a large scatter due to the stochasticity introduced by the luminosity differences between stars.

Since an IFU observation gives intrinsically luminosity-weighted kinematic measurements, it is already evident that it will suffer from a stochasticity effects driven by the presence of a few bright stars that can completely dominate certain spaxels. A procedure often employed to minimize the stochasticity consists in excluding from the kinematic analysis the spaxels that are found contaminated by a bright star. A simple strategy consists in masking the brightest spaxels in our IFU field-of-view and then recompute the velocity dispersion profile. We present in Fig. 2 the profiles obtained in the IFU field-of-view without masking, and after applying a medium and a strong masking (eliminating respectively 10% and 30% of the brightest spaxels). The figure shows that masking can significantly change the observed central velocity dispersion.

4. Conclusion

We have presented a procedure to simulate IFU observations of Galactic GCs. Our procedure allows to obtain a data cube of spectra in a selected field-of-view starting from any GC simulation. We used a mock IFU observation of a realistic 10^6 particle Monte Carlo cluster simulation to investigate what are the effects that can bias the kinematic measurements. Our analysis shows that IFU kinematic measurements can be strongly biased by the presence of a few bright stars, that can dominate certain spaxels. Moreover IFU kinematics gives a luminosity-weighted information, and therefore can in principle give different outcomes from what obtained from resolved line-of-sight kinematics. These stochasticity effects can prevent to obtain sound measurements of the central velocity dispersion of GCs and therefore must be carefully investigated before interpreting any signatures of the presence of IMBHs. We will use our procedure to study in detail the stochasticity connected to luminosity-weighted kinematics and to determine an efficient masking technique to recover unbiased results. Moreover, we will compare mock observations of models with and without IMBHs to help understanding if the presence of an IMBH can be definitively recovered using common dynamical modeling techniques.

References

Bahcall, J. N. & Wolf, R. A. 1976, *ApJ*, 209, 214

Bianchini, P., Varri, A. L., Bertin, G., & Zocchi, A. 2013, *ApJ*, 772, 67

Bianchini, P., Norris, M., van de Ven, G., & Schinnerer, E. in preparation

Cappellari, M. & Emsellem, E. 2004, *PASP*, 116, 138

den Brok, M., van de Ven, G., van den Bosch, R., *et al.* 2014, *MNRAS*, 438, 487

Downing, J. M. B., Benacquista, M. J., Giersz, M., & Spurzem, R. 2010, *MNRAS*, 407, 1946

Fabricius, M. H., Noyola, E., Rukdee, S., *et al.* 2014, *ApJ* (Letters) 787, L26

Ferrarese, L. & Merritt, D. 2000, *ApJ* (Letters) 539, L9

Gebhardt, K., Pryor, C., O'Connell, R. D., Williams, T. B., *et al.* 2000, *AJ*, 119, 1268

Gustafsson, B., Edvardsson, B., Eriksson, K., *et al.* 2008, *A&A*, 486, 951

Kacharov, N., Bianchini, P., Koch, A., *et al.* 2014, *A&A*, 567, 69

Kotulla, R., Fritze, U., Weilbacher, P., & Anders, P. 2009, *MNRAS*, 396, 462

Kroupa, P. 2001, *MNRAS*, 322, 231

Lanzoni, B. Mucciarelli, A., *et al.* 2013, *ApJ*, 769, 107

Lützgendorf, N., Kissler-Patig, M., Noyola, E., *et al.* 2011, *A&A*, 533, 36

Lützgendorf, N., Kissler-Patig, M., Gebhardt, K., *et al.* 2013, *A&A*, 552, 49

Magorrian, J., Tremaine, S., Richstone, D., *et al.* 1998, *AJ*, 115, 2285

Noyola, E., Gebhardt, K., Kissler-Patig, M., *et al.* 2010, *ApJ* (Letters) 719, L60

Plummer, H. C. 1911, *MNRAS*, 71, 460

van den Bosch R., de Zeeuw, T. Gebhardt, K., *et al.* 2006, *ApJ*, 641, 852

van der Marel, R. P. & Anderson, J. 2010, *ApJ*, 710, 1063

Star clusters and black holes in galaxies across cosmic time
Proceedings IAU Symposium No. 312, 2014
Y. Meiron, S. Li, F.-K. Liu & R. Spurzem, eds.
© International Astronomical Union 2016
doi:10.1017/S1743921315007851

Expansion techniques for collisionless stellar dynamical simulations

Yohai Meiron

Kavli Institute for Astronomy and Astrophysics at Peking University, Beijing 100871, China
email: ymeiron@pku.edu.cn

Abstract. We present *ETICS*, a collisionless N-body code based on two kinds of series expansions of the Poisson equation, implemented for graphics processing units (GPUs). The code is publicly available and can be used as a standalone program or as a library (an AMUSE plugin is included). One of the two expansion methods available is the self-consistent field (SCF) method, which is a Fourier-like expansion of the density field in some basis set; the other is the multipole expansion (MEX) method, which is a Taylor-like expansion of the Green's function. MEX, which has been advocated in the past, has not gained as much popularity as SCF. Both are particle-field methods and optimized for collisionless galactic dynamics, but while SCF is a "pure" expansion, MEX is an expansion in just the angular part; thus, MEX is capable of capturing radial structure easily, while SCF needs a large number of radial terms.

Keywords. methods: numerical – stars: kinematics and dynamics

1. Introduction

A stellar system could naively be described as a set of $3N_\star$ coupled, second-order, non-linear ordinary differential equations, where $N_\star$ is the number of stars. Solving such an equation set numerically is practically only possible at the very low end of the realistic $N_\star$-range, and even so could be very challenging with current computer hardware. Thus, various techniques are used to simplify the mathematical description of the system; these are often designed to fit a particular problem in stellar dynamics and yield unphysical results when applied to another problem.

Direct N-body simulation is one of the main techniques used to study gravitational systems in general and galaxies in particular. In this technique, the distribution function is sampled at $N \ll N_\star$ points in a Monte-Carlo fashion. This N depends on the computational capabilities an may be orders of magnitudes smaller. This simplification can cause problems, as some dynamical processes depend on the number density rather than just the mass density. The most well-known of these processes is two-body relaxation. The relaxation time (the characteristic time for a particle's velocity to change by an order of itself due to encounters with other particles) scales with the crossing time roughly as $N/\ln N$. Thus, the ratio between the relaxation times in a real and a simulated system is of a similar order of magnitude as the undersampling factor.

Galaxies are often described as collisionless stellar systems, which means that the relaxation time is much longer than the timescale of interest (except perhaps at the very center). This property could be very useful. Since a particle's orbit is basically what it would be if it were moving in a smooth gravitational field, we could evaluate the field instead of calculating all of the stellar interactions, which is computationally cheaper. Another useful property is that galaxies are often spheroidal in shape. Even highly flattened galaxies will have a spherical dark halo component. Thus, a spherical shape could be used as a zeroth-order approximation for the gravitational field, and higher-order terms could be written using spherical harmonics.

These two facts are utilized by series expansion techniques such as the multipole expansion (MEX) and the self-consistent field (SCF) methods. They historically come from different ideas and are mathematically distinct. In the context of numerical simulations, however, they serve a similar function: to evaluate the gravitational force on all N particles generated by this same collection of particles in a way that discards spurious small-scale structure (in other words, smooths the field). In Meiron *et al.* (2014) we first presented the *ETICS*† code which encapsulates these techniques, fully describing the mathematics behind it, implementation and accuracy. In this proceeding we briefly summarize this and show basic performance benchmarks.

2. Formalism

Both the MEX and SCF methods are ways of solving the Poisson equation

$$\nabla^2 \Phi(\boldsymbol{r}) = 4\pi\rho(\boldsymbol{r}), \tag{2.1}$$

the formal solution of which is given by the following integral:

$$\Phi(\boldsymbol{r}) = -\int \frac{\rho(\boldsymbol{r}')\mathrm{d}^3 r'}{|\boldsymbol{r}-\boldsymbol{r}'|}. \tag{2.2}$$

The expression $|\boldsymbol{r}-\boldsymbol{r}'|^{-1}$ is the Green's function of the Laplace operator in three dimensions and in free space (no boundary conditions), and the integral is over the whole domain of definition of $\rho(\boldsymbol{r})$.

In both MEX and SCF, the integrand above is expanded as a series of terms, each of which is more easily numerically integrable; this is done in two different ways, lending the two methods quite different properties. In brief, MEX is a Taylor-like expansion of the Green's function, while SCF is a Fourier-like expansion of the density. Another way to look at it is that in both methods the integrand is written as a series of functions (of $\boldsymbol{r}$) with coefficients: in MEX, one uses the given density to evaluate the functions, while their coefficients are known in advance; in SCF, one evaluates coefficients, while the functions are known in advance. The standard form of MEX in three dimensions is

$$\Phi(\boldsymbol{r}) = -\sum_{l=0}^{\infty} \frac{4\pi}{2l+1} \sum_{m=-l}^{l} \left[q_{lm}(r)r^{-(l+1)} + p_{lm}(r)r^l \right] Y_{lm}(\theta,\phi) \tag{2.3}$$

$$q_{lm}(r) = \int_{r'<r} r'^l \rho(\boldsymbol{r}')Y_{lm}^*(\theta',\phi')\mathrm{d}^3 r' \tag{2.4}$$

$$p_{lm}(r) = \int_{r<r'} r'^{-(l+1)} \rho(\boldsymbol{r}')Y_{lm}^*(\theta',\phi')\mathrm{d}^3 r'. \tag{2.5}$$

Since in practice the density field is made of N discrete points, they must be sorted by r in order for the above integrals to be evaluated in one pass.

The standard form of SCF in three dimensions is

$$\Phi(\boldsymbol{r}) = \sum_{n=0}^{\infty}\sum_{l=0}^{\infty}\sum_{m=-l}^{l} A_{nlm}\Phi_{nl}(r)Y_{lm}(\theta,\phi) \tag{2.6}$$

$$A_{nlm} = \int \rho(\boldsymbol{r}')\Phi_{nl}(r')Y_{lm}^*(\theta',\phi')\mathrm{d}^3 r'. \tag{2.7}$$

† `https://github.com/sahmes/etics`

$\Phi_{nl}(r)$ are functions forming an orthogonal set, the radial basis, which is a key difference between MEX and SCF. The choice of basis is not unique, and the basis functions themselves need not represent physical potentials, but it is convenient to take the zeroth term ($n = l = 0$) to represent some physical system, and to construct the rest of the set by some orthogonalization method. Hernquist & Ostriker (1992) constructed a radial basis using Gegenbauer polynomials that at zeroth order was a Hernquist (1990) model, which they argued was well suited to study galaxies; this is the basis we adopt in *ETICS*.

3. Implementation

GPUs are powerful and cost-effective devices for high-performance parallel computing. They are used to accelerate many scientific calculations, especially in astrophysics, such as the dynamics of dense star clusters and galaxy centers (see review by Spurzem *et al.* 2012). The main issue when implementing a mathematical technique for GPUs is to make the algorithm parallel. In many cases there is already a parallel implementation of the method, but a GPU is different from a multi-CPU parallel environment, and the parallelization scheme is not always transferable. A full description of the implementation would be very long and technical, thus we only briefly overview it here; the full details are found in Meiron *et al.* (2014).

Both the MEX and SCF algorithms have two distinct steps: calculating the expansion from the particles, and using the expansion to calculate the forces on the particles. These two parts are done consecutively in *ETICS*, while each part is done in parallel on the GPU. While in SCF the full coefficient list is calculated before it is used for force calculation, in MEX, due to use of very large arrays in the global memory, the force is calculated in parts, i.e., first the zeroth (monopole) order force is calculated, then the $l = 2$ correction is added to it, and so forth.

The current implementation of the MEX method relies on *Thrust* (Bell & Hoberock 2011), a C++ template library of parallel algorithms which is part of the CUDA framework. The particles are first sorted by radius using a *Thrust* subroutine; this is followed by a loop over l, inside of which the spherical harmonics are calculated; these are used to calculate the contributions of the particle to q_{lm} and p_{lm}, which are saved in global memory. Finally, *Thrust* subroutines are dispatched to perform forward and backward cumulative sums.

For SCF, no sorting or cumulative summation are done. In principle, a similar algorithm could be used, i.e., the contribution of each particle to each coefficient is stored in global memory, and later the contributions are summed. The problem is that there are $\frac{1}{2}(n_{\max} + 1)$ more SCF coefficients than there are MEX functions. There is a better scheme that does not require such heavy use of the global GPU memory. Since a normal (rather than cumulative) summation is needed, it is fairly easy to use the very fast "shared memory" (another memory type in the GPU, available to all threads in a single block) to accumulate the particles' contributions, and then perform parallel reductions within the block (later the CPU finalizes by summing the contribution from each block). The parallel SCF algorithm of Hernquist *et al.* (1995) could not be used here due to the difference between how the GPU and CPU access and cache memory (it could however be used for the inter-node level parallelization).

The calculation of the forces (or potentials) at the last step is very similar between MEX and SCF (except, as mentioned before, the MEX force is calculated for each multipole level separately), as each particle only needs to know its own coordinates and the series expansion and there is no dependency between the threads.

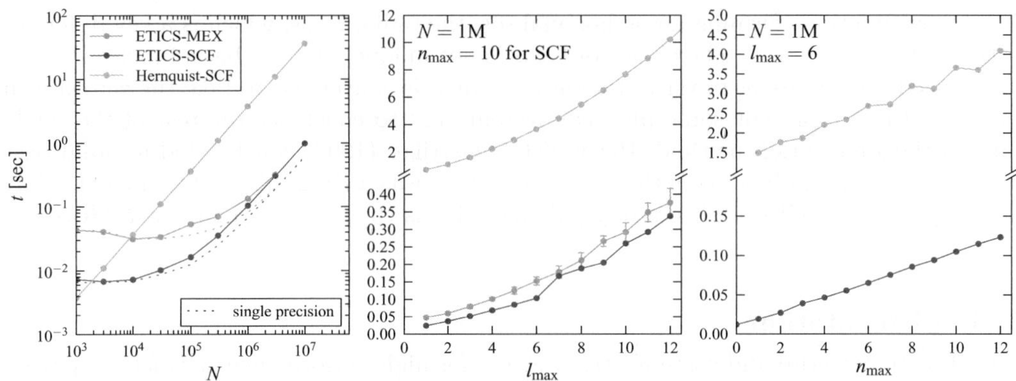

Figure 1. Scaling of one full force calculation time. Hernquist's SCF code (in green) is a CPU and *ETICS* is a GPU code with both MEX (red) and SCF (blue) methods; for the GPU codes, dotted lines show the performance in single-precision mode. The parameter which is held constant is indicated on each panel. The CPU code shows some erratic behavior due to compiler optimization. Note that the CPU and GPU tests use different hardware.

4. Performance

We tested the performance of *ETICS* (both MEX and SCF) on a single Nvidia Tesla K20 GPU on the Laohu supercomputer at the NAOC in Beijing. For comparison, we also tested the Fortran CPU SCF code by Lars Hernquist on the ACCRE cluster at Vanderbilt University in Nashville, Tennessee (we used a node with an Intel Xeon E5520 CPU). Figure 1 shows the time it takes to do one full force calculation as a function of N, l_{max}, and n_{max}. Note that the timing only depends on the number of particles (and expansion cutoffs) and not on their spatial distribution.

The CPU and GPU SCF codes are both theoretically $\mathcal{O}(l_{max}^2 n_{max} N)$. At low N, the GPU is not fully loaded and *ETICS* performance seems superlinear with N. *ETICS*-MEX is theoretically $\mathcal{O}(l_{max}^2 N \, logN)$, but this again is an asymptotic behavior which is not observed. The lack of good GPU load for $N \lesssim 10^6$ is much more evident than the $NlogN$ nature of the algorithm. The GPU global memory was the limiting factor in how many particles could be used with both methods. All codes should scale quadratically with l_{max}, but as the middle panel of Figure 1 shows, this behavior is not as clear for *ETICS*-MEX. This is due to the extensive memory access this code requires, which rivals the calculation time. Memory latency on GPUs is not easy to predict; due to caching and the way memory is copied in blocks, and the latency depends not only on the amount of memory accessed but also on the memory access pattern.

References

Bell, N. & Hoberock, J. 2011, in *GPU Computing Gems Jade Edition*, ed. W.-M W. Hwu (Waltham, MA: Morgan Kaufmann), 359

Hernquist, L. 1990, *ApJ*, 356, 359

Hernquist, L. & Ostriker, J. P. 1992, *ApJ*, 386, 375

Hernquist, L., Sigurðsson, S., & Bryan, G. L. 1995, *ApJ*, 446, 717

Meiron, Y., Li, B., Holley-Bockelmann, K., & Spurzem, R. 2014, *ApJ*, 792, 98

Spurzem, R., Berczik, P., Berentzen, I., *et al.* 2012, in *Large-Scale Computing Techniques for Complex System Simulations*, ed. W. Dubitzky, K. Kurowski, & B. Schott (Hoboken, NJ: John Wiley & Sons), 35

Star clusters and black holes in galaxies across cosmic time
Proceedings IAU Symposium No. 312, 2014
Y. Meiron, S. Li, F.-K. Liu & R. Spurzem, eds.

© International Astronomical Union 2016
doi:10.1017/S1743921315007863

The role of three-body stability in tidally interacting globular clusters

Gareth F. Kennedy[1,2]

[1]National Astronomical Observatories of China, Chinese Academy of Sciences,
Beijing 100012, China
email: `gareth.f.kennedy@gmail.com`

[2]Monash Centre for Astrophysics, Monash University, Clayton, Vic, Australia, 3800

Abstract. The role of stability in the general three-body problem is investigated with regard to the tidal radius of a globular cluster (GC) in a galactic potential. This proceedings is a summary of two papers which outline the stability method (Kennedy 2014a) and compare the predicted stability boundary radius to observations of velocity dispersion profiles in Milky Way GCs (Kennedy 2014b).

Keywords. gravitation, stellar dynamics, methods: analytical, stars: kinematics, globular clusters: general

1. Introduction

The discovery of flattening in the velocity dispersion profile for the galactic globular clusters (GCs) NGC 5139 and NGC 7078 (Scarpa *et al.* 2003), where dark matter is thought not to exist, has generated a lot of excitement with some studies claiming that this is direct evidence for a breakdown of Newton's laws at low accelerations. Explanations for the observed deviation broadly fit into three categories; tidal interactions, dark matter or a modified gravity theory (e.g. MOND). All of these theories can produce a flattening of the velocity dispersion profile for large radii, in the case of MOND this occurs at the critical acceleration of 1.2×10^{-10} m/s^2 Milgrom (1983). The premise of this proceeding is that the velocity dispersion profile will flatten in the outer regions of the cluster where stellar orbits become unstable.

A stability analysis, detailed in Kennedy (2014a), is used to predict the occurrence of unstable stellar orbits in the outermost region of a GC. Stars on unstable orbits around the cluster centre will random walk in orbital binding energy until their eventual escape from the cluster. The timescale for this random walk is currently being investigated using a high resolution N-body simulation of a tidally interacting star cluster on an orbit based on NGC 6341 and will be presented in the third paper in this series. This simulation has run for over 1 Gyr and consists of over 10^5 particles, resolving each star in the cluster. During the random walk stage stars have a different distribution of velocities when compared to stars in an isolated cluster. This manifests as a flattening of the velocity dispersion profile beyond the transition radius between stable and unstable orbits.

The transition radius from stable to unstable orbits was compared to observational data of the velocity dispersion profiles for 15 Milky Way GCs with approximately known orbital parameters in Kennedy (2014b). It was found that the transition radius predicts where the velocity dispersion flattens and that there is no need for any MOND type theories to explain the observations.

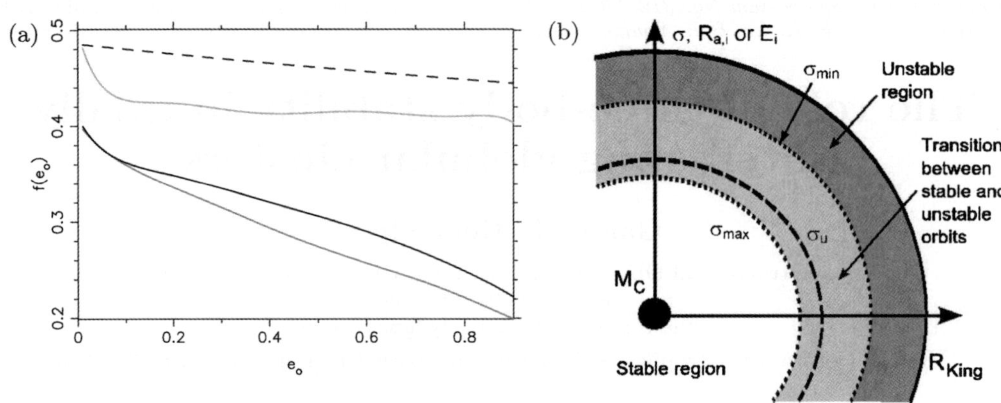

Figure 1. Left panel shows the eccentricity dependence for the transition from stable inner orbits to unstable exterior orbits. The indicative value for the stability boundary (r_c) is shown as a black curve, while the minimum and maximum extents of the partially stable region are the red and green curves respectively. The King radius (Equation 2.2) is shown as the dashed black curve for comparison. Right panel shows a conceptualisation of the stability of stellar orbits in a GC. The distance from the cluster centre associated with the transition from unstable (dark shading) to stable orbits (unshaded inner region) is indicated by the ratio of outer to inner periods σ_u. The region where orbits can be found in either unstable or stable configurations is shown as a light shading between $\sigma_{\min}$ and $\sigma_{\max}$.

2. Stability theory

The method used for determining the stability boundary is described in detail in the first paper in this series Kennedy (2014a) which was based on the stability of the general three-body problem as derived in Mardling (2008) and Mardling (2013).

To describe this stability transition radius the functional form

$$r_t = R_p \left(\frac{M_C}{M_G} \right)^{1/3} f(e) \tag{2.1}$$

is adopted, where M_G is the effective point mass for the Milky Way galaxy, e is the eccentricity of the cluster orbit around the galaxy, and R_p is the distance of closest approach to the galaxy. This form allows direct comparison between the stability boundary and other tidal radii estimates, for example the classical King (1962) tidal radius is given by

$$f(e) = 0.7 \left(3 + e \right)^{-1/3}. \tag{2.2}$$

The tidal radius (r_t) will be used to denote the maximum theoretical tidal radius of a GC by using Equations 2.1 and 2.2.

In Kennedy (2014a) the stability boundary was found to be well approximated by three polynomials indicating the minimum and maximum limits of the partial stability region and an indicative value referred to as the chaos radius and denoted by r_c. The three polynomials for $f(e)$ are shown in Figure 1 (a) with the eccentricity function associated with r_c shown as a solid black curve, the reader is referred to Kennedy (2014a) for details. For radii $r < r_{\min}$ all stars are expected to be on stable orbits, whereas for $r > r_{\max}$ they are expected to be on unstable orbits. A schematic diagram of the cluster stability is shown in Figure 1 (b) where the ratio of GC-galaxy orbital period to the orbital period of the star in the GC (σ) is used as a proxy for radius.

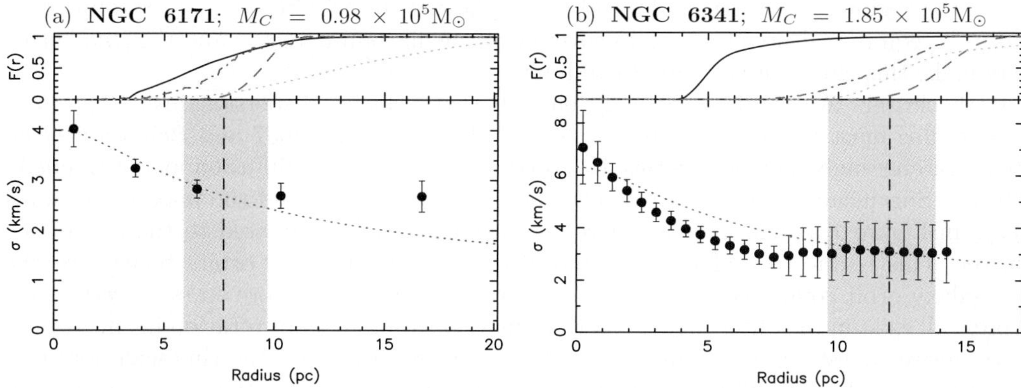

Figure 2. The observed velocity dispersion profile for two GCs. The vertical lines show the best fit to the flattening of the velocity dispersion with errors denoted by the shaded region and the red dotted curve indicates the theoretical velocity dispersion profile for the quoted cluster mass. The cumulative distributions above each velocity dispersion profile show the chaos radius (solid black curve), MOND radius (red dashed curve), tidal radius (green dotted curve) and best fit flattening radius (dot-dashed blue curve).

3. Implications for observations

The bottom panels in Figure 2 show the velocity dispersion as a function of distance from the cluster centre for two of the 15 GCs presented in Kennedy (2014b). References for the velocity dispersion data and basis for the GC-galaxy orbital parameters are given in Kennedy (2014b). The equilibrium velocity dispersion as a function of cluster radius, r, for a Plummer sphere is Dejonghe (1987)

$$\sigma^2(R) = \frac{3\pi G M_C}{64 r_h} \left(1 + \frac{r^2}{r_h^2}\right)^{-1/2} \tag{3.1}$$

where M_C is the cluster mass, r_h is the half-mass radius and G has its usual value. This function is shown as the dotted curves for the quoted cluster masses in Figure 2.

The key requirements for predicting the flattening of GCs is the cluster mass and the GC-galaxy orbital parameters. Both of which have large observational uncertainties. Uncertainties in the orbital elements (R_p and e) and cluster mass (M_C) are treated in Kennedy (2014b) using 10^6 realisations of the GC-galaxy orbit for each GC in a realistic galactic potential taking the observed GC proper motions as initial conditions. For each set of orbital elements and cluster mass the predicted radii for each model are determined and the resulting cumulative distributions for each radii are shown in the top panels of Figure 2 for NGC 6171 and NGC 6341.

The MOND radius r_m as calculated for each cluster from where the acceleration from the cluster potential goes beneath the MOND limit of $a_0 = 1.2 \times 10^{-10}$ m/s^2 is shown in the top panels of Figure 2 as a dashed red line. The transition from stable to unstable orbits r_c (Figure 1 a) is shown as a solid black line, the tidal radius r_t (Equation 2.2) is shown as a dotted green line and the flattening radius that best fits the data is shown as a dot-dashed blue line. The region of best fit for the flattening radius is also shown as the shaded region in the bottom panels.

Both NGC 6171 and NGC 6341 are promising candidates for distinguishing Newtonian and MOND models since these clusters have small relative errors for both the cluster mass and orbital parameters (Kennedy 2014b). For NGC 6341 the existing data is quite good, although further observations of the velocity dispersion at large radii would better probe the region close to the tidal radius. In the case of NGC 6171 more resolution in radius is required, although the data coverage of the flattening region and the small error bars make this cluster a very good candidate.

Both clusters are rotating strongly enough that velocity dispersion inside the half-mass radius must be modelled in more detail (Drukier *et al.* 2007 and Bellazzini *et al.* 2012). Interestingly increased rotation is predicted by the chaos diffusion model Kennedy (2014a). Specifically, the region between the chaos radius (r_c) and the maximum radius ($f_{\max}$ in Figure 1 a) will have a different rotational profile compared to the rest of the cluster. This is due to preferential removal of stars on prograde orbits relative to the GC-galaxy orbit compared to stars on retrograde orbits. Thus more stars on retrograde orbits will exist in the outer regions on the cluster, leading to a net rotation in this region.

To answer questions regarding the emergence of rotation and the timescale for stars to escape via chaotic diffusion, an N-body simulation based on the mass and orbital parameters of NGC 6341 is currently in progress, the results from which will be presented in an upcoming publication.

4. Conclusions

Flattening of the velocity dispersion of GCs is predicted to occur beyond the stability radius by consideration of Newtonian three-body stability. Kennedy (2014a) presented an easy to use stability radius which has a much stronger dependence on the eccentricity of the cluster-galaxy orbit than the classical tidal radius. As the stability radius depends on the GC-galaxy orbit and not just on the cluster mass, further observations of the motion of GCs will provide a way of distinguishing these predictions from MOND models.

At present the orbital determinations are not quite sufficient to definitively rule out those MOND models that predict flattening for all 15 GCs (see Kennedy 2014b). However NGC 6171 and NGC 6341 have been identified as promising candidates for distinguishing Newtonian and MOND models, requiring only moderate improvement in the observations. In particular, NGC 6171 may already be showing evidence of chaotic diffusion of stars leading to flattening at the chaos radius as seen in the overlap between the black and blue curves in Figure 2 (a). Clarification of this phenomenon will require more resolution in the velocity dispersion and better orbital determination.

References

Bellazzini, M., Bragaglia, A., Carretta, E., Gratton, R. G., Lucatello, S., Catanzaro, G., & Leone, F. 2012, *A&A*, 538, A18

Dejonghe, H. 1987, *MNRAS*, 224, 13

Drukier, G. A., Cohn, H. N., Lugger, P. M., Slavin, S. D., Berrington, R. C., & Murphy, B. W. 2007, *AJ*, 133, 1041

Kennedy, G. F. 2014, *MNRAS*, 444, 3328

Kennedy, G. F. 2014, *MNRAS*, 445, 4443

King, I. 1962, *AJ*, 67, 471

Mardling, R. A. 2008, *Lecture Notes in Physics, Vol. 760: The Cambridge N-body Lectures*, 59

Mardling, R. A. 2013, *MNRAS*, 435, 2187

Milgrom, M. 1983, *ApJ*, 270, 365

Scarpa, R., Marconi, G., & Gilmozzi, R. 2003, *A&AL*, 405, L15

Star clusters and black holes in galaxies across cosmic time
Proceedings IAU Symposium No. 312, 2014
Y. Meiron, S. Li, F.-K. Liu & R. Spurzem, eds.
© International Astronomical Union 2016
doi:10.1017/S1743921315007875

Planetary systems in star clusters

Maxwell Xu Cai[1,2], **Rainer Spurzem**[1,2,3] **and M. B. N. Kouwenhoven**[2]

[1]National Astronomical Observatories, Chinese Academy of Sciences,
Beijing 100012, 20A Datun Road, Chaoyang District, Beijing, P.R. China
email: maxwell@nao.cas.cn

[2]Kavli Institute for Astronomy and Astrophysics, Peking University
5 Yi He Yuan Road, Haidian District, Beijing

[3]Astronomisches Rechen-Institut, Zentrum für Astronomie, University of Heidelberg,
Mönchhofstrasse 12-14, 69120 Heidelberg, Germany

Abstract. In the solar neighborhood, where the typical relaxation timescale is larger than the cosmic age, at least 10% to 15% of Sun-like stars have planetary systems with Jupiter-mass planets. In contrast, dense star clusters, characterized by frequent close encounters, have been found to host very few planets. We carry out numerical simulations with different initial conditions to investigate the dynamical stability of planetary systems in star cluster environments.

Keywords. planetary system, star cluster, stability, free-floating planets

1. Introduction

Recent studies suggest that most stars are formed in groups and clusters (Lada & Lada 2003). The evolution of star clusters is driven by internal (e.g. two-body relaxation, mass loss by stellar evolution), as well as external processes (e.g. tidal forces of a galaxy) (Spitzer 1987, Elson *et al.* 1987). Since these mechanisms finally lead to the dissolution of clusters they have a finite life time. So, if our sun has been born in an open star cluster, the solar planetary system may also have formed before the dissolution of its parent star cluster.

Observations of exoplanets have been possible since the early 1990s. As of December 2014, more than 1,800 exoplanets have been discovered, among which are 473 multi-planet systems like our own solar system (http://exoplanet.eu). Nevertheless, most exoplanets are found in the solar neighborhood; very few exoplanets (e.g. Kepler-66, Kepler-67) are found in star clusters. Star clusters, especially cores of globular clusters, are characterized by high density of stars and high velocity dispersions, which results in frequent close encounters. Current observations suggest that planetary systems in dense regions of star clusters have difficulties to maintain their orbital stability. This could be due to instability against external gravitational perturbations, though there may also be some instrumental biases due to detection limits.

Previous studies include direct *N*-body simulation (Spurzem *et al.* 2009) and Monte-Carlo simulations (Hao *et al.* 2013). We present numerical simulations to investigate the dynamical stability of planetary systems in star clusters. Coupling planetary system with star cluster dynamics is a multi-scale problem, due to time scales ranging from $\sim 1 - 100$ years for planetary orbits to $\sim 10^4 - 10^6$ years for stars orbiting in the cluster. Full direct coupled simulations are expensive, though not impossible (see e.g. Spurzem *et al.* 2009); but frequently Monte Carlo approaches are used (see e.g. Hao *et al.* 2013). The AMUSE framework is a convenient toolbox to study such a problem (Portegies Zwart *et al.*, 2013; Portegies Zwart *et al.*, 2009). Within AMUSE we integrate star cluster dynamics directly with NBODY6++ (Spurzem 1999), while the planets are integrated with MERCURY6

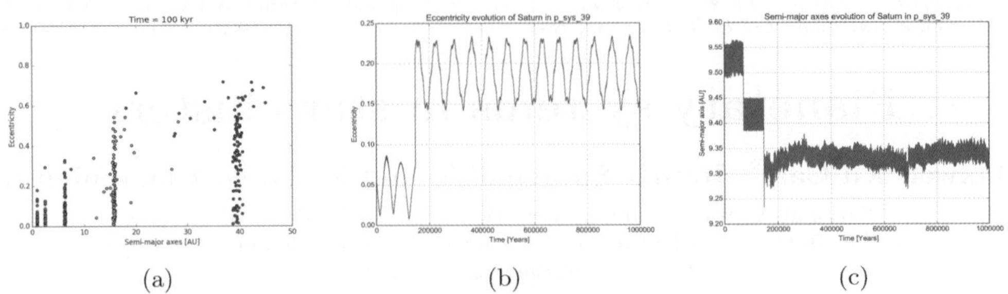

Figure 1. (a) Diffusion of the $a-e$ space at $t = 100$ kyr of a simulation; (b) Eccentricity evolution of Saturn in one of the planetary systems; (c) Corresponding semi-major axis evolution of Saturn in the same planetary system as (b).

(Chambers 1999). The perturbations are implemented with the **BRIDGE** scheme (Fujii *et al.*, 2007; Pelupessy *et al.*. 2013). All coupling is done within **AMUSE** framework.

2. Results and conclusions

For the initial setup of our system we follow the previously cited papers (Plummer sphere for stellar density, $N = 2$k and $N = 10$k stars, virial radius $R_v = 1$ pc; (A) current gas giant configuration or (B) equal mass and mutual Hill radii separation for the planets). For case (B) the inner planet has $M = 1M_J$ and $a = 1$ AU. We distribute 50 (100) planetary systems randomly across the $N = 2$k ($N = 10$k) cluster, to sample different positions and environments of planetary systems in star clusters.

Figure 1a shows the $a - e$ space (semi-major axis verses eccentricity) at the end of the $N = 10$k simulation (corresponding to $T = 0.1$ Myr). The evolution of semi-major axes and eccentricities of Saturn in the $N = 2$k simulation are shown in Figures 1b and 1c, respectively. Our simulations show that generally planetary systems have difficulty surviving within dense cluster environments, but some compact ones are sheltered in the potential well of their hosts from disruption.

We acknowledge support by NAOC CAS through the Silk Road Project and (RS) through the Chinese Academy of Sciences Visiting Professorship for Senior International Scientists, Grant Number 2009S1-5. We are grateful for the supports from the **AMUSE** team, especially Simon Portegies Zwart, Inti Pelupessy, Arjen van Elteren and Nathan de Vries for various useful discussions. MXC and MBNK acknowledge the **AMUSE** team for supporting their visits to Leiden, and Michiko Fujii for useful discussions.

References

Chambers, J. E. 1999, *MNRAS*, 304, 793
Elson, R., Hut, P., & Inagaki, S. 1987, *ARAA*, 25, 565
Fujii, M., Iwasawa, M., Funato, Y., & Makino, J. 2007, *PASJ*, 59, 1095
Hao, W., Kouwenhoven, M. B. N., & Spurzem, R. 2013, *MNRAS*, 433, 867
Lada, C. J. & Lada, E. A. 2003, *ARAA*, 41, 57
Pelupessy, F. I., van Elteren, A., de Vries, N., *et al.* 2013, *A&A*, 557, AA84
Portegies Zwart, S. *et al.* 2013, *Computer Physics Communications*, 183, 456
Portegies Zwart, S., McMillan, S., Harfst, S., *et al.* 2009, *New Astronomy*, 14, 369
Spitzer, L. 1987, *Princeton, NJ, Princeton University Press*, 1987, 191 p.,
Spurzem, R. 1999, *Journal of Computational and Applied Mathematics*, 109, 407
Spurzem, R., Giersz, M., Heggie, D. C., & Lin, D. N. C. 2009, *ApJ*, 697, 458

Star clusters and black holes in galaxies across cosmic time
Proceedings IAU Symposium No. 312, 2014
Y. Meiron, S. Li, F.-K. Liu & R. Spurzem, eds.
© International Astronomical Union 2016
doi:10.1017/S1743921315007887

Dynamics of primordial binary stars in multiple-population globular clusters

Jongsuk Hong[1], Enrico Vesperini[1], Antonio Sollima[2], Steve McMillan[3], Franca D'Antona[4] and Annibale D'Ercole[2]

[1]Department of Astronomy, Indiana University, Bloomington, IN, 47401, USA
[2]INAF – Osservatorio Astronomico di Bologna, via Ranzani 1, I-40127 Bologna, Italy
[3]Department of Physics, Drexel University, Philadelphia, PA 19104, USA
[4]INAF – Osservatorio Astronomico di Roma, via di Frascati 33, I-00040 Monteporzio, Italy

Abstract. We have performed a survey of N-body simulations to explore the dynamics of primordial binaries in multiple-population globular clusters. We show that, as a consequence of the initial differences between the spatial distribution of first-generation (FG) and second-generation (SG) stars, SG binaries are disrupted more efficiently than FG binaries. The effects of dynamical evolution on the surviving binaries produces a difference between the SG and the FG binary binding energy distribution with the SG population characterized by a larger fraction of high binding energy (more bound) binaries. We also explore the evolution of the radial variation of the SG-to-FG binary number ratio and find that although the global binary fraction decreases more rapidly for the SG population, the local binary fraction measured in the cluster inner regions may still be dominated by SG binaries.

Keywords. globular clusters : general – stars : chemically peculiar

---

1. Introduction

Theoretical models for the formation of multiple-population globular clusters predict that second-generation (SG) stars form in a compact subsystem in the cluster central regions (see e.g. D'Ercole *et al.* 2008; Bekki 2011) and several observational studies have indeed found clusters in which SG stars are more centrally concentrated than the first-generation (FG) population (see e.g. Sollima *et al.* 2007; Bellini *et al.* 2009; Lardo *et al.* 2011). In a previous study (Vesperini *et al.* 2011), we have shown that the initial differences in the structural properties of FG and SG stars can have significant implications for the evolution of the binary star population . We present here a brief summary of the results of an extended survey of N-body simulations aimed at exploring the evolution of binary stars in multiple-population globular clusters (see Hong *et al.* 2015 for a complete presentation of our results).

2. Results

We have followed the evolution of systems with different structural parameters, binary fraction, and binary binding energy. In Fig.1 we show the results of two of the simulations of our survey. For the results shown in Fig. 1, the initial multiple-population system is characterized by a ratio of the FG to the SG half-mass radius equal to 5 and an initial binary fraction equal to 10 per cent. We compare the results of our simulation of the multiple-population cluster with those of simulations of a single-population cluster with the same total mass and radius. We present here the results for two values of the binary binding energy corresponding to values of the hardness parameter (as measured in the reference single-population cluster) equal to $x_{g,0} = 5$ and $x_{g,0} = 20$.

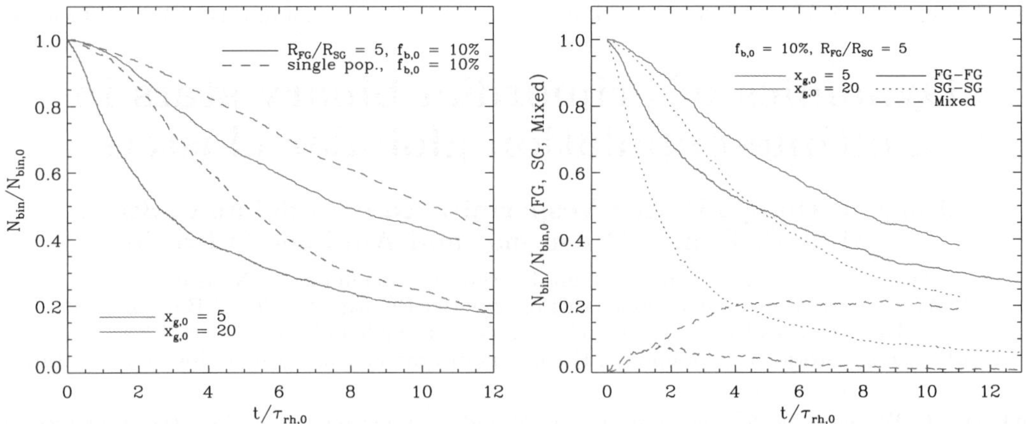

Figure 1. Left Time evolution (time is normalized to the initial half-mass relaxation time) of the total number of binaries (normalized to the initial total number of binaries) for multiple-population and single-population systems for binaries with different initial hardness. **Right** Time evolution of the number of FG, SG and mixed binaries (binaries with one FG and one SG component) for a multiple-population cluster and binaries with different initial hardness parameter.

The two panels of Fig.1 illustrate some of the general conclusions reached in our investigation. In particular the left panel shows that binary disruption is in general enhanced in multiple-population clusters compared to single-population systems. The right panel shows the preferential disruption of SG binaries. This is a consequence of the fact that the SG population is more spatially concentrated than the FG population.

Some binaries are not disrupted and can undergo an exchange event during which mixed binaries composed of one FG and one SG component are produced.

Although SG binaries are disrupted at a larger rate than FG binaries and the global number of FG binaries is always larger than that of SG binaries, the local binary fraction in the cluster inner regions may be larger for SG binaries. We have shown that the SG-to-FG binary number ratio measured at a distance from the cluster center equal to about 1–2.5 times r_{h} (where r_{h} is the cluster half-mass radius) is approximately equal to the global value of this ratio. Our simulations also show that the effects of dynamical evolution produce a difference between the SG and the FG binary binding energy distribution of the surviving binaries with the SG population characterized by a larger fraction of high binding energy (more bound) binaries.

Acknowledgements

JH acknowledges support from AAS International Travel Grant.

References

Bekki, K., 2011, MNRAS, 412, 2241

Bellini, A., Piotto, G., Bedin, L. R., *et al.* 2009, *A&A*, 507, 1393

D'Ercole, A., Vesperini, E., D'Antona, F., McMillan, S. L. W., & Recchi, S., 2008, MNRAS, 391, 825

Hong, J. Vesperini, E., Sollima, A., McMillan, S. L. W., D'Antona, F., & D'Ercole, A., 2015, submitted to MNRAS

Lardo, C., Bellazzini, M., Pancino, E., *et al.* 2011, *A&A*, 525, AA114

Sollima, A., Ferraro, F. R., Bellazzini, M., *et al.* 2007, *ApJ*, 654, 915

Vesperini, E., McMillan, S. L. W., D'Antona, F., & D'Ercole, A., 2011, MNRAS, 416, 355

Star clusters and black holes in galaxies across cosmic time
Proceedings IAU Symposium No. 312, 2014
Y. Meiron, S. Li, F.-K. Liu & R. Spurzem, eds.

© International Astronomical Union 2016
doi:10.1017/S1743921315007899

Evolutionary models of rotating dense stellar systems: challenges in software and hardware

Jose Fiestas

National Astronomical Observatories of China, Chinese Academy of Sciences, Datun Lu 20A,
Chaoyang District, Beijing 100012, China
email: `fiestas@nao.cas.cn`

Abstract. We present evolutionary models of rotating self-gravitating systems (e.g. globular clusters, galaxy cores). These models are characterized by the presence of initial axisymmetry due to rotation. Central black hole seeds are alternatively included in our models, and black hole growth due to consumption of stellar matter is simulated until the central potential dominates the kinematics in the core. Goal is to study the long-term evolution ($\sim$ Gyr) of relaxed dense stellar systems, which deviate from spherical symmetry, their morphology and final kinematics. With this purpose, we developed a 2D Fokker-Planck analytical code, which results we confirm by detailed N-Body techniques, applying a high performance code, developed for GPU machines. We compare our models to available observations of galactic rotating globular clusters, and conclude that initial rotation modifies significantly the shape and lifetime of these systems, and can not be neglected in studying the evolution of globular clusters, and the galaxy itself.

Keywords. methods: numerical – gravitation – stellar dynamics – black hole – globular clusters: general - galactic nuclei

---

1. Systems without black hole

As initial models, we use rotating King Models. We select a set of galactic Globular Clusters which are flattened by rotation, and apply the technique of Fiestas et.al. (2006) in order to find initial configurations for these systems. Figure 1 shows simulation data (symbols without error bars), observational data (symbols with error bars), together with curves of isotropic rotating spheroids (solid lines).

Observations show in many cases clear deviations from isotropic rotators from which our models do not deviate significantly. Note that final configurations are a consequence of dynamical relaxation, and angular momentum transfer. It shows that only in a few cases, ellipticity is determined by rotation, while in other cases anisotropy and/or tidal effects modify the cluster shape. More studies are being done in order to rule out observation uncertainties, which will be target of our future simulations.

2. Systems with black hole

The gravitational dominance of the central BH of mass $M_{\rm bh}$ over the surrounding stars with velocity dispersion $\sigma_{\rm c}$ vanishes at a radius defined by

$$r_{\rm h} = \frac{GM_{\rm bh}}{\sigma_{\rm c}^2}$$

Our simulations show that rotation grows inside $r_{\rm h}$, what can be explain through capture of strong rotating massive stars by the BH, and angular momentum transfer happening from the lighter to the massive component as a consequence of dynamical relaxation.

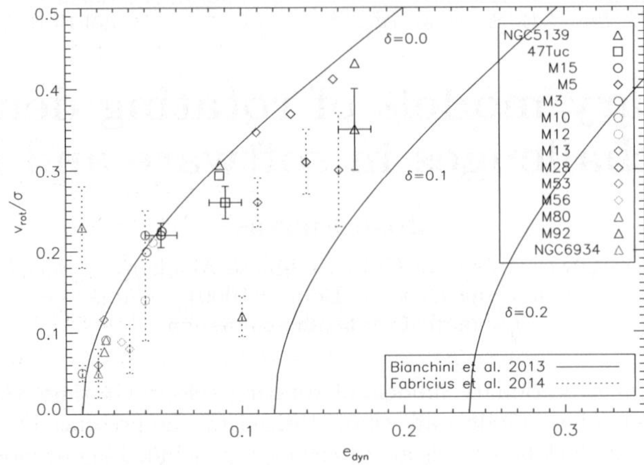

Figure 1. v_{rot}/σ against ellipticity for selected galactic globular clusters

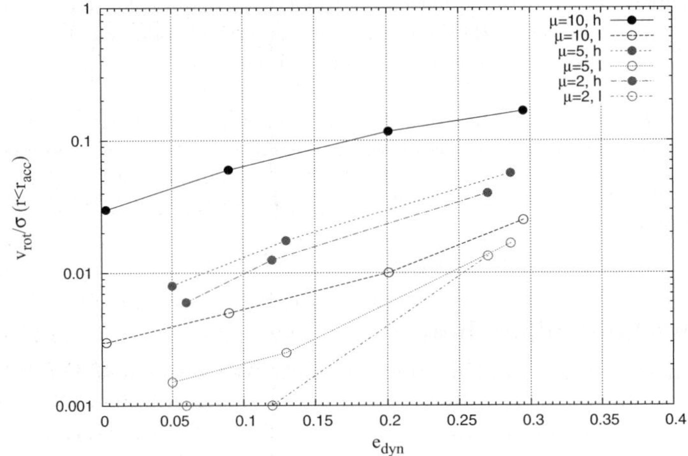

Figure 2. v_{rot}/σ for all BH models

Figure 2 shows final rotation of simulation data inside rh vs. dynamical ellipticity for different individual mass rates $\mu = m_h/m_l$ for heavy (filled circles) and light (open circles) stars. Symbols are connected with lines, along which initial rotation increases from left to right. This correlation is a consequence of angular momentum transfer, enhanced by relaxation processes, which maintain v_{rot} increasing towards the center. Although the obtained v_{rot}/σ in our set of models is difficult to detect by current observational techniques, ellipticity in galactic cores or globular star clusters is a measurable quantity, which can be used to detect this feature, and give conclusions about the formation and evolution of systems harboring central BHs.

References

Bahcall J. N. & Wolf R. A. 1976, *ApJ*, 209, 214
Fiestas J., Spurzem R. & Kim E. 2006, *MNRAS*, 373, 677
Fiestas J. & Spurzem R. 2010, *MNRAS*, 405, 194
Fiestas J., Porth O. & Spurzem R. 2012, *MNRAS* 419, 57

Star clusters and black holes in galaxies across cosmic time
Proceedings IAU Symposium No. 312, 2014
Y. Meiron, S. Li, F.-K. Liu & R. Spurzem, eds.

© International Astronomical Union 2016
doi:10.1017/S1743921315007905

Dynamical modeling of the Arches cluster using Fokker-Planck calculations

Joowon Lee[1] and Jihye Shin[2]

[1] School of Space Research, Kyung Hee University, Yongin, Gyeonggi 446-701, Korea
email: joowon.lee@khu.ac.kr

[2] Kavli Institute for Astronomy and Astrophysics, Peking University, 5 Yiheyuan Road,
Haidian District, Beijing 100871, China

Abstract. The Arches cluster is a young, compact, and massive star cluster located in $\sim$30pc away from the Galactic Center in projection. The cluster is located in the extreme environment of the Galactic Center, making it an excellent target for understanding the effects of star-forming environment on the mass function of star clusters. In this study, we estimate the initial condition (mass, concentration parameter, and galactocentric radius) of the Arches cluster by comparing Fokker-Planck calculations with observed velocity dispersion, surface density and mass function data.

Keywords. Galaxy: globular clusters: individual(Arches), Galaxy: kinematics and dynamics

---

1. Introduction

Arches cluster, one of the starburst clusters near the Galactic Center (GC), is known to be young (2–2.5 Myr; Najarro *et al.* 2004, Martins *et al.* 2008), massive ($\sim 10^4$ $M_\odot$; Figer *et al.* 1999, 2002), and compact ($\sim 10^5$ $M_\odot pc^{-3}$; Espinoza *et al.* 2009). Arches cluster is located in $\sim$30 pc away from the GC in projection, thus it is expected to be formed under the extreme environments of high stellar and gas densities, turbulent motion, tidal forces, strong magnetic and radiation field (Stolte *et al.* 2002). Therefore, Arches cluster has been an excellent target for understanding massive star formation, stellar mass function, and dynamical evolution of the cluster near the GC.

2. Observation and models

Fokker-Planck (FP) results are compared with the observations of velocity dispersion (Clarkson *et al.* 2012), surface density (Espinoza *et al.* 2009), and mass function (Kim *et al.* 2006). We use the FP model that solves the two dimensional orbit-averaged FP equation and considers multi-mass clusters, three-body heating, tidal boundary condition, and stellar evolution (Takahashi 1997). For a mass spectrum, we use Kroupa initial mass function (Kroupa 2002) whose mass range is 0.1 - 150 $M_\odot$. King profile (King 1966) is adopted to assign the initial velocity and density distribution. We perform a parametric survey for concentration parameter W_0, galactocentric radius r_g, and initial mass of the cluster $M_{\rm init}$ and trace the best parameters for the current observations of the Arches cluster.

3. Results

The figure of merits for velocity dispersion, surface density, and mass function are calculated to quantify the similarity between the observations and FP results (Figure 1).

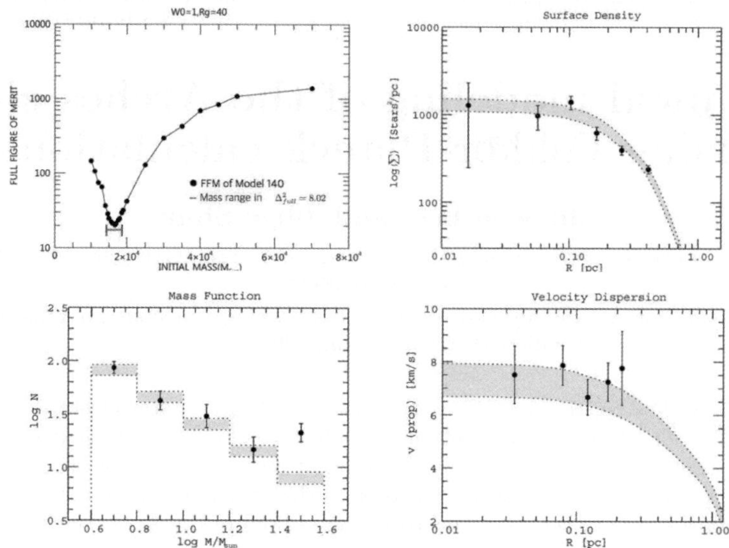

Figure 1. Left-top : Results of the full figure of merit of Model 140 (W_0=1 and r_g=40pc). **Right-top and bottom** : Comparisons of observations with FP results. Surface density (Espinoza *et al.* 2009), mass function (Kim *et al.* 2006), and velocity dispersion (Clarkson *et al.* 2012) data are used. Blue areas correspond to the results for 95.4% confidence level (2σ) of the best model.

Observed velocity dispersion, surface density and mass function are compared with the FP results in Figure 1 (right-top and two bottom panels). We calculate the *full figure of merit* (left-top panel of Figure 1) which is the total of three figure of merits, and find that the best model has $W_0 = 1$ and $r_g = 40$pc (Model 140). The initial mass range of the Arches cluster from our best model, which is $M_{\rm init} = (1.5-1.8) \times 10^4$ M$_\odot$ within the 95.4% confidence level.

References

Clarkson, W. I., Ghez, A. M., Morris, M. R., Lu, J. R., Stolte, A., McCrady, N., Do. T., & Yelda, S. 2012, *ApJ*, 751, 132

Espinoza, P., Selman, F. J., & Melnick, J. 2009, *A&A*, 501, 563

Figer, D. F., Najarro, F., Gilmore, D., Morris, M., Kim, S. S., & Serabyn, E. 2002, *ApJ*, 581, 258

Figer, D., Kim, S. S., Morris, M., Serabyn, E., Rich, R. M., & McLean, I. S. 1999, *ApJ*, 525, 750

King, I. R. 1966, *AJ*, 71, 64

Kim, S. S., Figer, D. F., Kudritzki, R. P., & Najarro. F. 2006, *ApJ*, 653, L113

Kroupa, P. 2002, *Science*, 295, 82

Martins, F., Hillier, D. J., Paumard, T., Eisenhauer, F., Ott, T., & Genzel, R. 2008, *A&A*, 478, 219

Najarro, F., Figer, D. F., Hillier, D. J., & Kudritzki, R. P. 2004, *ApJ*, 611, L105

Stolte, A., Grebel, D. K., Brandner W., & Figer, D. F. 2002, *A&A*,394, 459

Takahashi, K. 1997, *PASJ*, 49, 547

Star clusters and black holes in galaxies across cosmic time
Proceedings IAU Symposium No. 312, 2014
Y. Meiron, S. Li, F.-K. Liu & R. Spurzem, eds.
© International Astronomical Union 2016
doi:10.1017/S1743921315007917

The grey extinction curve in NGC 3603

Xiaoying Pang[1], Anna Pasquali[2] and Eva K. Grebel[2]

[1] Shanghai Institute of Technology, 100 Haiquan Road, Fengxian district, Shanghai 201418,
China
email: xypang@bao.ac.cn
[2] Astronomisches Rechen-Institut, Zentrum für Astronomie der Universität Heidelberg,
Mönchhofstr. 12–14, 69120 Heidelberg, Germany

Abstract. We use photometry in the *F220W*, *F250W*, *F330W*, *F435W* filters from the High Resolution Channel of the Advanced Camera for Surveys and photometry in the *F555W*, *F675W*, and *F814W* filters from the Wide Field and Planetary Camera 2 aboard the Hubble Space Telescope to derive individual stellar reddenings and extinctions for member stars in the HD 97950 cluster in the giant H II region NGC 3603. Within the standard deviation associated with $E(\lambda - \mathrm{F555W})/E(\mathrm{F435W} - \mathrm{F555W})$ in each filter, the cluster extinction curve at ultraviolet wavelengths tends to be greyer than the average Galactic extinction laws from Cardelli *et al.* (1989) and Fitzpatrick *et al.* (1999). It is closer to the extinction law derived by Calzetti *et al.* (2000) for starburst galaxies, where the 0.2175 μm bump is absent.

Keywords. star cluster, extinction, reddening

1. Extinction Curve in NGC 3603

We have obtained extinctions in the *F220W*, *F250W*, *F330W*, *F435W*, *F555W*, *F675W*, and *F814W* filters for the a hundred cluster main-sequence (MS) member stars in the HD 97950 cluster in NGC 3603. The membership of these stars was determined via relative proper motions in Pang *et al.*(2013). We adopt the foreground reddening from Pandey *et al.* (2000), with an assumption of a Fitzpatrick (1999) extinction law for the foreground ($R_V = 3.1$). After correcting for foreground reddening, the total to selective extinction ratio is $R_{F555W} = 3.75 \pm 0.87$ in the cluster.

In Figure 1, we plot the mean extinction curves for the MS members (black line) already corrected for the foreground extinction, and compare it with Cardelli *et al.*'s ($R_V = 3.1$, 1989), Fitzpatrick's ($R_V = 3.1$, 1999), and Calzetti *et al.*'s ($R_V = 4.05$, 2000) extinction laws (red dashed lines). We normalize the extinction curves to both the visual color excess $E(F435W - F555W)$ (upper panels) and extinction A_λ/A_{F555W} (lower panels). Within the standard deviation associated with $E(\lambda - \mathrm{F555W})/E(\mathrm{F435W} - \mathrm{F555W})$ (error bars), the extinction curve (black line) we obtained for the HD 97950 cluster (upper panels) tends to be greyer than the Galactic extinction laws with $R_V = 3.75$ (red solid lines) and $R_V = 3.1$ (red dashed lines, Cardelli *et al.* 1989; Fitzpatrick 1999) in the near-UV. The cluster extinction curve is closer to the extinction law of Calzetti *et al.* (2000), especially in the UV pass bands (black box). This indicates an anomalous extinction in NGC 3603, which may due to the clumpy dust distribution within the cluster, and the size of dust grains being larger than the average Galactic ISM.

Acknowledgements

This work is funded by Shanghai Key Lab for Astrophysics, No: SKLA1301; and by National Natural Science Foundation of China, No: 11347205.

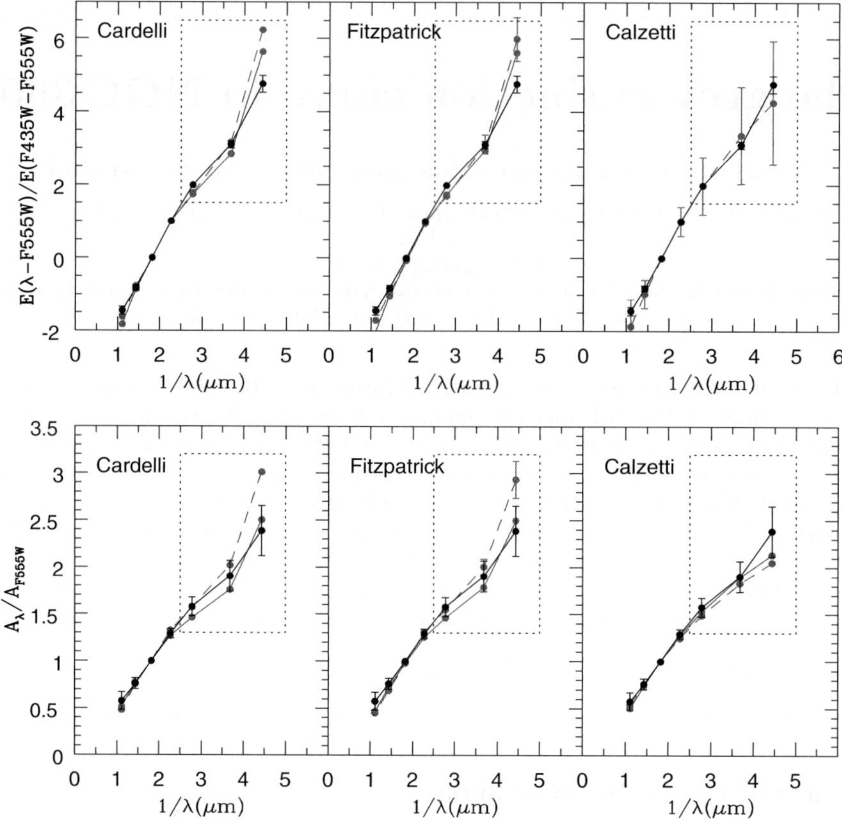

Figure 1. Mean extinction curve $E(\lambda - \mathrm{F555W})/E(\mathrm{F435W} - \mathrm{F555W})$ (upper panels) and A_λ/A_{F555W} (lower panels) for the HD 97950 cluster in NGC 3603 (black lines) already corrected for foreground extinction. The red solid lines trace Cardelli *et al.*'s (1989), Fitzpatrick's (1999) and Calzetti *et al.*'s (2000) extinction laws with the same value of R_V as the cluster ($R_V = 3.75$). The red dashed lines are the average Galactic extinction laws ($R_V = 3.1$) of Cardelli *et al.* (1989) and Fitzpatrick (1999), and Calzetti *et al.*'s (2000) extinction law for starburst galaxies ($R_V = 4.05$). The error bars are the standard deviation associated with $E(\lambda-\mathrm{F555W})/E(\mathrm{F435W}-\mathrm{F555W})$ and A_λ/A_{F555W} in each filter. Uncertainties of Fitzpatrick's (1999) and Calzetti *et al.*'s (2000) extinction laws are also indicated with error bars. For Fitzpatrick's (1999) extinction law, its uncertainty approaches zero for wavelengths larger than 4000 Å owning to the normalization to visual band. No such information is available for Cardelli *et al.*'s (1989) extinction law. The dashed black boxes cover the UV filters where the cluster extinction curve differs most dramatically from the reference extinction laws.

References

Calzetti, D., Armus, L., Bohlin, R. C., *et al.* 2000, *ApJ*, 533, 682
Cardelli, J. A., Clayton, G. C., & Mathis, J. S. 1989, *ApJ*, 345, 245
Fitzpatrick, E. L. 1999, *PASP*, 111, 63
Pandey, A. K., Ogura, K., & Sekiguchi, K. 2000, *PASJ*, 52, 847
Pang, X., Grebel, E. K., Allison, R. J., *et al.* 2013, *ApJ*, 764, 73

Star clusters and black holes in galaxies across cosmic time
Proceedings IAU Symposium No. 312, 2014
Y. Meiron, S. Li, F.-K. Liu & R. Spurzem, eds.

© International Astronomical Union 2016
doi:10.1017/S1743921315007929

Evolution of a globular cluster with a two-component BH mass spectrum

**Dawoo Park[1], Chunglee Kim[1], Hyung Mok Lee[1,2]
and Yeong-Bok Bae[1]**

[1]Astronomy Program, Department of Physics and Astronomy, Seoul National University,
1 Gwanak-ro, Gwanak-gu, Seoul 151-742, Korea.
email:dawoo@astro.snu.ac.kr

[2]Centre for Theoretical Physics, Seoul National University, 1 Gwanak-ro, Gwanak-gu, Seoul
151-742, Korea

Abstract. Globular clusters (GCs) is dominated by behaviors of high-mass components such as
neutron stars (NSs) and black holes (BHs). In this work, we perform direct N-body simulations
of a GC assuming two BH masses. We used the BH masses of 10 and 20 $M_\odot$ with total math
ratio between those two populations is assumed to be 2:1 and 5:1. Our results show that the
heavier BHs (20 $M_\odot$) are depleted in the early stage of cluster evolution. The existence of heavier
BH components increase the retention rate of lighter BHs during the cluster evolution. About
30% of ejected BHs are in binaries.

Keywords. Globular Clusters, Black Holes, Gravitational wave.

---

1. Introduction

During an evolution of GCs, massive components such as BHs and NSs quickly segregate into the cluster core due to dynamical friction. At the center of the cluster, the
enhancement of the BH number density causes dynamical interactions between those
massive components to form binaries. Some BHs are ejected from the cluster through
three-body interactions. Among those ejected, BH-BH binaries in tight orbits are expected to merge with strong GW emission. A global network of GW detectors on Earth
will be capable of detecting BH-BH binary mergers.

Recently, Bae *et al.* (2014) investigated formation and ejection NS-NS and BH-BH
binaries, fixing BH masses to be 10 $M_\odot$. They suggested that about 30% of ejected NSs
and BHs are in a form of binaries.

As an extension of Bae *et al.* (2014), we study the evolution of a GC and ejection of
BH-BH binaries assuming two mass components for the black holes: 10 and 20 $M_\odot$. This
is a simplification of models suggested by Tanikawa (2013) and Belczynski *et al.* (2006).

2. Models and Results

GC model. We use NBODY6 code for our simulation (Aarseth 2003). We employed the
King model with $W_0 = 6$ and central velocity dispersion of 5 km/s as the initial model.
We assumed static tidal field by the Galaxy. A cluster is composed of 3 mass components,
20 $M_\odot$, 10 $M_\odot$, and 0.7 $M_\odot$ particles, where the BH mass fraction is fixed to be 2.5%
of the total mass. We investigated three models (A, B and C) with different BH mass
ratios. Stellar evolution and primordial binaries are not included in our models. More
specific model information is given in Table 1.

Table 1. Models used in this work.

Model[1]	$N_{\text{total}}(\times 10\text{k})^2$	BH mass ratio[3]	$N_{\text{run}}{}^4$
A	5(10)	2:1	13(1)
B	5(10)	5:1	13(1)
C	5(10)	10 M$_\odot$ BH only	13(1)

Notes: [1] Model names, [2] the total number of particles used in each run (N_{total}) in a unit of 1000, [3] the BH mass ratio ($\equiv$ total mass of 10 M$_\odot$ BHs/total mass of 20 M$_\odot$ BHs), [4] the number of simulations performed.

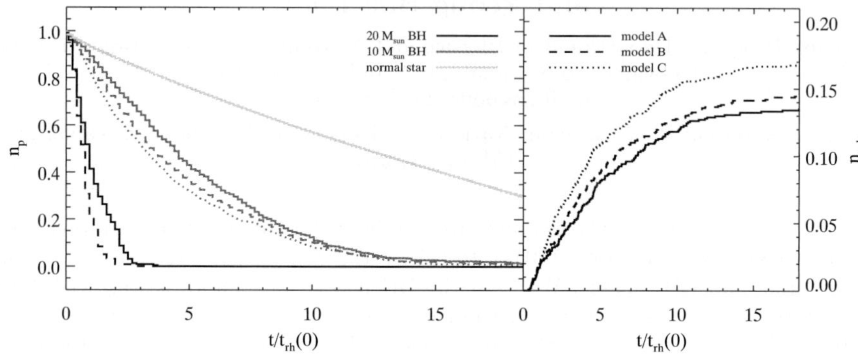

Figure 1. *Left :* Number fraction of retained BHs normalized by their initial number. *Right :* Cumulative number of ejected BH-BH binaries, regardless of binary mass.

BHs retained in the cluster. The left panel of Figure 1 shows the number of stars and BHs in a cluster normalized by their initial number (n_{p}) with time. During the cluster evolution, heavier 20 M$_\odot$ BHs are quickly depleted before $t/t_{\text{rh}}(0) \sim 3$, while a number of 10 M$_\odot$ BHs decreases slower. After enough evolution time ($t/t_{\text{rh}}(0) \sim 10$), about 10% of initial 10 M$_\odot$ BHs can be retained in the cluster.

BH-BH binaries ejected from the cluster. The right panel of Figure 1 shows the cumulative number of ejected BH-BH binaries ($n_{\text{e,b}}$), normalized. The result shows about 30% of initial BHs are ejected in binaries. $n_{\text{e,b}}$ is inversely proportional to the initial portion of 20 M$_\odot$ BHs. We find 10 M$_\odot$ − 10 M$_\odot$ binaries are ejected only after the depletion of 20 M$_\odot$ BHs. Unequal mass binaries (10 M$_\odot$ − 20 M$_\odot$) are rare (less than 3% of the total number of ejected BHs).

Acknowledgements

This work was supported by the National Research Foundation of Korea (NRF) grant NRF-2006-0093852 funded by the Korean government.

References

Aarseth S. J. 2003, *Gravitational N-body Simulations*, Cambridge Univ. Press, Cambridge
Bae, Y.-B., Kim, C., & Lee, H. M. 2014, *MNRAS*, 440, 2714
Belczynski, K., Sadowski, A., Rasio, F. A., & Bulik, T. 2006, *ApJ*, 650, 303
Tanikawa, A. 2013, *MNRAS*, 435, 1358

Star clusters and black holes in galaxies across cosmic time
Proceedings IAU Symposium No. 312, 2014
Y. Meiron, S. Li, F.-K. Liu & R. Spurzem, eds.

© International Astronomical Union 2016
doi:10.1017/S1743921315007930

The coupling of a disk corona and a jet for the radio/X-ray correlation in black hole X-ray binaries

Erlin Qiao

National Astronomical Observatories, Chinese Academy of Sciences, Beijing 100012, China
email: `qiaoel@nao.cas.cn`

Abstract. We interpret the radio/X-ray correlation of $L_{\rm R} \propto L_{\rm X}^{\sim 1.4}$ for $L_{\rm X}/L_{\rm Edd} \gtrsim 10^{-3}$ with a detailed disk corona-jet model, in which the accretion flow and the jet are connected by a parameter, η, describing the fraction of the matter in the accretion flow ejected outward to form the jet. We calculate $L_{\rm R}$ and $L_{\rm X}$ at different $\dot{M}$, adjusting η to fit the observed radio/X-ray correlation of the black hole X-ray transient H1743-322 for $L_{\rm X}/L_{\rm Edd} > 10^{-3}$. It is found that the value of η for this radio/X-ray correlation for $L_{\rm X}/L_{\rm Edd} > 10^{-3}$, is systematically less than that of the case for $L_{\rm X}/L_{\rm Edd} < 10^{-3}$, which is consistent with the general idea that the jet is often relatively suppressed at the high luminosity phase in black hole X-ray binaries.

Keywords. accretion, accretion disk, X-rays: individual: H1743-322, X-rays: binaries, radio continuum: stars

1. Introduction

Recently, a growing number of black hole X-ray binaries (BHBs) have been discovered with a steeper radio/X-ray correlation of $L_{\rm R} \propto L_{\rm X}^{\sim 1.4}$ for $L_{\rm X} \gtrsim 10^{-3} L_{\rm Edd}$, which is suggested to be explained within the framework of a radiatively efficient accretion flow-jet model. Observationally, there is evidence of a coupling of the disk corona and jet at high mass accretion rates and a coupling of the radiatively inefficient accretion flow (RIAF) and jet at low mass accretion rates (Wu *et al.* 2013). Meanwhile, theoretically, for $\dot{M} \gtrsim \alpha^2 \dot{M}_{\rm Edd}$ (with α the viscosity parameter, $\dot{M}_{\rm Edd} = 1.39 \times 10^{18} M/M_\odot$ g s^{-1}), the accretion flow will transit from a RIAF to a disk corona system (Qiao & Liu 2009, 2010, 2013; Narayan & Yi 1995b). In this work, we propose a disk corona-jet model to explain the radio/X-ray correlation of $L_{\rm R} \propto L_{\rm X}^{\sim 1.4}$ for $L_{\rm X}/L_{\rm Edd} \gtrsim 10^{-3}$. We briefly introduce the disk corona-jet model and the results in Section 2. Section 3 is the conclusion.

2. The coupled disk corona-jet model and the result

In the model, the disk and corona are radiatively and dynamically coupled (Liu *et al.* 2002, 2003). Thus, the energy fraction dissipated respectively in the corona and disk, and coronal density and temperature can all be self-consistently determined for given black hole mass and accretion rate. Then the spectrum emitted by the disk and corona can be calculated by Monte Carlo simulation. Assuming a fraction of matter in the accretion flow, $\eta \equiv \dot{M}_{\rm jet}/\dot{M}$, is ejected to form the jet, we can also calculate the emergent spectrum from the jet. So far, the theoretical understanding of the jet formation is poor. Specifically, it is difficult to put constraints on the dependence of η on $\dot{M}$ in our model, so we set η as an independent parameter on $\dot{M}$ to fit the observations.

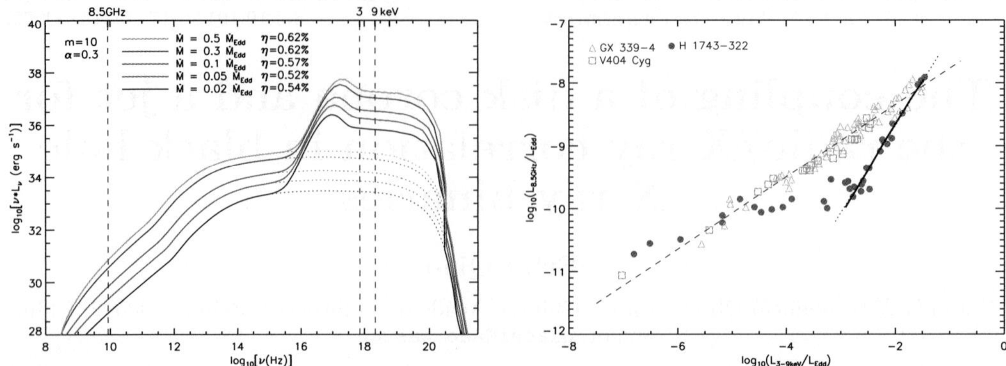

Figure 1. Left panel: Emergent spectra of the disk corona-jet model around a black hole with $M = 10 M_\odot$ assuming $\alpha = 0.3$ for modeling the radio/X-ray correlation of the black hole X-ray transient H1743-322 for $L_{3-9\mathrm{keV}} / L_{\mathrm{Edd}} > 10^{-3}$. From the bottom up, the solid lines are the combined emergent spectra of the disk corona-jet model for $\dot{M} = 0.02, 0.05, 0.1, 0.3$ and $0.5\ \dot{M}_{\mathrm{Edd}}$, and the corresponding dotted lines are the emergent spectra from the jet with $\eta = 0.54\%, 0.52\%, 0.57\%, 0.62\%$ and 0.62% respectively. Right panel: $L_{8.5\mathrm{GHz}}/L_{\mathrm{Edd}}$ as a function of $L_{3-9\mathrm{keV}}/L_{\mathrm{Edd}}$. The red $\bullet$ symbols are the observations for H1743-322, the green $\triangle$ symbols are the observations for GX 339-4, and orange $\square$ symbols are the observations for V404 Cyg. The dotted line is the best-fitting linear regression of H1743-322 for $L_{3-9\mathrm{keV}} / L_{\mathrm{Edd}} > 10^{-3}$. The dashed line is the best-fitting linear regression of GX 339-4 and V404 Cyg. The thick solid line is the model line, and the model spectra are shown in the left panel. The date are originally from Coriat *et al.* (2011).

We calculate L_{R} and L_{X} at different $\dot{M}$, adjusting η to fit the observed radio/X-ray correlation of the black hole X-ray transient H1743-322 for $L_{\mathrm{X}}/L_{\mathrm{Edd}} > 10^{-3}$. It is found that always the X-ray emission is dominated by the disk corona and the radio emission is dominated by the jet. See Figure 1 for the details. We noted that the value of η for the radio/X-ray correlation of $L_{\mathrm{R}} \propto L_{\mathrm{X}}^{\sim 1.4}$ for $L_{\mathrm{X}}/L_{\mathrm{Edd}} > 10^{-3}$, is systematically less than that of the case for $L_{\mathrm{X}}/L_{\mathrm{Edd}} < 10^{-3}$, which is consistent with the general idea of jets often relatively suppressed at the high luminosity phase in black hole X-ray binaries.

3. Conclusion

We investigate the radio/X-ray correlation of $L_{\mathrm{R}} \propto L_{\mathrm{X}}^{\sim 1.4}$ for $L_{\mathrm{X}}/L_{\mathrm{Edd}} \gtrsim 10^{-3}$ within the framework of a disk corona-jet model. We note an interesting result, i.e., the fraction of the ejected matter η ($\sim 0.57\%$) for the radio/X-ray correlation of $L_{\mathrm{R}} \propto L_{\mathrm{X}}^{\sim 1.4}$ for $L_{\mathrm{X}}/L_{\mathrm{Edd}} \gtrsim 10^{-3}$, is systematically less than that of the case for $L_{\mathrm{X}}/L_{\mathrm{Edd}} < 10^{-3}$ (at least $\eta \gtrsim 1\%$), which may put some constraints on the jet formation, i.e., by suggesting that the strength of the jet power is relatively suppressed during the high luminosity phase in BHBs.

References

Coriat, M. *et al.*, 2011, *MNRAS*, 414, 677
Liu, B. F. & Mineshige, S., Ohsuga K., 2003, *ApJ*, 587, 571
Liu, B. F. & Mineshige, S., Shibata K., 2002, *ApJ*, 572, L17
Narayan, R. & Yi, I., 1995b, *ApJ*, 452, 710
Qiao, Erlin, Liu, B. F., 2009, *PASJ*, 61, 403
Qiao, Erlin, Liu, B. F., 2010, *PASJ*, 62, 661
Qiao, Erlin, Liu, B. F., 2013, *ApJ*, 764, 2
Wu, Q., Cao, X., Ho, Luis C., & Wang, Ding-Xiong, 2013, *ApJ*, 770, 31

Star clusters and black holes in galaxies across cosmic time
Proceedings IAU Symposium No. 312, 2014
Y. Meiron, S. Li, F.-K. Liu & R. Spurzem, eds.

© International Astronomical Union 2016
doi:10.1017/S1743921315007942

The X-ray spectral evolution and radio–X-ray correlation in radiatively efficient black-hole sources

Ai-Jun Dong[1], Qingwen Wu[1] and Xiao-Feng Cao[2]

[1]School of Physics, Huazhong University of Science and Technology, Wuhan 430074, China

[2] School of Physics and Electronics Information, Hubei University of Education, 430205, Wuhan, China

Abstract. We explore X-ray spectral evolution and radio–X-ray correlation simultaneously for four X-ray binaries (XRBs). We find that hard X-ray photon indices, Γ, are anti- and positively correlated to X-ray fluxes when the X-ray flux, $F_{3-9\mathrm{keV}}$, is below and above a critical flux, $F_{\mathrm{X,crit}}$, which may be regulated by ADAF and disk-corona respectively. We find that the data points with anti-correlation of $\Gamma - F_{3-9\mathrm{keV}}$ follow the universal radio–X-ray correlation of $F_{\mathrm{R}} \propto F_{\mathrm{X}}^{b}$ ($b \sim 0.5 - 0.7$), while the data points with positive X-ray spectral evolution follow a steeper radio–X-ray correlation ($b \sim 1.4$, the so-called 'outliers track'). The bright active galactic nuclei (AGNs) share similar X-ray spectral evolution and radio–X-ray correlation as XRBs in 'outliers' track, and we present a new fundamental plane of $\log L_{\mathrm{R}} = 1.59^{+0.28}_{-0.22} \log L_{\mathrm{X}} - 0.22^{+0.19}_{-0.20} \log M_{\mathrm{BH}} - 28.97^{+0.45}_{-0.45}$ for these radiatively efficient BH sources.

Keywords. accretion, accretion disks - galaxies: jets - X-rays:binaries - galaxies:active

1. Introduction

Quasi-simultaneous radio and X-ray fluxes of X-ray binaries (XRBs) in low hard (LH) state roughly follow a 'universal' non-linear correlation of $F_{\mathrm{R}} \propto F_{\mathrm{X}}^{b}$ ($b \sim 0.5 - 0.7$). By taking into account the mass of black hole (BH), the relation was extended to active galactic nuclei (AGNs), which is called "fundamental plane" of BH activity. The plane can be described by Merloni *et al.* (2003),

$$\log L_{\mathrm{R}} = 0.60^{+0.11}_{-0.11} \log L_{\mathrm{X}} + 0.78^{+0.11}_{-0.09} \log M_{\mathrm{BH}} + 7.33^{+4.05}_{-4.07}, \tag{1.1}$$

However, in recent years, more and more XRBs were found to lie well outside the scatter of the former universal radio–X-ray correlation. These outliers roughly form a different 'outliers' track, which follow a steeper radio–X-ray correlation with an index of $b \sim 1.4$ as initially found in H1743−322(Coriat *et al.* 2011).

2. Results

We explore hard X-ray spectral evolution in four XRBs (GX 339-4, H1743-322, XTE J1752-223 and Swift J1753-0127) with multiple, quasi-simultaneous radio and X-ray observations. We find that hard X-ray photon indices, Γ, are anti- and positively correlated to X-ray fluxes when the X-ray flux, $F_{3-9\mathrm{keV}}$, is below and above a critical flux, $F_{\mathrm{X,crit}}$, and the critical Eddington ratio for the transition of anti- and positive $\Gamma - F_{3-9\mathrm{keV}}$ correlation is $L_{\mathrm{bol}}/L_{\mathrm{Edd}} \sim 1\%$ (L_{bol} and L_{Edd} are bolometric luminosity and Eddington luminosity respectively), which are consistent with prediction of advection dominated accretion flow (ADAF) and disk-corona model respectively (Wu & Gu 2008, Cao *et al.* 2014). We further find that the radio–X-ray correlations are also clearly different when

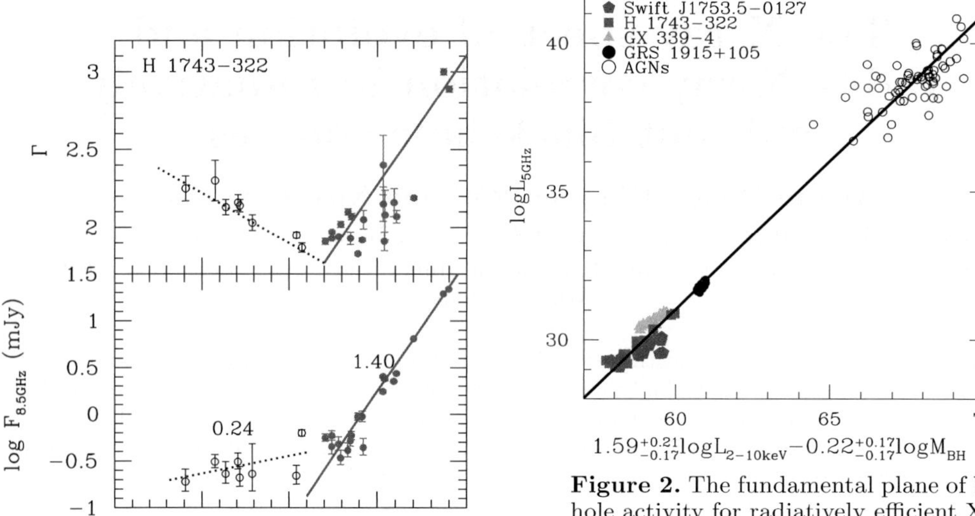

Figure 1. The radio–X-ray correlation and the X-ray spectral evolution for H1743-322.

Figure 2. The fundamental plane of black hole activity for radiatively efficient XRBs and AGNs. The solid line is best fit for the whole sample.

the X-ray fluxes are higher and lower than the critical flux that defined by X-ray spectral evolution. The data points with $F_{3-9\mathrm{keV}} \gtrsim F_{\mathrm{X,crit}}$ have a steeper radio–X-ray correlation ($F_\mathrm{X} \propto F_\mathrm{R}^b$ and $b \sim 1.1 - 1.4$), which roughly form the 'outliers' track. However, the data points with anti-correlation of $\Gamma - F_{3-9\mathrm{keV}}$ either stay in the universal track with $b \sim 0.6$ or stay in transition track (from the universal to 'outliers' tracks or vice versa). Therefore, our results support that the universal and 'outliers' tracks of radio–X-ray correlations are regulated by radiatively inefficient and radiatively efficient accretion model respectively(see Figure 1 for an example of H1743-322 and more details can be found in Cao *et al.* 2014).

We further compile a sample of bright radio-quiet AGNs and find that their hard X-ray photon indices and Eddington ratios are positively correlated, which is similar to that of 'outliers' of XRBs, where both bright AGNs and 'outliers' of XRBs have $L_{\mathrm{bol}}/L_{\mathrm{Edd}} \gtrsim 1\%$. The Eddington-scaled radio–X-ray correlation of these AGNs is also similar to that of 'outliers' of XRBs, which has a form of $L_{5\mathrm{GHz}}/L_{\mathrm{Edd}} \propto (L_{2-10\mathrm{keV}}/L_{\mathrm{Edd}})^c$ with $c \simeq 1.59$ and 1.53 for AGNs and XRBs respectively. Both the positively correlated X-ray spectral evolution and the steeper radio–X-ray correlation can be regulated by a radiatively efficient accretion flow (e.g., disk-corona). Based on these similarities, we further present a new fundamental plane for 'outliers' of XRBs and bright AGNs in black-hole (BH) mass, radio and X-ray luminosity space,

$$\log L_\mathrm{R} = 1.59^{+0.28}_{-0.22} \log L_\mathrm{X} - 0.22^{+0.19}_{-0.20} \log M_\mathrm{BH} - 28.97^{+0.45}_{-0.45}, \tag{2.1}$$

with a scatter of $\sigma_\mathrm{R} = 0.51$dex. This fundamental plane is suitable for radiatively efficient BH sources(more details can be found in Dong *et al.* 2014), while the fundamental plane of Merloni *et al.* (2003) is most suitable for radiatively inefficient BH sources.

References

Cao, X.-F., Wu, Q., & Dong, A.-J. 2014, *ApJ*,788,52
Coriat M., Corbel S., Prat L., *et al.* 2011, *MNRAS*, 414, 677
Dong, A.-J., Wu, Q., & Cao, X.-F., 2014, *ApJL*, 787, 20
Merloni, A., Heinz, S., & Matteo, T. D., 2003, *MNRAS*, 345, 1057
Wu, Q. & Gu, M., 2008, *ApJ*, 682, 212

Star clusters and black holes in galaxies across cosmic time
Proceedings IAU Symposium No. 312, 2014
Y. Meiron, S. Li, F.-K. Liu & R. Spurzem, eds.

© International Astronomical Union 2016
doi:10.1017/S1743921315007954

GraviDy: a modular, GPU-based, direct-summation N-body code

Cristián Maureira-Fredes and Pau Amaro-Seoane

Max-Planck-Institut für Gravitationsphysik (Albert Einstein Institut Potsdam-Golm, Germany
email: `cristian.maureira.fredes@aei.mpg.de`

Abstract. The direct-summation of N gravitational forces is a complex problem for which there is no analytical solution. Dense stellar systems such as galactic nuclei and stellar clusters are the loci of different interesting problems. In this work we present a new GPU, direct-summation N-body integrator written from scratch and based on the Hermite scheme. The first release of the code consists of the Hermite integrator for a system of N bodies with softening. We find an acceleration factor of about ≈ 90 of the GPU version in a single node as compared to the Serial-Single-CPU one. We additionally investigate the impact of using softening in the dynamics of a dense cluster. We study how it affects the two body relaxation, as compared with another code, NBODY6, which uses KS regularization, so as to understand the role of softening in the evolution of the system. This initial release is the first step towards more and more realistic scenarios, starting for a proper treatment for binary evolution, close encounters and the role of a massive black hole.

Keywords. N-body, Hermite integrator, GPU

---

1. Introduction

The dynamical evolution of a dense stellar system such as e.g. a globular cluster or a galactic nucleus has been addressed extensively by a number of authors. For Newtonian systems compounded by more than two stars we must rely on numerical approaches in the general case which provide us with solutions which are more or less accurate. A very well known example is the family of direct-summation N-body integrators of Aarseth for dense stellar systems (see Aarseth 1999; Spurzem 1999; Aarseth 2003† or also KIRA Portegies Zwart *et al.* 2001‡). The progress in both software and hardware has put us now in a position in which we start to get closer and closer to simulate realistic systems. There has been recently an effort at porting existing codes to graphics processing units (GPUs), like e.g. the work of Portegies Zwart *et al.* (2007); Hamada & Iitaka (2007); Belleman *et al.* (2008) on single nodes or using large GPU clusters Berczik *et al.* (2011); Nitadori & Aarseth (2012); Capuzzo-Dolcetta *et al.* (2013).

2. Integrator

To numerically integrate the system of equations we adopt the widely known 4th-order Hermite integrator (H4 henceforth) presented in (Makino & Aarseth 1992; and see also Aarseth 1999, 2003), which is a scheme based on a predictor-corrector scenario, in other words, the extrapolation and interpolation of the equations of motion. An advantage of the choice for H4 is that we can use the family of Aarseth's codes as a test for our implementation.

† All versions of the code are publicly available at the URL
`http://www.ast.cam.ac.uk/ sverre/web/pages/nbody.htm`
‡ `http://www.sns.ias.edu/ starlab/`

3. Code structure

We present here the structure of our new N-body code developed purely in C/C++, using CUDA/MPI/OpenMP for parallelization. One of the main concerns of GRAVIDY is to treat our N-body code as a piece of software, being written using an "Iterative and incremental development", which is a software development methodology similar to the development APOD cycle (CUDA Best Practices, NVIDIA 2012; Assess, Parallelize, Optimize and Deploy). Another key point in the development of GRAVIDY is its legibility. We have focused on making it easy to read and modify by other users or potential future developers without compromising the computational performance of the algorithm. The code organization was thought to be highly modular, in other words, every critical process of the integrator is represented by a separated class or chain of functions. Because maintainability is one of our design goals, its documentation is also a critical factor (using Doxygen) to explain the role of every function.

4. Testing the code

We performed a number of tests to measure the performance and the accuracy of GRAVIDY using different amount of particles, which agreed with the previous work.

5. On-going and future work

We are currently working on the implementation of a binary treatment which is based on the time-symmetric Hermite 4th order proposed by Kokubo *et al.* (1998) and implemented by Konstantinidis & Kokkotas (2010). On the other hand, we are testing the adaptation of an Hermite 6th order scheme, as a alternative for the current 4th order. And finally we are working on the multi-GPU implementation of GRAVIDY. A detailed version of this work is under development and will be published soon.

References

Aarseth, S. J. 1999, *The Publications of the Astronomical Society of the Pacific*, 111, 1333-1346.
Spurzem, R. 1999, *Journal of Computational and Applied Mathematics*, 109, 407-432.
Aarseth, S. J. 2003, ISBN 0521432723.
Portegies Zwart, S. F., McMillan, S. L. W., Hut, P., & Makino, J. 2001, *MNRAS*, 321, 199-226.
Portegies Zwart, S. F., Belleman, R. G., & Geldof, P. M. 2007, *New A*, 12, 641-650.
Hamada, T. & Iitaka, T. 2007, *New A*.
Belleman, R. G., Bédorf, J., &Portegies Zwart, S. F. 2008, *New A*, 13, 103-112.
Berczik, P., Nitadori, K., Zhong, S., Spurzem, R., Hamada, T., Wang, X., Berentzen, I., Veles, A., & Ge, W. 2011, *International conference on High Performance Computing*, 8-18.
Nitadori, K. & Aarseth, S. J. 2012, *MNRAS*, 424, 545-552.
Capuzzo-Dolcetta, R., Spera, M., & Punzo, D. 2013, *Journal of Computational Physics*, 236, 580-593.
Kokubo, E., Yoshinaga, K., & Makino, J. 1998, *MNRAS*, 297, 1067-1072.
Konstantinidis, S. & Kokkotas, K. D. 2010, *A&A*, 522, A70.
Makino, J. & Aarseth, S. J. 1992, *PASJ*, 44, 141-151.

Star clusters and black holes in galaxies across cosmic time
Proceedings IAU Symposium No. 312, 2014
Y. Meiron, S. Li, F.-K. Liu & R. Spurzem, eds.

© International Astronomical Union 2016
doi:10.1017/S1743921315007966

Performance evaluation of the Hermite scheme on many-core accelerators

Naohito Nakasato

Department of Computer Science and Engineering, University of Aizu,
Aizu-Wakamatsu, Fukushima, 965-8580, Japan
email: **nakasato@u-aizu.ac.jp**

Abstract. We are developing a software library to calculate gravitational interaction for the Hermite scheme on parallel computing systems supported by OpenCL API. Our library is partly compatible with a standard GRAPE-6A interface and is easily usable in existing N-body codes. Since our library is based on OpenCL standard API, our library is working on many parallel computing systems such as a multi-core CPU, a GPU, and a many-core architecture. We report the performance evaluation of our library on computing platforms from various vendors.

Keywords. methods: N-body simulations

1. Introduction

Recent development of parallel computing architecture enable us to use multi/many-core architecture such as CPUs, GPUs and Xeon Phi(MIC) for modeling star clusters (e.g. Gaburov *et al.* 2009; Tanikawa *et al.* 2012; Nitadori & Aarseth 2012; Wang *et al.* 2015). In this paper, we report a new library to speed-up the Hermite scheme (Makino & Aarseth 1992) on the emerging parallel architectures. Our library utilizes the OpenCL API that is a standard API working on many parallel architectures while previous techniques/libraries rely on a specific computing platform. We make our library compatible to the GRAPE-6A library (Fukushige *et al.* 2005) so that it is usable as a replacement to the GRAPE library. In the following section, we present optimization techniques used in our library to gain high performance on GPUs and report the performance evaluation on different GPU architectures.

2. Performance optimization

Fundamentally, there are I- and J- parallelism in GRAPE-like evaluation of N-body kernels. "I-particle" is a particle where we want to compute the force. "J-particles" are particles that exert the force on the I-particle. In our current implementation, each thread running on a GPU first loads N_i I-particles and loop-over all other particles (J-particles) to compute particle interaction such as potential energy, acceleration and jerk for the I-particles (Makino & Aarseth 1992). In this loop, the thread loads N_j particles at a time. A simplest case is $(N_i, N_j) = (1, 1)$ where each thread compute one-to-one interaction on every iteration.

A combination of (N_i, N_j) is a key to gain optimal performance on GPUs since GPU architectures favor more floating-point operations per memory access operations. Larger N_i and N_j in an N-body kernel, the number of arithmetic operations per unit memory access is larger so that the performance of the kernel is better. In Figure 1, we compare the performance in GFLOPS with a different combination of (N_i, N_j). For this comparison,

Table 1. Evaluated Architectures and Optimal (N_i, N_j)

Architecture	SP peak	DP peak	Kernel D	Kernel DS
FirePro W9100	5337	2619	(1,2)/987	(2,2)/1300
GeForce TITAN	5376	1570	(1,2)/333	(1,1)/793
Radeon 7970	3789	947	(1,1)/626	(2,2)/1063
MIC 5110P	1888	994	(1,2)/188	(2,2)/255

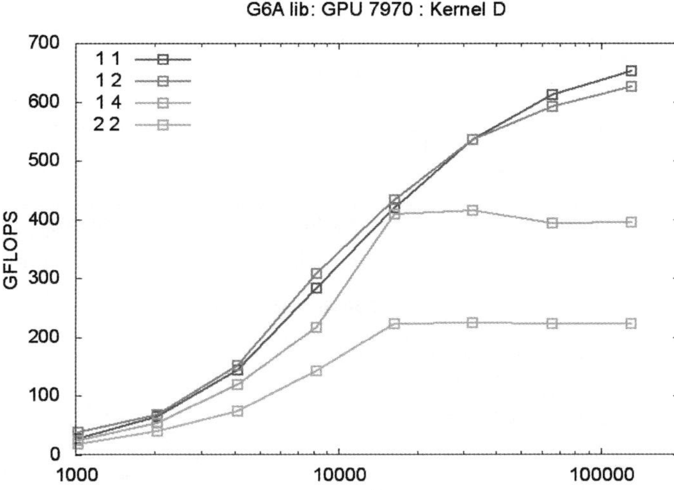

Figure 1. Performance of our library as a function of N with different combinations of (N_i, N_j)

we have used Radeon 7970 GPU and N-body kernels are implemented with double-precision (DP) operations. In this particular case, $(N_i, N_j) = (1,1)$ and $(1,2)$ show the comparable performance. We did similar evaluations on other parallel architectures and found an optimal combination for a given architecture as shown in Table 1. In this table, each row shows GPUs and MIC. The second and third columns show the peak GFLOPS in single-precision (SP) and DP operations, respectively. The forth and fifth columns present the optimal combination of (N_i, N_j) and the performance in GFLOPS for $N = 131072$. We have tested two variants of kernels as explained below. Roughly speaking, the performance of "Kernel D" that uses only DP operations is proportional to the peak performance in DP operations except GeForce TITAN. TITAN GPU favors a proprietary CUDA programming API and other authors have reported much better performance than our result on CUDA-enabled GPUs (Gaburov *et al.* 2009; Nitadori & Aarseth 2012).

Also, we have appropriately used SP and DP operations in particle interactions since GPUs are much faster on SP operations as shown in Table 1. Accordingly, we have two variant of N-body kernels as Kernel D and Kernel DS. "Kernel D" uses only DP operations while "Kernel DS" adopts a mixed-precision technique that is similar to Gaburov *et al.* (2009). Note that we use actual DP operations instead of emulated DP operations by combining 2 SP variables to represent emulated DP variable. In addition, we use vector type variables defined in OpenCL language such as float4, double2 to explicitly express vectorized evaluation of the particle interaction. In Figure 2, we present the performance of our library as a function of N. "Kernel DS" shows better performance

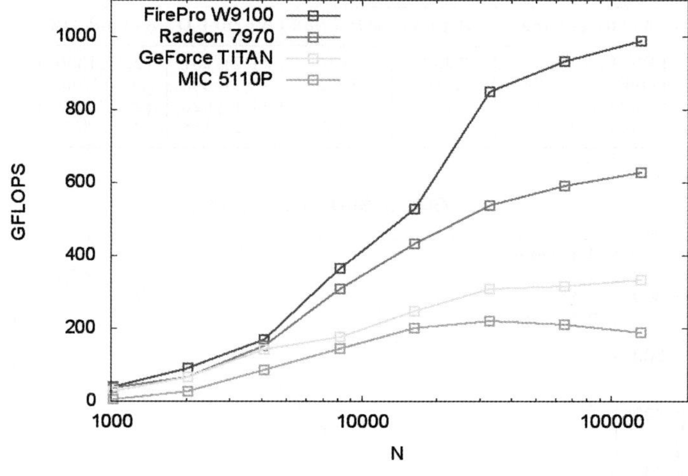

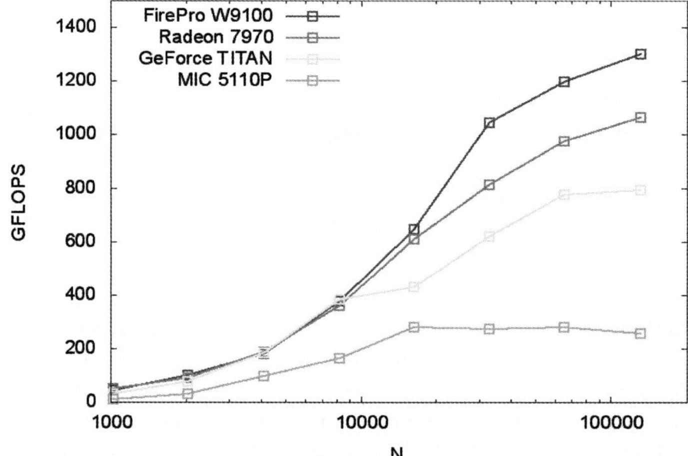

Figure 2. Performance of two variants of kernels as a function of N

than "Kernel D" over 1 TFLOPS on W9100 and 7970 GPUs. Our library also works on multi-core CPUs thanks to the standard OpenCL API.

Finally, we have further optimized Kernel DS by adopting different data storage. Performance of compute kernels on GPUs is sensitive to memory access pattern of the kernels. In "Kernel DS", we use separate arrays to store the position, velocity and mass of particles. This usage of data storage is a so-called structure of arrays (SoA). When a thread load the data of I- or J-particles, the memory access is stride access in SoA. Alternative way to store particles data is to use a structure for each particle that holds the the position, velocity and mass of the particle. This is a so-called array of structures (AoS). In Figure 3, we present the comparison of the performance with the data storage as SoA and AoS on 7970 GPU. The two integers are the combination of (N_i, N_j). AoS is 20 - 40 % better performance than SoA at $N = 131072$.

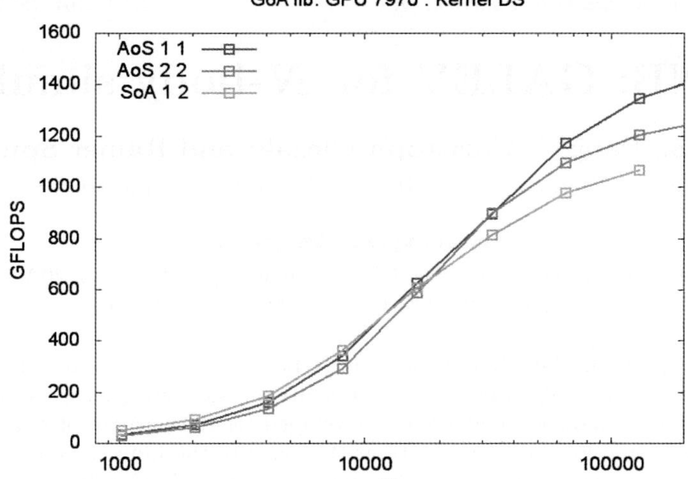

Figure 3. Performance of two variants of kernels as a function of N

3. Summary

Our GRAPE-6A compatible Hermite Scheme library shows fairly good performance on various GPU/MIC architectures. Since the OpenCL standard works on many platforms, our approach is a very effective way to implement a similar library for N-body integrations.

References

Toshiyuki, F., Junichiro, M., & Atsushi, K., 2005, *Publications of the Astronomical Society of Japan*, 57, 6, 1009–1021

Gaburov, E., Harfst, S., & Portegies Zwart, S., 2009, *New Astronomy*, 14, 7, 630–637

Makino, J., & Aarseth, S. J., 1992, *Publications of the Astronomical Society of Japan*, 44, 2, 141–151

Nitadori, K., & Aarseth, S. J., 2012, *Monthly Notices of the Royal Astronomical Society*, 424, 1, 545–552

Tanikawa, A., Yoshikawa, K., Okamoto, T., & Nitadori, K., 2012, *New Astronomy*, 17, 2, 82–92

Wang, L., Spurzem, R., Aarseth, S. J., Nitadori, K., Berczik, P., Kouwenhoven, M. B. N., & Naab, T., 2015, *Monthly Notices of the Royal Astronomical Society*, 450, 4, 4070–4080

Star clusters and black holes in galaxies across cosmic time
Proceedings IAU Symposium No. 312, 2014
Y. Meiron, S. Li, F.-K. Liu & R. Spurzem, eds.
© International Astronomical Union 2016
doi:10.1017/S1743921315007978

GalevNB: GALEV for N-body simulations

Xiaoying Pang[1,2], Christoph Olczak[1] and Rainer Spurzem[2]

[1]Shanghai Institute of Technology, 100 Haiquan Road, Fengxian district, Shanghai 201418,
China
email: xypang@bao.ac.cn

[2]National Astronomical Observatories, Chinese Academy of Sciences, 20A Datun Road,
Chaoyang District, Beijing 100012, P.R. China

Abstract. We report on GalevNB (Galev for N-body simulations), an integrated software solution that provides N-body users direct access to the software package GALEV (GALaxy EVolutionary synthesis models). GalevNB is developed for the purpose of a direct comparison between N-body simulations and observations. It converts the fundamental stellar properties of N-body simulations, i.e., stellar mass, temperature, stellar luminosity and metallicity, into observational magnitudes for a variety of filters of widely used instruments/telescopes (HST, ESO, SDSS, 2MASS), and into spectra that span from far-UV (90 Å) to near-IR (160 μm).

Keywords. N-body, magnitude, spectra

1. Introduction

The output parameters of `NBODY6++` (Aarseth 1999) simulations are mostly theoretical values. To make a direct comparison between N-body simulation data and observations, we combine GALEV (GALaxy EVolutionary synthesis models; Kotulla *et al.* 2009), a flexible algorithm to combine astrophysical colors in many filters and spectra of stars (Lejeune, Cuisinier & Buser 1997, 1998) or sets of stars, with `NBODY6++` simulations. In this paper, we present the structure of this new code: GalevNB (Galev for N-body simulations). Adapting subroutines from GALEV, GalevNB can produce spectra spanning the range from far-UV at 90 Å to far IR at 160 μm, with a spectral resolution of 20 Å in the UV-optical and 50-100 Å in the near IR range. Given a list of requested filters in HST, ESO, SDSS, 2MASS etc., GalevNB convolves the spectra with the filter response functions and applies the chosen zero-points (Vegamag, ABmag, and STmag) to yield absolute magnitudes. GalevNB bridges theoretical parameters and their observed values, thus allows us to understand the color and spectra evolution of star clusters, and to determine the initial conditions and parameters of star cluster simulations with a direct comparison to observations.

2. GalevNB structure and execution

The main program of GalevNB is `GalevNB.f90`, which parses single snapshot files (stellar evolution only) generated by `NBODY6(++)`. It uses seven subroutines (`startomaginit`, `specint_initialize`, `reset_weights`, `startomag`, `add_star`, `spec2mag`, `spec_output`) of GALEV package to convert effective temperature, stellar luminosity, metallicity, and mass into observational magnitudes and spectra. The functions of these routines are presented in Table 1. The GalevNB package contains four folders: 1) `spectral_templates`, in which locate all the spectral template files from the BaSeL library of model atmospheres (Lejeune, Cuisinier & Buser 1997, 1998); 2) `standard_filters`, contains a large set of filter response functions (FUV, NUV, U, B, V, R, I, J, H, K) that are used as

Table 1. Functions of subroutines computing magnitudes and spectra

Subroutine	Function
specint_initialize	initialize the stellar spectra
reset_weights	reset the weight of stellar spectra
add_star	integrate the flux of all stars in the cluster
spec_output	output spectra
startomaginit	initialize the stellar magnitude
spec2mag	convolve the stellar spectra with the filter response function
startomag	compute magnitudes for stars

Table 2. Column contents for the filter information file: `filterlist.dat`

Column	Content	ID of zero point
1	Filter name	
2	Corresponding path of the filter response function	
3	ID of selected zero point (default value is 1)	
4	Standard zero point in the Vega magnitude system	1
5	Standard zero point in the AB magnitude system	2
6	Standard zero point in the ST magnitude system	3
7	Optional user-defined zero point	4

standard reference filters; 3) `filter_response_curves`, includes filter response functions from magnitude systems of HST, ESO instruments, 2MASS, SDSS, Johnson, and Cousins in separate subfolders. We also provide a choice of user-specify filter response functions. Information about the entire set of available filters is included in the file `filterlist.dat`. Please be aware that `filterlist.dat`, in which the user specify their own choice of magnitude system by uncommenting the line of chosen filter, MUST be presented in the same directory as the `NBODY6(++)` snapshot files. The content of the file, `filterlist.dat`, is presented in Table 2.

To compile GalevNB, the user should have C++ and Fortran compilers installed. The input file of GalevNB should be a single snapshot output from `NBODY6(++)` simulations. In case of a file containing all snapshots (called `sev.83` in `NBODY6++` and `fort.83` in `NBODY6`), we provide the user with a shell script `generate_snapshots.sh` in the folder, `scripts`, for retrieving single snapshot data out of `sev.83` and `fort.83`. The user can select his/her preferred filters (maximum 20) by uncommenting the row of the corresponding filter in `filterlist.dat`, and choose his/her desired magnitude system (Table 2). Magnitudes of individual stars and the whole cluster, and spectra of the cluster or chosen stellar types are produced, respectively.

Acknowledgements

This work is funded by National Natural Science Foundation of China, No: 11443001.

References

Aarseth, S. J. 1999, *PASP*, 111, 1333
Kotulla, R., Fritze, U., Weilbacher, P., & Anders, P. 2009, *MNRAS*, 396, 462
Lejeune, T., Cuisinier, F., & Buser, R. 1997, *A&A Supplement*, 125, 229
Lejeune, T., Cuisinier, F., & Buser, R. 1998, *A&A Supplement*, 130, 65

Star clusters and black holes in galaxies across cosmic time
Proceedings IAU Symposium No. 312, 2014
Y. Meiron, S. Li, F.-K. Liu & R. Spurzem, eds.
© International Astronomical Union 2016
doi:10.1017/S174392131500798X

Acceleration of hybrid MPI parallel NBODY6++ for large N-body globular cluster simulations

Long Wang[1], Rainer Spurzem[2,1], Sverre Aarseth[3], Keigo Nitadori[4], Peter Berczik[2], M.B.N. Kouwenhoven[1] and Thorsten Naab[5]

[1]Kavli Institute for Astronomy and Astrophysics, Peking University, Beijing, China
email: long.wang@pku.edu.cn

[2]National Astronomical Observatories and Key Laboratory of Computational Astrophysics,
Chinese Academy of Sciences, Beijing, China

[3]Institute of Astronomy, University of Cambridge, Cambridge, UK

[4]RIKEN Advanced Institute for Computational Science, Kobe, Japan

[5]Max-Planck Institut für Astrophysik, Garching, Germany

Abstract. Previous research on globular clusters (GCs) dynamics is mostly based on semi-analytic, Fokker-Planck, Monte-Carlo methods and on direct N-body (NB) simulations. These works have great advantages but also limits since GCs are massive and compact and close encounters and binaries play very important roles in their dynamics. The former three methods make approximations and assumptions, while expensive computing time and number of stars limit the latter method. The current largest direct NB simulation has $\sim$ 500k stars (Heggie 2014). Here, we accelerate the direct NB code NBODY6++ (which extends NBODY6 to supercomputers by using MPI) with new parallel computing technologies (GPU, OpenMP + SSE/AVX). Our aim is to handle large N (up to 10^6) direct NB simulations to obtain better understanding of the dynamical evolution of GCs.

Keywords. methods: numerical, globular clusters: general

1. The features of NBODY6/NBODY6++

Star clusters are dense groups in which the relaxation timescale is short. Thus, the frequent close encounters and binaries play an important role in star cluster dynamics. NBODY6 (Aarseth 2003) is a state-of-the-art direct NB simulation code specifically designed for star clusters. It uses the 4th order Hermite integrator with a block time step and neighbor scheme which improve the integration speed. Kustaanheimo & Stiefel (1965) (KS) and Chain regularization (Mikkola & Aarseth 1993) are used to ensure the highly accurate treatment of close encounter and binaries.

The direct NB simulation is very costly. The cost per relaxation timescale can be $O(N^{10/3}/\ln N)$ (Makino & Hut 1988). Thus parallelization is necessary if we want to simulate large particle system like GCs. We present our new version of NBODY6++GPU by combining the massively parallel multi-node code NBODY6++ (Spurzem 1999 and Hemsendorf *et al.* 2003) with the hybrid parallelization libraries (GPU + OpenMP + AVX/SSE) from NBODY6-GPU (Nitadori & Aarseth 2012). The parallelization structure of NBODY6++GPU is shown in Figure 1. One cycle of integration can be separated to three hierarchical time step blocks: (1) Regular block: particles belong to the current regular time step blocks will obtain forces calculated from all other particles by using GPU acceleration. (2) Irregular block: particles belong to the irregular time step blocks only cumulate forces from their neighbors. AVX/SSE together with OpenMP is used to

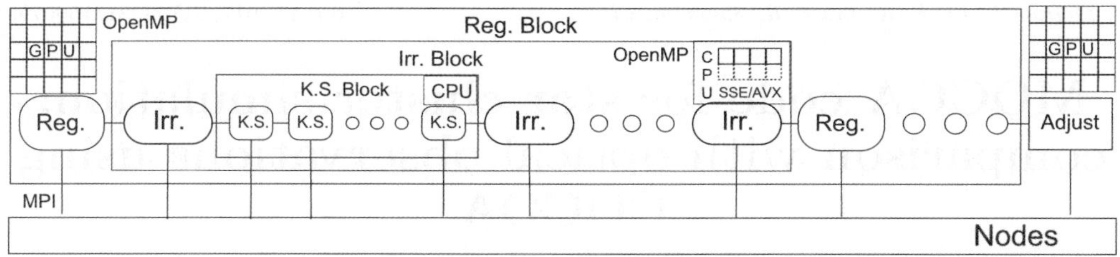

Figure 1. The NBODY6++GPU code parallelization structure. The MPI, GPU, OpenMP and AVX/SSE are used for different components.

accelerate position and velocity prediction and force calculations. (3) KS block: KS pairs will be integrated inside KS time step blocks with MPI parallelization support.

2. Benchmarks and resources

The detailed performance analysis will be published soon (Wang *et al.* 2015 and Huang *et al.* 2015). For 10^6 particles with Plummer (1911) sphere and a Kroupa, Tout & Gilmore (1993) initial mass function (IMF), we get ~ 800 seconds computing time per NB time unit with 8 nodes (160 Intel Ivy Bridge cores (2.8 GHz) and 16 NVIDIA K20X GPU) on the "Hydra" Cluster, Max-Planck-Supercomputing center, Garching. For a realistic globular cluster simulation with one million stars including 5% primordial binaries, King (1966) profile and Kroupa (2001) IMF, the same configuration of hardware results in a stable computing time of ~ 1500 seconds per NB time unit.

Future improvements will speed up the KS regularization performance. With the current NBODY6++GPU we can already handle the million-body GC direct NB simulations. Future developments of hardware with faster bandwidth and latency and optimizations of communication and data management will open the window for simulations of larger system like nuclear star clusters.

The code is free for downloading by two links:

- Subversion: svn co http://silkroad.bao.ac.cn/repos/betanb6
- GitHub: git clone https://github.com/lwang-astro/betanb6pp.git

The user manual is also provided in downloading resource with detailed descriptions of code features, input and outputs.

References

Aarseth S. J., 1963, *MNRAS*, 126, 223
Heggie D. C., 2014, *MNRAS*, 445, 3435
Hemsendorf M., Khalisi E., Omarov C. T., & Spurzem R., 2003, *High Performance Computing in Science and Engineering.* Springer Verlag, 71, 388
Huang, S., Spurzem, R., & Berczik, P. 2015, *RAA*, in press (arXiv:1508.02510)
King I. R., 1966, *AJ*, 71, 64
Kroupa P., Tout C. A., & Gilmore G., 1993, *MNRAS*, 262, 545
Kroupa P., 2001, *MNRAS*, 322, 231
Kustaanheimo P. & Stiefel E., 1965, *J. Reine Angew. Math.*, 218, 204
Makino J. & Hut P., 1988, *ApJS*, 68, 833
Mikkola S. & Aarseth S. J., 1993, *CeMDA*, 57, 439
Nitadori K. & Aarseth S. J., 2012, *MNRAS*, 424, 545
Spurzem R., 1999, *JCoAM*, 109, 407
Plummer H. C., 1911, *MNRAS*, 71, 460
Wang L., Spurzem R., Aarseth S., *et al.* 2015, *MNRAS*, 450, 4070

Star clusters and black holes in galaxies across cosmic time
Proceedings IAU Symposium No. 312, 2014
Y. Meiron, S. Li, F.-K. Liu & R. Spurzem, eds.

© International Astronomical Union 2016
doi:10.1017/S1743921315007991

MOCCA code for star cluster simulation: comparison with optical observations using COCOA

Abbas Askar[1], Mirek Giersz[1], Wojciech Pych[1], Arkadiusz Olech[1] and Arkadiusz Hypki[1,2]

[1]Nicolaus Copernicus Astronomical Centre, Polish Academy of Sciences,
ul. Bartycka 18, 00-716 Warsaw, Poland
email: askar@camk.edu.pl

[2]Leiden Observatory, Leiden University, P.O. Box 9513, 2300 RA Leiden, The Netherlands

Abstract. We introduce and present preliminary results from COCOA (Cluster simulatiOn Comparison with ObservAtions) code for a star cluster after 12 Gyr of evolution simulated using the MOCCA code. The COCOA code is being developed to quickly compare results of numerical simulations of star clusters with observational data. We use COCOA to obtain parameters of the projected cluster model. For comparison, a FITS file of the projected cluster was provided to observers so that they could use their observational methods and techniques to obtain cluster parameters. The results show that the similarity of cluster parameters obtained through numerical simulations and observations depends significantly on the quality of observational data and photometric accuracy.

Keywords. stellar dynamics, globular clusters: general, methods: numerical

1. Introduction

Due to advancements in computational technology over the past couple of decades, there has been extensive work done in modeling the dynamical evolution of star clusters using numerical simulation codes like direct N-body and Monte Carlo codes. In order to be able to get feedback from observers and vice versa, we need to extend numerical simulations for direct comparisons with observations. For this purpose, we are actively developing the COCOA (Cluster simulatiOn Comparison with ObservAtions) code that can create observational data from the output of numerical simulations of star clusters. COCOA has been developed in Python and it combines various tools to enable quick comparisons between simulation and observational data. As an input, the code uses the extended snapshot from the MOCCA code (Hypki & Giersz 2013; Giersz *et al.* 2013). We are working on developing additional features in COCOA. The current version of the code can:

- Project numerical data from star cluster simulations on to the plane of the sky;
- Give complete projected snapshot with magnitudes and details of binary systems;
- Simulate observations of stars in binary systems inside star clusters;
- Generate FITS file and photometric data from projected snapshot;
- Compute surface brightness and velocity dispersion profiles (using either projected center of mass velocities or velocities of individual stars in binary systems);
- Fit projection to the King model and determine cluster parameters

In these proceedings, we present initial results from COCOA for a test star cluster after 12 Gyr of evolution, simulated using MOCCA. We determined the surface brightness, velocity dispersion profiles and cluster parameters using COCOA. For comparison, a FITS

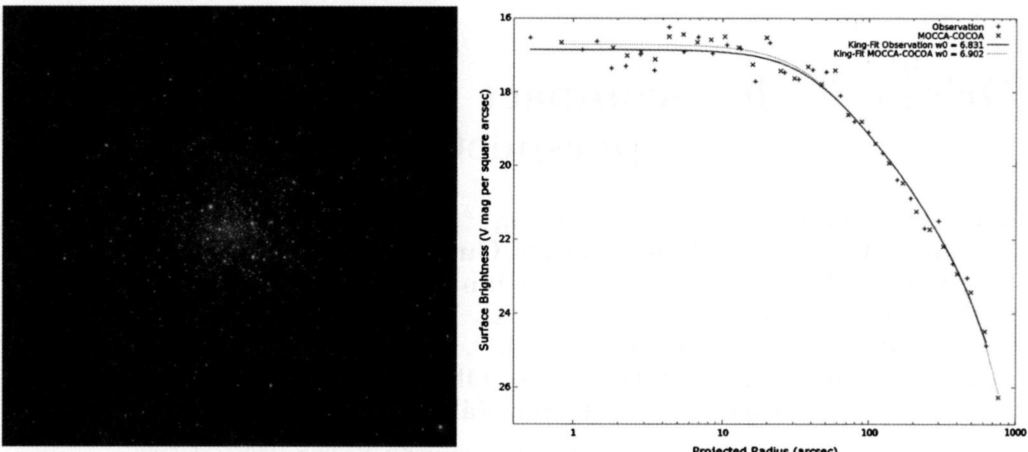

Figure 1. The left panel shows the grayscale synthetic image of the model cluster created using COCOA. The right panel shows the surface brightness profile and King fit of the cluster using projected MOCCA-COCOA data and observational data.

Table 1. Comparison of cluster parameters obtained through projected simulation and observational data. These results may vary with different projections of simulation data.

Data	King Scale Radius (pc)	King Parameter W_0	Half-light Radius (pc)
MOCCA-COCOA	0.91	6.90	2.06
Observation	0.98	6.83	2.07

file of the projection was provided to observers so that they could use their observational methods and techniques to obtain cluster photometry.

2. Comparing MOCCA-COCOA results & observational data

The test star cluster evolved using MOCCA for 12 Gyr was tidally underfilled, had a very high primordial binary fraction (95%) and initially contained 553k objects. Using the sim2obs tool in COCOA, we generated a FITS image (Figure 1) from the MOCCA snapshot after 12 Gyr of evolution. The simulated observation of the model cluster was made with a 2.5m optical telescope and the distance to the cluster was set to 5 kpc. The FITS files (2×2 mosaic of the cluster) generated by sim2obs were used by observers to obtain photometric data for 49253 objects extending up to a radius of 465 arcsec. We compared the the surface brightness profile from the photometry done by the observers with the SBP from MOCCA-COCOA results (Figure 1). Photometric data from observers was limited and the observational SBP extends to a smaller radius compared to all the data from MOCCA simulations. For this comparison, we limited the MOCCA-COCOA projected data to ∼700 arcsec. After photometric corrections, the SBP from the observational data and projected simulation data match closely. With COCOA we are able to fit the SBP to the King Model. Table 1 shows a comparison of cluster parameters obtained from the SBP and King fit for projected simulation and observation data.

References

Giersz, M., Heggie, D. C., Hurley, J. R., & Hypki, A. 2013, *MNRAS*, 431, 2184
Hypki, A. & Giersz, M. 2013, *MNRAS*, 429, 1221

Star clusters and black holes in galaxies across cosmic time
Proceedings IAU Symposium No. 312, 2014
Y. Meiron, S. Li, F.-K. Liu & R. Spurzem, eds.
© International Astronomical Union 2016
doi:10.1017/S1743921315008005

Origin of ultra-compact dwarfs: a dynamical perspective

Hong-Xin Zhang[1,2,3,4], **Eric W. Peng**[1,2], **Patrick Côté**[5], **Chengze Liu**[6,7], **Laura Ferrarese**[5], **Jean-Charles Cuillandre**[8], **Nelson Caldwell**[9], **Stephen D. J. Gwyn**[5], **Andrés Jordán**[10], **Ariane Lançon**[11], **Biao Li**[1,2], **Roberto P. Muñoz**[10,11], **Thomas H. Puzia**[10], **Kenji Bekki**[12], **John Blakeslee**[5], **Alessandro Boselli**[13], **Michael J. Drinkwater**[14], **Pierre-Alain Duc**[15], **Patrick Durrell**[16], **Eric Emsellem**[17,18], **Peter Firth**[14] and **Ruben Sánchez-Janssen**[5]

[1] Department of Astronomy, Peking University, Beijing 100871, China
email: hongxin@pku.edu.cn

[2] Kavli Institute for Astronomy and Astrophysics, Peking University, Beijing 100871, China

[3] CAS-CONICYT Fellow

[4] Chinese Academy of Sciences South America Center for Astronomy, Camino EI Observatorio #1515, Las Condes, Santiago, Chile

[5] National Research Council of Canada, Herzberg Astronomy and Astrophysics Program, 5071 West Saanich Road, Victoria, BC V9E 2E7, Canada

[6] Center for Astronomy and Astrophysics, Department of Physics and Astronomy, Shanghai Jiao Tong University, Shanghai 200240, China

[7] Shanghai Key Lab for Particle Physics and Cosmology, Shanghai Jiao Tong University, Shanghai 200240, China

[8] Canada–France–Hawaii Telescope Corporation, Kamuela, HI 96743, USA

[9] Harvard-Smithsonian Center for Astrophysics, Cambridge, MA, 02138

[10] Instituto de Astrofísica, Facultad de Física, Pontificia Universidad Católica de Chile, Av. Vicuña Mackenna 4860, 7820436 Macul, Santiago, Chile

[11] Observatoire astronomique de Strasbourg, Université de Strasbourg, CNRS, UMR 7550, 11 rue de l'Universite, F-67000 Strasbourg, France

[12] School of Physics, University of New South Wales, Sydney 2052, NSW, Australia

[13] Aix Marseille Université, CNRS, LAM (Laboratoire d'Astrophysique de Marseille) UMR 7326, F-13388 Marseille, France

[14] School of Mathematics and Physics, University of Queensland, Brisbane, QLD 4072, Australia

[15] Laboratoire AIM Paris-Saclay, CNRS/INSU, Université Paris Diderot, CEA/IRFU/SAp, F-91191 Gif-sur-Yvette Cedex, France

[16] Department of Physics & Astronomy, Youngstown State University, Youngstown, OH 44555

[17] Université de Lyon 1, CRAL, Observatoire de Lyon, 9 av. Charles André, F-69230 Saint-Genis Laval; CNRS, UMR 5574; ENS de Lyon, France

[18] European Southern Observatory, Karl-Schwarzschild-Str. 2, D-85748 Garching, Germany

Abstract. Discovery of ultra-compact dwarfs (UCDs) in the past 15 years blurs the once thought clear division between classic globular clusters (GCs) and early-type galaxies. The intermediate nature of UCDs, which are larger and more massive than typical GCs but more compact than typical dwarf galaxies, has triggered hot debate on whether UCDs should be considered galactic in origin or merely the most extreme GCs. Previous studies of various scaling relations, stellar populations and internal dynamics did not give an unambiguous answer to the primary origin of UCDs. In this contribution, we present the first ever detailed study of global dynamics of 97

UCDs ($r_\mathrm{h} \gtrsim 10$ pc) associated with the central cD galaxy of the Virgo cluster, M87. We found that UCDs follow a different radial number density profile and different rotational properties from GCs. The orbital anisotropies of UCDs are tangentially-biased within ~ 40 kpc of M87 and become radially-biased with radius further out. In contrast, the blue GCs, which have similar median colors to our sample of UCDs, become more tangentially-biased at larger radii beyond ~ 40 kpc. Our analysis suggests that most UCDs in M87 are not consistent with being merely the most luminous and extended examples of otherwise normal GCs. The radially-biased orbital structure of UCDs at large radii is in general agreement with the scenario that most UCDs originated from the tidally threshed dwarf galaxies.

Keywords. galaxies: clusters, globular clusters, galaxies: nuclei, galaxies: elliptical and lenticular, cD, galaxies: kinematics and dynamics

---

1. Data and samples

This work is devoted to a comparative study of the dynamical properties of M87 UCDs, GCs and surrounding dwarf ellipticals. Our samples of 97 confirmed UCDs and 911 confirmed GCs within the central 1.5° of M87 are collected from both literature (e.g. Hanes *et al.* 2001; Strader *et al.* 2011) and our new observations. For new observations, we selected UCD and GC candidates for spectroscopic followup with Hectospec/MMT and 2dF/AAT based on the high-sensitivity ($g_\mathrm{lim} = 25.9$ mag at 10σ for point sources) and high-resolution (FWHM $\sim 0.6''$ in i band) u^*griz imaging data from the Next Generation Virgo Cluster Survey (NGVS, Ferrarese *et al.* 2012).

The readers are referred to Zhang *et al.* (2015) for details of the samples and new observations. Briefly, our spectroscopic surveys of Virgo UCDs and GCs have been highly efficient, thanks to an unprecedentedly clean sample of Virgo UCD and GC candidates selected from NGVS. Our surveys covered nearly all of the area encompassed by one scale radius of the NFW dark matter halo toward the Virgo A subcluster ($2°.143 = 0.617$ Mpc; McLaughlin 1999). Because the half-light radius $r_\mathrm{h,NGVS}$ measurement for relatively faint sources is subject to large uncertainties, we only select UCDs with $g \leqslant 21.5$ mag and $r_\mathrm{h,NGVS} \geqslant 11$ pc. In addition, sources with HST size measurement $r_\mathrm{h,HST} > 9.5$ pc are also included as UCDs, irrespective of their brightness. All the other confirmed Virgo compact clusters are regarded to be GCs. Overall, our sample of UCDs is expected to be $\sim 60\%$ complete at $g_0 < 21.5$ mag. The median $(g - i)_0$ color of our samples of UCDs and blue GCs are 0.75 and 0.74 respectively, and about 92% of our UCDs fall into the color range of the blue GCs. So we place additional emphasis on a comparison between properties of UCDs and blue GCs in this work.

2. Results and discussion

Surface number density profiles. Our UCD sample is $\sim 98\%$ complete at $g_0 < 20.5$, which corresponds to $M_g < -10.6$. In Figure 1, we show the radial number density profiles of the 59 UCDs with $g_0 < 20.5$ mag, together with profiles of the photometrically selected blue GCs, red GCs ($g_0 < 24$ mag, Durrell *et al.* 2014) and the surrounding dE galaxies.

We adopted the Sérsic function to quantify (curves in Figure 1) the radial profiles of UCDs and GCs. Details about the profile fitting is described in Zhang *et al.* (2015). UCDs have a radial number density profile that is shallower than GCs in the inner $\sim 15'$ and as steep as the red GCs at large radii. The surrounding dE galaxies have much flatter and extended number density profiles than UCDs and GCs.

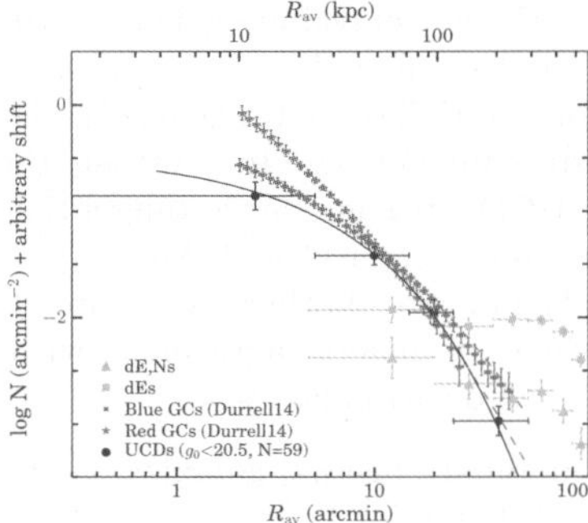

Figure 1. Radial number density profiles of UCDs (black), blue GCs (blue), red GCs (red), dE, Ns (cyan triangles) and all dE (cyan squares) galaxies. Overplotted on the data are the best-fit Sérsic profiles for UCDs and GCs. Note that radial profiles of the GCs have been vertically shifted down arbitrarily (2.1 for the blue and 1.7 for the red GCs) for comparison purpose.

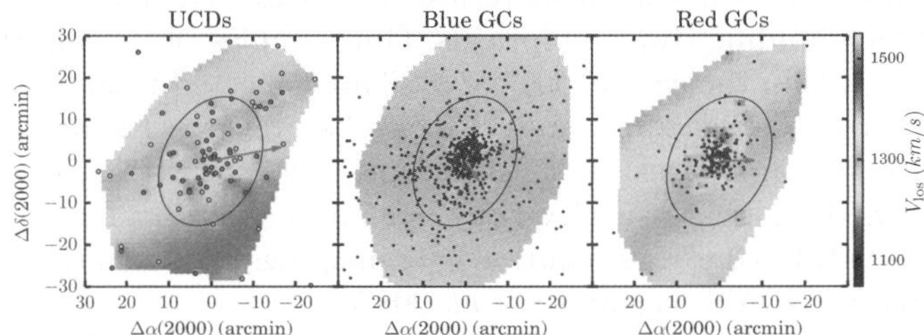

Figure 2. Spatial distribution of the UCDs (*left*), blue GCs (*middle*) and red GCs (*right*) are over plotted on their respective surface fitting (*the color background*) to the spatial distribution of line-of-sight velocities with the Kriging technique for the inner $30'$ of M87. The *black ellipses* represent the stellar isophotes of M87 at $10R_{\rm e}$, and the *black solid line* marks the photometric minor axis of M87 in each panel. The *red arrows*, with the length being proportional to the rotation amplitude, mark the direction of rotation axis from our global kinematics fitting of $v_{\rm los}$ vs. PA to the inner $30'$. The global kinematics fitting, which is primarily driven by the inner regions that contain most of the data points, matches the central velocity field from Kriging surface fitting. Among the three populations, the blue GCs seem to have an overall velocity field more aligned with the photometric major axis than the other two populations.

Velocity field. The UCDs and blue GCs have similar intrinsic velocity dispersion. The rotation amplitude of UCDs is more than 3 times stronger than that of the blue and red GCs. In addition, the overall velocity field of blue GCs is aligned with the photometric major axis of M87 much better than UCDs and the red GCs. Our test suggests that the probability of finding a rotation amplitude greater than or equal to the best-fit value for UCDs purely by chance is $\sim 2\%$, the probability for blue GCs is $\sim 27\%$, and $\sim 12\%$ for the red GCs. The significantly stronger rotation of UCDs as compared to GCs suggests that UCDs are kinematically distinct from GCs.

Orbital anisotropies. To constrain the orbital anisotropy profiles β_r of UCDs, blue and red GCs, we turn to the spherically symmetric Jeans equation. In determining β_r, we adopt the most recent determination of M87 mass profile by Zhu *et al.* (2014) based on made-to-measure modeling of over 900 M87 GCs. In addition, the surface number density profiles are de-projected to derive the 3D number density profiles which are used in solving the Jeans equation. Details of the Jeans modeling, including the functional form of β_r, can be found in Zhang *et al.* (2015). Basically, we used a maximum likelihood method to fit Jeans models to the observed $v_{\rm los,i}$ as a function of geometric average radius R_i, by assuming that $v_{\rm los,i}$ at R_i follows a Gaussian distribution. The most probable model β_r

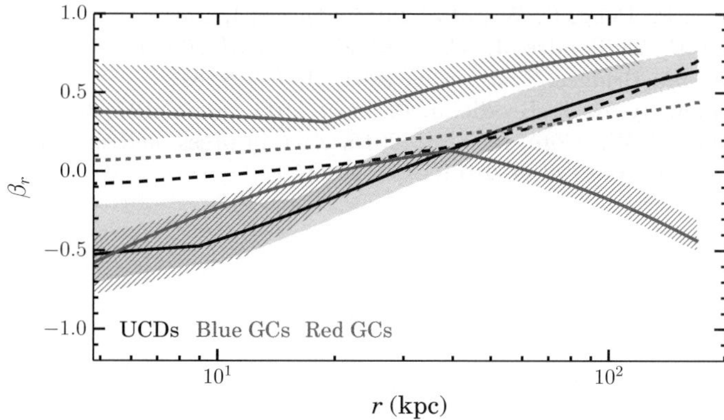

Figure 3. Variation of the anisotropy parameters as a function of the 3D radius. The profiles for UCDs, blue GCs, and red GCs are represented as black, blue, and red solid curves respectively. Following the same color code, the hatched regions of different styles mark the 68% confidence intervals for blue GCs, and red GCs. The grey shaded region marks the 68% confidence interval for the UCDs. The short dashed curves (*black* for UCDs, *blue* for blue GCs) represent the anisotropy profiles predicted by a universal relation between the number density slope and β for relic high-σ density peaks as found in cosmological simulations by Diemand *et al.* (2005).

profile for each population is taken as the fiducial one, and the 68% confidence intervals are determined by randomly resampling the real data sets, with $\sim 10\%$ of data points being left out for each resample.

The determined β_r profiles for UCDs, blue and red GCs are shown in Figure 3. The UCD system has an anisotropy profile that becomes more radial with radius, with β_r being negative within the inner ~ 40 kpc and being positive beyond. The blue GCs have a radially increasing β_r profile in the inner ~ 40 kpc but a radially decreasing profile at larger radii. Among the three populations, the red GCs exhibit the largest radially-biased orbital structure across the explored radius range.

Although being more negative in the innermost radii, the β_r profile of UCDs determined from Jeans analysis is more or less consistent with the cosmological simulations of Diemand *et al.* (2005). In contrast, the blue GCs exhibit large deviation from the cosmological simulations. The significantly tangentially-biased orbital structure of UCDs at small radii can be partly attributed to a strong tidal transformation. The finding that the blue GCs are tangentially-biased, rather than radially-biased, at large radii may indicate that the blue GCs in the outer halo of M87 have not yet established an equilibrium state and is still in an early and active stage of assembly by continuously accreting the surrounding dwarf galaxies (e.g. Côté, Marzke & West 1998).

3. Conclusion

We conclude that most UCDs in M87 are not consistent with being merely the most luminous and extended examples of otherwise normal GCs. The radially-biased orbital structure of UCDs at large radii is in general agreement with the scenario that UCDs are tidally stripped nuclei of dwarf galaxies. The distinct rotational properties of UCDs, as compared to GCs and the surrounding dE galaxies, indicate that the primary UCD progenitors do not necessarily resemble the present-day surviving dwarf galaxies.

References

Côté, P., Marzke, R. O., & West, M. J. 1998, *ApJ*, 501, 554
Diemand, J., Madau, P., & Moore, B. 2005, *MNRAS*, 364, 367

Durrell, P. R., Côté, P., Peng, E. W., *et al.* 2014, *ApJ*, 794, 103
Ferrarese, L., Côté, P., Cuillandre, J. -C., *et al.* 2012, *ApJS*, 200, 4
Hanes, D. A., Côté, P., Bridges, T. J., *et al.* 2001, *ApJ*, 559, 812
Strader, J., Romanowsky, A. J., Brodie, J. P., *et al.* 2011, *ApJS*, 197, 33
Zhang, H. -X., Peng, E. W., Côté, P., *et al.* 2015, *ApJ*, in press, arXiv: 1501.03167
Zhu, L., Long, R. J., Mao, S., *et al.* 2014, *ApJ*, 792, 59

Star clusters and black holes in galaxies across cosmic time
Proceedings IAU Symposium No. 312, 2014
Y. Meiron, S. Li, F.-K. Liu & R. Spurzem, eds.

© International Astronomical Union 2016
doi:10.1017/S1743921315008017

Black hole and nuclear cluster scaling relations: $M_{\mathrm{bh}} \propto M_{\mathrm{nc}}^{2.7 \pm 0.7}$

Alister W. Graham

Centre for Astrophysics and Supercomputing, Swinburne University of Technology, Hawthorn, Victoria 3122, Australia.
email: AGraham@swin.edu.au

Abstract. There is a growing array of supermassive black hole and nuclear star cluster scaling relations with their host spheroid, including a bent (black hole mass)–(host spheroid mass) M_{bh}–M_{sph} relation and a different (massive compact object mass)–(host spheroid velocity dispersion) M_{mco}–σ relations for black holes and nuclear star clusters. By combining the observed $M_{\mathrm{bh}} \propto \sigma^{5.5}$ relation with the observed $M_{\mathrm{nc}} \propto \sigma^{1.6-2.7}$ relation, we derive the expression $M_{\mathrm{bh}} \propto M_{\mathrm{nc}}^{2-3.4}$, which should hold until the nuclear star clusters are eventually destroyed in the larger core-Sérsic spheroids. This *new* mass scaling relation helps better quantify the rapid evolutionary growth of massive black holes in dense star clusters, and the relation is consistently recovered when coupling the observed $M_{\mathrm{nc}} \propto M_{\mathrm{sph}}^{0.6-1.0}$ relation with the recently observed quadratic relation $M_{\mathrm{bh}} \propto M_{\mathrm{sph}}^{2}$ for Sérsic spheroids.

Keywords. galaxies, black holes, nuclear star clusters.

1. Introduction

Over the past two decades there has been wide-spread interest in the scaling relations connecting supermassive black holes (SMBHs) with their host galaxy, and in particular with their host bulge. This has been, in part, due to observations which suggested that they grow in tandem, with feedback from the black hole (previously) thought to establish a near constant 0.1–0.2% mass ratio with the host spheroid. Over the last decade there has been a quieter realisation that the nuclear star clusters (NSCs)† at the centres of most Sérsic galaxies also correlate with the properties of their host spheroid. This connection continues until the disappearance / destruction of the clusters in the (massive) core-Sérsic galaxies with partially depleted cores (Bekki & Graham 2010, and references therein). Given the coexistence of SMBHs within dense star clusters (e.g. Seth *et al.* 2008; González Delgado *et al.* 2008; Graham & Spitler 2009; Graham 2012b; Leigh *et al.* 2012; Neumayer & Walcher 2012; Scott & Graham 2013) one may wonder if massive black holes might also be intimately connected with their host star cluster, in addition to their host spheroid, or perhaps SMBHs and dense NSCs merely inevitable co-inhabitants at the bottom of each galaxy's gravitational potential well.

In this review talk I briefly present the latest scaling relations between both SMBHs and NSCs with their host spheroid's (*i*) velocity dispersion (Section 2) and (*ii*) stellar mass (Section 3). After then reminding ourselves what Sérsic and core-Sérsic galaxies are (Section 4), these relations are consistently brought together in a way that eliminates the spheroid and yields, for the first time, the mass relation between SMBHs and their host NSCs (Section 5).

† Nuclear star clusters are so-named because of their location at the nuclei of galaxies.

2. The $M_{\rm mco}$-σ relations

Massive black holes and dense nuclear star clusters† are collectively referred to here as massive compact objects (mco). In the (massive compact object mass)–(host spheroid velocity dispersion) $M_{\rm mco}$-σ diagram, SMBHs and NSCs follow different tracks.

The $M_{\rm bh} \propto \sigma^X$ relation has a logarithmic slope X of around 5.5±0.3 (Graham & Scott 2013; McConnell & Ma 2013). Galaxies with bars have also been observed to display an apparent offset to lower black hole masses in the $M_{\rm bh}$–σ diagram (Graham 2008; Hu 2008; Graham & Li 2009; Graham *et al.* 2011). As was noted by Hu (2008) and Graham (2008), this may be due to under-massive black holes in what might be pseudobulges (an idea preferred by Greene *et al.* 2010 and Kormendy & Bender 2011), or instead it may be due to the occurrence of higher velocity dispersions. Hartmann *et al.* (2013) have recently shown that the dynamics associated with bars are indeed fully capable of explaining the observed offset in the $M_{\rm bh}$–σ diagram in terms of elevated velocity dispersions (see also Brown *et al.* 2013 and Debattista *et al.* 2013), and Graham & Scott (2013) have found no offset between barred and unbarred galaxies in the $M_{\rm bh}$–$L_{\rm sph}$ diagram, disfavouring the pseudobulge idea suggested 7 years ago.

The $M_{\rm nc}$–σ^Y relation has a much shallower slope than the $M_{\rm bh}$–σ^X relation. Excluding nuclear disks, Graham (2012b) reported a logarithmic slope Y of 1.57 ± 0.24, while Scott & Graham (2013) reported a value of 2.11 ± 0.31 having over-lapping error bars. Leigh *et al.* (2012) have however reported a steeper value of 2.73 ± 0.29, attributed to their inclusion of nuclear disks which can be an order of magnitude more massive than the biggest nuclear star clusters.

3. Sérsic versus core-Sérsic galaxies

Sérsic galaxies contain spheroids, either bulges or the main elliptical galaxy itself, whose projected light is well described by Sérsic's $R^{1/n}$ (1963) model. These Sérsic spheroids may additionally contain NSCs. In contrast, core-Sérsic galaxies display a partially depleted core, not due to dimming by dust and typically less than a few hundred parsec in radius, relative to the inward extrapolation of their outer Sérsic profile (Graham *et al.* 2003; Trujillo *et al.* 2004). The Sérsic versus core-Sérsic divide built on but differs from the "core" versus "power-law" galaxy divide (Lauer *et al.* 1995) in that "core" galaxies do not always have a partially depleted core relative to their outer profile (see Dullo & Graham 2014, and references therein). The core-Sérsic galaxies are thought to have formed from the dry merger of Sérsic (and/or core-Sérsic) galaxies, wherein the SMBHs sink to the centre via the gravitational ejection of stars from the core of the newly formed galaxy.

With Sérsic indices from less than 1 to ∼4, Sérsic spheroids follow a log-linear luminosity-(central surface brightness) relation ($L - \mu_0$) and a log-linear $L - n$ relation (e.g. Graham & Guzmán 2003). Due to the non-homology in their light profiles, i.e. the fact that they do not all have the same ($R^{1/4}$, for example) light profiles, this systematic change in n with luminosity produces a non-linear luminosity-dependent difference between μ_0 and $\langle\mu\rangle_{\rm e}$ (the mean surface brightness within the effective half light radius, $R_{\rm e}$). This results in a strongly curved $L - \langle\mu\rangle_{\rm e}$ relation. Given that $L = 2\pi\langle I\rangle_{\rm e} R_{\rm e}^2$, where $\langle I\rangle_{\rm e}$ is the average intensity associated with the average surface brightness, the $L - R_{\rm e}$ relation is also strongly curved. These relations are in fact so curved that the faint ($n \lesssim 2$) and bright ($n \gtrsim 2$) arms of the relations have, before the consequences of structural non-homology

---

† By this term we mean to exclude (obvious) nuclear discs, which can be much more extended than compact nuclear star clusters (see Balcells *et al.* 2007).

were known, been mistakenly heralded as evidence for a dichotomy between faint and bright early-type galaxies (see Graham *et al.* 2013 for an extended review).

Due to the depleted cores in the core-Sérsic spheroids (typically $M_B < -20.5 \pm 0.75$ mag), they branch off from the $L - \mu_0$ relation toward lower central surface brightnesses. Core-Sérsic and Sérsic spheroids/galaxies do however follow the same steep $M_{\mathrm{bh}}-\sigma$ relation (e.g. Graham & Scott 2013).

4. The M_{mco}-M_{sph} relations

Before getting to observations of the M_{mco}-M_{sph} relations, one can already predict the general behavior in the case of the M_{bh}-M_{sph} relation. This is done by noting a transition or bend in the $L - \sigma$ relation found by Davies *et al.* (1983) such that low-luminosity early-type galaxies (not pseudobulges) follow the relation $L-\sigma^2$ while the high-luminosity galaxies ($M_B < -20.5$ mag) follow a steeper relation. Matković & Guzmán (2005) explained the bend in terms of Sérsic versus core-Sérsic galaxies. This bend was recently shown again as a bend in the $M_{\mathrm{sph}}-\sigma$ relation by Davies and his collaborators in Cappellari *et al.* (2013).

Coupled with the log-linear $M_{\mathrm{bh}}-\sigma$ relation noted in Section 2, the bent $M_{\mathrm{sph}}-\sigma$ relation necessitates a bent $M_{\mathrm{bh}}-M_{\mathrm{sph}}$ relation. As was pointed out in Graham (2012a), for things to be consistent there *must* be a bent relation rather than the log-linear relation which had been assumed and claimed for well over a decade. This of course introduces a huge change to our understanding of the physical relation between SMBHs and their host spheroid.

While the bent $M_{\mathrm{bh}}-M_{\mathrm{sph}}$ relation was first presented in Graham (2012a) with actual data, the black hole masses did not probe very far down the mass function, making the discovery somewhat hard to see (although still statistically significant). However in Graham & Scott (2013), see also Scott, Graham & Schombert (2013), they were able to include data down to $M_{\mathrm{bh}} \approx 10^6 M_\odot$, and in Graham & Scott (2014, in prep.) it reaches down to $10^5 M_\odot$ through the inclusion of over 100 active galactic nuclei with indirectly measured black hole masses. What the two papers in 2013 confirmed is that Sérsic galaxies follow a near-quadratic $M_{\mathrm{bh}}-M_{\mathrm{sph}}$ relation, i.e. a power-law with a slope close to 2, as predicted in Graham (2012b). It is only the core-Sérsic galaxies, built from simple additive mergers, which branch off and follow a near-linear $M_{\mathrm{bh}}-M_{\mathrm{sph}}$ relation. As a result, fitting a single log-linear relation to samples of Sérsic and core-Sérsic galaxies produces a slope greater than 1 and a relation which is not optimal for either population.

The $M_{\mathrm{bh}}/M_{\mathrm{sph}}$ mass ratio for core-Sérsic galaxies was found by Graham (2012a) to be 0.36%, double the previously assumed constant value for all galaxy types, and it was increased to 0.49% in Graham & Scott (2013). However, due to the quadratic relation for the Sérsic galaxies, their $M_{\mathrm{bh}}/M_{\mathrm{sph}}$ mass ratio can be far lower.

The $M_{\mathrm{nc}}-M_{\mathrm{sph}}$ relation was found by Balcells *et al.* (2003) among the bulges of disk galaxies, and later by Graham & Guzmán (2003) using a sample of predominantly elliptical galaxies. The slope of this relation has since been measured many times, most recently by den Brok *et al.* (2014) who reports $L_{\mathrm{nc}} \propto L_{\mathrm{sph}}^{0.57\pm0.05}$ ($F814W$), in fair agreement with the value of 0.60 ± 0.10 from Scott & Graham (2013) for the $M_{\mathrm{nc}}-L_{\mathrm{sph}}$ (K-band) relation for early-type galaxies. Previous works have claimed slopes around 0.75 but as high as 1 when including nuclear disks (e.g. Grant *et al.* 2005; Wehner & Harris 2006; Côté *et al.* 2006; Balcells *et al.* 2007).

5. The (new) $M_{\rm bh}$-$M_{\rm nc}$ relation

Coupling $M_{\rm nc} \propto M_{\rm sph}^{0.6-1.0}$ with $M_{\rm bh} \propto M_{\rm sph}^2$ for the Sérsic spheroids gives $M_{\rm bh} \propto M_{\rm nc}^{2-3.3}$.

Coupling $M_{\rm bh} \propto \sigma^{5.5}$ with $M_{\rm nc} \propto \sigma^{1.6-2.7}$ from Section 2 gives $M_{\rm bh} \propto M_{\rm nc}^{2.0-3.4}$.

Depending on which precise slopes one adopts from the wedded pair of relations above, one ends up with a different slope for the new relation between black hole mass and host nuclear star cluster mass. While the author's past work would favour a steeper exponent (3.4), encompassing the wider literature suggests something like $M_{\rm bh} \propto M_{\rm nc}^{2.7\pm0.7}$ given the range of slopes for the initial relations. It is hoped that further observations and analysis, combined with theory, will be able to refine and explain this steep (non-linear) relation which may not simply be a consequence of the relations from which it was derived here.

As noted in the Introduction, not all galaxies with SMBHs have NSCs, and as such a certain degree of common sense is required in application of this new relation. For instance, core-Sérsic galaxies do not have a NSC, which was likely eroded away prior to the formation of their partially-depleted cores. Given this, as one approaches the high-mass end of the Sérsic galaxy sequence (from lower masses), it is expected that the NSCs will flay and some Sérsic galaxies would no longer house any significant NSC (e.g., NGC 5831, Graham *et al.* 2003). At the low-mass end, and as with the $M_{\rm bh}-\sigma$ relation and the $M_{\rm bh}-M_{\rm sph}$ relation, the frequency of massive black holes below $10^5 M_\odot$ is not yet known. The occurrence of NSCs is however known to tailor off, or at least become harder to identify, in early-type galaxies fainter than $M_{F814W} = -15$ mag (den Brok *et al.* 2014), which may then reflect some kind of lower bound to the relation. Finally, it is remarked that massive BHs in globular clusters may be better matched to the new $M_{\rm bh}$-$M_{\rm nc}$ relation than the $M_{\rm bh}$-$M_{\rm sph}$ relation.

6. Conclusions

SMBHs grow rapidly relative to their stellar nurseries, i.e. the nuclear cluster of stars which still enshroud many. This is not to say that we know if the SMBHs were born in these nurseries; although once they become one hundred million solar mass grown-ups their nursery is gone. The growth of the BH relative to the NSC is extremely rapid: $M_{\rm bh} \propto M_{\rm nc}^{2.7\pm0.7}$, with the author favouring higher values for the exponent, especially if new data steepens the $M_{\rm bh}-\sigma$ relation, and if the $M_{\rm bh}$-$M_{\rm sph}$ relation *is* super-quadratic for the Sérsic galaxies.

Acknowledgements

The author thanks the conference organizers for bringing together researchers of massive black holes and star clusters, and for the opportunity to present this invited review/update which resulted in the formulation of the steep (black hole mass)–(nuclear star cluster mass) relation given here.

References

Balcells, M., Graham, A. W., Domínguez-Palmero, L., & Peletier, R. F. 2003, *ApJ Lett.*, 582, L79

Balcells, M., Graham, A. W., & Peletier, R. F. 2007a, *ApJ*, 665, 1084

Bekki, K. & Graham, A. W. 2010, *ApJ Lett.*, 714, L313

Brown, J. S., Valluri, M., Shen, J., & Debattista, V. P. 2013, *ApJ*, 778, 151

Cappellari, M., McDermid, R. M., Alatalo, K. *et al.* 2013, *MNRAS*, 432, 1862

Côté, P., Piatek, S., Ferrarese, L., *et al.* 2006, *ApJS*, 165, 57

Davies, R. L., Efstathiou, G., Fall, S. M., Illingworth, G., & Schechter, P. L. 1983, *ApJ*, 266, 41

Debattista, V. P., Kazantzidis, S., & van den Bosch, F. C. 2013, *ApJ*, 765, 23

den Brok, M., Peletier, R. F., Seth, A., *et al.* 2014, *MNRAS*, accepted (arXiv:1409.4766)

Dullo, B. T. & Graham, A. W. 2014, *MNRAS*, 444, 2700

González Delgado, R. M., Pérez, E., Cid Fernandes, R., & Schmitt, H. 2008, *AJ*, 135, 747

Graham, A. W. 2008, *ApJ*, 680, 143

Graham, A. W. 2012a, *ApJ*, 746, 113

Graham, A. W. 2012b, *MNRAS*, 422, 1586

Graham, A. W. 2013, in "Planets, Stars and Stellar Systems", Vol. 6, p.91-140, T.D.Oswalt & W.C Keel (eds.), Springer Publishing (arXiv:1108.0997)

Graham, A. W., Erwin, P., Trujillo, I., & Asensio Ramos, A. 2003, *AJ*, 125, 2951

Graham, A. W. & Guzmán, R. 2003, *AJ*, 125, 2936

Graham, A. W., Onken, C. A., Athanassoula, E., & Combes, F. 2011, *MNRAS*, 412, 2211

Graham, A. W. & Li, I-H. 2009, *ApJ*, 698, 812

Graham, A. W. & Scott, N. 2013, *ApJ*, 764, 151

Graham, A. W. & Spitler, L. R. 2009, *MNRAS*, 397, 2148

Grant, N. I., Kuipers, J. A., & Phillipps, S. 2005, *MNRAS*, 363, 1019

Greene, J. E., Peng, C. Y., Kim, M., *et al.* 2010, *ApJ*, 721, 26

Hartmann, M., Debattista, V. P., Cole, D. R., *et al.* 2013, *MNRAS*, 441, 1243

Hu, J. 2008, *MNRAS*, 386, 2242

Kormendy, J. & Bender, R. 2011, *Nature*, 469, 377

Lauer, T. R., Ajhar, E. A., Byun, Y.-I., *et al.* 1995, *AJ*, 110, 2622

Leigh, N., Böker, T., & Knigge, C. 2012, *MNRAS*, 424, 2130

Matković, A., & Guzmán, R. 2005, *MNRAS*, 362, 289

McConnell, N. J. & Ma, C.-P. 2013, *ApJ*, 764, 184

Neumayer, N. & Walcher, C. J. 2012, *Advances in Astronomy*, 2012, (arXiv:1201.4950)

Scott, N. & Graham, A. W. 2013, *ApJ*, 763, 76

Scott, N., Graham, A. W., & Schombert, J. 2013, *ApJ*, 768, 76

Sérsic, J.-L. 1963, *Boletin de la Asociacion Argentina de Astronomia*, vol.6, p.41

Seth, A., Agüeros, M., Lee, D., & Basu-Zych, A. 2008, *ApJ*, 678, 116

Trujillo, I., Erwin, P., Asensio Ramos, A., & Graham, A. W. 2004, *AJ*, 127, 1917

Wehner, E. H. & Harris, W. E. 2006, *ApJ Lett.*, 644, L17

Star clusters and black holes in galaxies across cosmic time
Proceedings IAU Symposium No. 312, 2014
Y. Meiron, S. Li, F.-K. Liu & R. Spurzem, eds.
© International Astronomical Union 2016
doi:10.1017/S1743921315008029

The Milky Way's nuclear star cluster and massive black hole

Rainer Schödel

Instituto de Astrofísica de Andalucía (CSIC), Glorieta de la Astronomía s/n, ES-18008,
Granada, Spain
email: `rainer@iaa.es`

Abstract. Because of its nearness to Earth, the centre of the Milky Way is the only galaxy nucleus in which we can study the characteristics, distribution, kinematics, and dynamics of the stars on milli-parsec scales. We have accurate and precise measurements of the Galactic centre's central black hole, Sagittarius A*, and can study its interaction with the surrounding nuclear star cluster in detail. This contribution aims at providing a concise overview of our current knowledge about the Milky Way's central black hole and nuclear star cluster, at highlighting the observational challenges and limitations, and at discussing some of the current key areas of investigation.

Keywords. Nuclear star clusters, massive black holes, Galactic centre

---

1. Introduction

The paradigm that massive black holes (MBHs) are located in the centres of all major galaxies has been firmly established over the past decades (e.g., Gültekin *et al.* 2009). Moreover, the recent detection of an MBH in an ultra-compact dwarf galaxy by Seth *et al.* (2014) indicates that these objects may be present in almost all galaxy types. In addition to MBHs, the majority of galaxies of all types contain stellar nuclei in the form of nuclear star clusters (NSCs). NSCs have similar sizes as globular clusters, but contain several times their mass. They thus belong to the densest stellar systems in the Universe (for an overview, see, e.g., Böker 2010, or Schödel *et al.* 2014b). The study of galactic nuclei is of great interest for astrophysics in general because it touches a wide range of topics, such as General Relativity (GR), dense N-body dynamics, star formation in extreme environments or the accretion history of MBHs. Unfortunately, the large distances of extragalactic nuclei mean that we can resolve only relatively large physical scales in them (at best a few 0.1 pc in the nearest systems). This means that the observed light per resolution element arises from tens of thousands to millions of stars and the radius of influence, where the MBH dominates stellar dynamics, is barely resolved.

The Milky Way appears to be a relatively normal barred spiral galaxy. Since its nucleus is located at a distance of only about 8 kpc from Earth, it provides us with a unique opportunity to study galactic stellar nuclei and MBHs. In the case of the Galactic centre (GC), we can resolve linear scales on the order of milli-parsecs (mpc) in the near-infrared (NIR) and thus examine individual stars and their kinematics and dynamics. Hence, the GC is of fundamental importance for investigating questions such as the validity of GR near an MBH, the interaction of stars with an MBH, the initial mass function in galaxy nuclei, or the existence of stellar cusps around MBHs through accurate and precise quantitative measurements. For further in-depth reading about most of the topics discussed here, I recommend the recent detailed review article about the GC by Genzel *et al.* (2010), as well as the shorter review by Schödel *et al.* (2014b), that is mainly focused on the Milky Way's NSC.

2. Observational constraints

While the Milky Way offers a unique template for the study of galactic nuclei on the one hand, there exist, on the other hand, significant constraints for observational studies of the GC. Our line-of-sight through the Galactic disc implies that interstellar extinction toward the Milky Way's nucleus is extreme. With $A_V \gtrsim 30$ mag, studies at visual wavelengths are all but impossible. Even in the NIR, at wavelengths around $2.2\,\mu$m (the so-called K-band), extinction still amounts to 2-5 mag (see Nishiyama *et al.* 2008 or Fritz *et al.* 2011). The presence of molecular clouds in the central few hundred parsecs of the Milky Way induces the additional difficulty that interstellar extinction varies significantly on angular scales of only a few arcseconds (Schödel *et al.* 2010). This makes even rough stellar classification through broad-band photometry, e.g., distinguishing between cool giants and massive main sequence stars, very challenging.

The sheer number of stars results in high surface number densities of at least a few (at $\sim$100 pc from the center) up to several tens (in the central parsec) of stars per square arcsecond. As a consequence, crowding limits the completeness of star counts to relatively bright magnitudes ($K \lesssim 15$) and can induce significant astrometric and photometric bias in seeing-limited (resolution $\sim 0.5''$ at $2.2\,\mu$m) imaging from the ground. For the central parsec of the GC, even the resolution offered by the Hubble Space Telescope (HST) is insufficient to overcome crowding. To overcome crowding, the resolution offered by an 8-10m-class telescope supported by adaptive optics (resolution $\sim 0.06''$) is needed. In any case, the 50% completeness limit for source detection lies at magnitudes as bright as $K \approx 18 - 19$ in the central parsec of the GC. The detection of solar mass main sequence stars in this region ($K \approx 21$, taking into account distance and extinction) will require the angular resolution of telescopes of the 30m-class.

3. Sagittarius A*: the Milky Way's central black hole

Stellar proper motion and line-of-sight velocity measurements carried out since the mid-1990s (see Eckart & Genzel 1996 or Ghez *et al.* 1998) have resulted in the accurate measurement of a large number of individual orbits of stars around the radio source Sagittarius A* (Gillessen *et al.* 2009 and Ghez *et al.* 2008). The currently existing data can be fitted accurately with Keplerian orbits and require a mass of $\sim 4 \times 10^6\ M_\odot$ to be concentrated within a radius of $\lesssim 0.6$ mpc of Sagittarius A* (Sgr A*). The resulting high mass density in combination with radio-to-X-ray measurements of the (very low) emission from this location make the black hole hypothesis the only one that can currently satisfyingly explain all observational data. Although the term Sgr A* refers, strictly speaking, to the electromagnetic radiation released by hot plasma close to the Milky Way's central MBH, it is frequently used as a name for the putative black hole itself.

Stellar orbits have also allowed us to measure the distance of Sgr A*, which is $\sim$8 kpc. The combined statistical and systematic uncertainties of the mass and distance of the black hole are currently already $< 10\%$ and $< 5\%$, respectively, and will further improve with continued monitoring of stellar orbits in the future.

Stars with orbital periods less than 20 yr are termed short-period stars. They are of special importance because their orbital parameters can be determined with high accuracy within reasonable time. They are therefore of paramount importance to probe the gravitational potential around Sgr A* and thus to determine the amount of extended mass around the black hole, that may be present in the form of stellar remnants, and, at the same time, to test the validity of general relativity. Several short-period stars are needed to break model degeneracies (see Meyer *et al.* 2012). The star with the currently shortest-known period orbits Sgr A* in just 11.5 years (Meyer *et al.* 2012). The star with the most accurately and precisely measured orbit is S2/S0-2. It will pass through pericentre again in 2018 and will then provide us with an opportunity to detect the transverse-Doppler effect and gravitational redshift terms of special and general relativity (for an overview of this topic, see, e.g., Genzel *et al.* 2010 or Schödel *et al.* 2014a)

4. Nuclear star cluster: morphology and kinematics

One of the most complete existing works on the stellar structures at the GC is the study by Launhardt *et al.* (2002). They show that the central hundreds of parsecs are dominated by the so-called *nuclear bulge* (NB), that is composed of a stellar disc (scale height $\lesssim 45$ pc, radius ~ 230 pc) and a compact NSC. The total stellar mass of the NB is about $1.4 \times 10^9 \, M_\odot$. Strong UV-radiation and the stellar luminosity function indicate significant recent star formation, in particular toward the NSC. As concerns the properties of the NSC, a limitation in this study and other, previous and later studies is, however, that they are significantly affected by low angular resolution and/or the strong and variable interstellar extinction. The shape of the NSC was assumed as spherical without being able to test this assumption. Two recent works have made substantial progress in this respect, but with different methodologies. Fritz *et al.* (2014) have corrected stellar number counts for extinction and corrected areas with extreme extinction by assuming symmetry with respect to the Galactic plane and axis. Schödel *et al.* (2014a) used IRAC/Spitzer mid-infrared images around $3 - 5\,\mu$m because interstellar extinction reaches a minimum in this region (Fritz *et al.* 2011). They thus did not have to assume any intrinsic symmetry but could instead directly demonstrate the point-symmetry of the NSC with respect to Sgr A*. Both studies agree in their findings, which is encouraging, given the completely different methodologies applied. The NSC is found to be intrinsically flattened. It is aligned with the Galactic plane, although a small misalignment ($\lesssim 10\,°$) cannot be excluded.

Both mentioned studies generally agree in the measured parameters of the Milky Way's NSC. According to Schödel *et al.* (2014a), the nuclear cluster of the Milky Way

(a) is precisely centred on Sgr A* (uncertainty < 0.2 pc on large scales; actually, higher angular resolution measurements of a smaller area show that the uncertainty is only on the order of 0.02 pc, see Schödel *et al.* 2007);

(b) has a ratio of minor to major axis of 0.71 ± 0.02;

(c) can be described adequately by a Sérsic law with an index $n = 2.0 \pm 0.2$;

(d) has a half light radius of 4.2 ± 0.4 pc; and

(e) has a luminosity of $4.1 \pm 0.4 \times 10^7 \, L_\odot$ and a mass of $2.5 \pm 0.4 \times 10^7 \, M_\odot$.

The most recent studies of the overall kinematics of the Milky Way's NSC were carried out by Feldmeier *et al.* (2014) and Chatzopoulos *et al.* (2015) (see also Fritz *et al.* 2014). Also these studies use completely different methodologies. While Chatzopoulos *et al.* (2015) use the proper motions and line-of-sight velocities of thousands of individual stars, obtained from observations with high angular resolution, Feldmeier *et al.* (2014) analyse a seeing-limited spectroscopic slit drift-scan map of the NSC. Encouragingly, the basic results are again in agreement.

(a) The kinematics of the NSC are consistent with its flattening.

(b) The NSC rotates in parallel to Galactic rotation. Its rotation velocity approaches asymptotically a values of $30 - 40\,\mathrm{km\,s^{-1}}$ at distances $r \gtrsim 4\,\mathrm{pc}$.

(c) The NSC can be described approximately by an isotropic rotator model (Chatzopoulos *et al.* 2015).

(d) The kinematically derived mass is consistent with the photometrically derived mass by Schödel *et al.* (2014a).

(e) The stellar mass-to-light ratio is $0.76 \pm 0.18\,M_\odot/L_\odot$ at $\sim2.2\,\mu\mathrm{m}$ (Chatzopoulos *et al.* 2015) and $0.56 \pm 0.26\,M_\odot/L_\odot$ at $\sim4.5\,\mu\mathrm{m}$ (Feldmeier *et al.* 2014).

(f) The kinematic analysis suggests a misalignment of the NSC major axis by $\sim9°$ with respect to the Galactic plane.

(g) There are indications of a separate kinematic component at distances $r = 1 - 2\,\mathrm{pc}$ from Sgr A*, whose position angle appears to be offset by $\sim90°$ from the overall cluster rotation.

The kinematic substructure and the overall misalignment of the NSC kinematic axis with the Galactic plane may indicate the residuals of distinct accretion events. Investigating this evidence further may provide hints to the formation history of the NSC.

Finally, it is important to note that kinematic modelling of the NSC almost always consistently underestimates the mass of the central MBH by factors of $30 - 50\%$. The studies by Chatzopoulos *et al.* (2015) and, to a lesser degree, by Schödel *et al.* (2009) and Do *et al.* (2013) are the only ones that derive BH masses largely consistent with the estimates from stellar orbits. However, it must be pointed out that these publications were produced when the correct solution was already known. Fritz *et al.* (2014) list and discuss and analyse quantitatively possible sources of bias, among them the effects of anisotropy and the flat stellar core of the cluster. They conclude that no effect by itself appears to be sufficient to explain the low derived BH mass in the Jeans modeling. A combination of various effects and/or the necessity for improved modeling may provide the answer. Data sets such as the pseudo integral-field slit-scan data of Feldmeier *et al.* (2014) are of great value to study the possible systematic effects that lead to erroneous BH masses. This is of particular importance with respect to spectroscopic studies of extragalactic MBHs, which may suffer similar biases.

5. Is there a stellar cusp around Sgr A*?

The formation of a stellar cusp in a relaxed stellar cluster around a MBH is a robust result of theoretical stellar dynamics, which predicts a stellar density increase in the form of a power-law within the sphere of influence of the black hole. The latter is generally defined by the radius within which the stellar mass corresponds to twice the black hole mass. Depending on the properties of the stellar cluster, in particular the number ratio between heavy and light stars, the stellar density within the sphere of influence will settle to a density distribution that can be described by $\rho \propto r^{-\gamma}$, with different values of γ for stars of different mass. Alexander & Hopman (2009) show that in the so-called weak mass segregation regime $\gamma \approx 7/4$ is valid for the heavy stars and $3/2 < \gamma < 7/4$ for the light stars. In the strong segregation regime, which is considered probable for the Milky Way's NSC, the density distribution of the rare massive stellar objects can be described by $2 \lesssim \gamma \lesssim 11/4$ and the one of the lighter stars by $3/2 \lesssim \gamma \lesssim 7/4$ (see also Preto & Amaro-Seoane 2010).

Searching for stellar cusps around extragalactic MBHs is a very difficult task because the related small angular scales of their radius of influence mean that one can only study the light density profile. The latter can, however, easily be biased by the presence of a small number of bright giants/supergiants as well as by interstellar extinction and recent star formation events near the MBH. Conclusive tests require number density counts or very careful removal of the influence of bright stars (see discussion in Schödel *et al.* 2007). With current technology, the GC is therefore the only reliable target to test the predictions of theoretical stellar dynamics on cusp formation.

Within the radius of influence of Sgr A*, about 2-4 pc (Feldmeier *et al.* 2014, Chatzopoulos *et al.* 2015), the projected stellar number density follows a density law of about $\rho \propto R^{-0.8}$, where R is the projected radius. This is consistent with a three-dimensional stellar cusp of $\rho \propto r^{-1.8}$, i.e., close to the predicted values. However, the number density of the stars old enough to be dynamically relaxed has been found to be almost flat within a projected radius of roughly 0.5 pc around Sgr A* (Buchholz *et al.* 2009; Do *et al.* 2009; Bartko *et al.* 2010). This means that the NSC appears to be characterised by a core instead of a cusp around Sgr A*. This surprising finding is inconsistent with the theoretical predictions.

There exist a range of models that try to explain the observed absence of a stellar cusp in the immediate vicinity of Sgr A*. The first class of explanations assumes that the density distribution of the observed stars is representative for the entire population of the NSC. In that case, the relaxation time in the NSC may simply be too long for the cusp to have formed (Merritt 2010) or the cusp may have been destroyed (e.g., Merritt & Szell 2006). The second class assumes that the cusp is invisible, i.e., that the observable giant stars are no adequate tracers of the underlying stellar distribution. This could happen if collisions destroy the envelopes of the giants and render them invisible (e.g., Dale *et al.* 2009; Amaro-Seoane & Chen 2014). Another explanation is provided by Löckmann *et al.* (2010): Continuous star formation over the Galaxy's lifetime results in the formation of stellar black holes within the NSC that migrate towards the centre due to dynamical friction and push out lighter stars to greater distances, thus turning the cusp of visible stars into a core.

At the moment it is not clear whether the absence of evidence for the stellar cusp is evidence for its absence. The observational constraints (see above) imply that we can currently only detect a small fraction of the stellar population of the NSC, that is, giants, supergiants, and massive, short-lived main and post-main sequence stars. The detection of main-sequence stars of (sub-)solar mass will require the sensitivity and, in particular, angular resolution of a 30m-class telescope. Also, future high-precision observations of stellar orbits may reveal the presence of an extended dark mass component around Sgr A* (see, e.g., Weinberg *et al.* 2005; Perets *et al.* 2009).

6. Star formation near Sgr A*

Several studies over the past decades have tried to infer the star formation history of the Milky Way's NSC, a difficult task given the challenging observational limitations. The most recent study was carried out by Pfuhl *et al.* (2011). It was mainly based on spectroscopy of 450 cool giant stars within a projected distance of 1 pc of Sgr A*, but also included some information on intermediate-mass main sequence stars from very sensitive spectroscopic observations. Pfuhl *et al.* (2011) find that about 80% of the stellar mass had already formed about 5 Gyr ago. About $1 - 2$ Gyr ago there was an apparent minimum in the star formation rate, which increased again during the last few hundred megayears. The data appear to indicate that the bulk of the NSC's stars formed with a canonical Chabrier/Kroupa initial mass function (see also Löckmann *et al.* 2010).

Much observational and theoretical effort has been focused on understanding the properties of the most recent star formation event in the GC. It is traced by roughly 180 O/B super giants and main sequence stars as well as Wolf-Rayet stars (e.g., Levin & Beloborodov 2003 Paumard *et al.* 2006; Bartko *et al.* 2009, Bartko *et al.* 2010, Lu *et al.* 2009, Lu *et al.* 2013). Almost all of these stars are located within 0.5 pc in projection from Sgr A* and a significant fraction of them rotate in a clockwise disk around the MBH. Their density increases toward the MBH and the disk-like pattern of their dynamics is consistent with their formation in a formerly existing dense gas disk around Sgr A*. The viability of this scenario has been confirmed by theoretical models (see, e.g., Bonnell & Rice 2008, and discussions in Levin & Beloborodov 2003 Lu *et al.* 2009, Bartko *et al.* 2010, or Genzel *et al.* 2010). The age of this star formation event (if it was a single one), is estimated to between 2.5 to 5.8 Myr (Lu *et al.* 2013). Intriguingly, this recent star formation event was almost certainly characterised by a top-heavy IMF (see, e.g., Nayakshin & Sunyaev 2005, Bartko *et al.* (2010), Lu *et al.* 2013), in contrast to the finding that most stars in the NSC formed with a standard IMF.

7. Summary

Because of extreme interstellar extinction and high stellar surface density the centre of the Milky Way poses, on the one hand, unique observational challenges. On the other hand, it is the only nucleus of a quiescent spiral galaxy nucleus that we can resolve observationally on scales of milli-parsecs and plays therefore a key role as a template, where to test many of our theoretical ideas. The existence of a central black hole at the GC has been established with high confidence through the measurements of stellar orbits and radio-to-X-ray observations of its electromagnetic counterpart, Sagittarius A*. Uncertainties on its mass and distance are already $\lesssim 10\%$ and can be expected to reach percent-level in the next decade. This will open the door to using the GC as a calibrator for cosmic distance measurements.

With a half-light radius of about 4 pc, a mass of $\sim 2.5 \times 10^7\ M_\odot$, and its complex stellar population, the Milky Way's NSC appears to be very similar to extragalactic nuclear clusters. It is truly central, both in morphology and kinematics, rotates in parallel to overall Galactic rotation, and is significantly flattened along the Galactic plane. Apparent kinematic substructures and the possible kinematic misalignment of the NSC with the Galactic plane may be remnants of individual accretion events, that may have contributed to building the cluster. A point of potential importance for measuring the masses of extragalactic MBHs is that the mass of Sgr A* has almost always been consistently under-estimated when applying Jeans modeling to kinematic data.

One of the key questions about galactic nuclei that we can test in the Milky Way's NSC is the presence of a stellar cusp around the MBH. Contrary to robust theoretical predictions, the NSC shows a core-like structure in the central parsec. It is not clear whether the cusp is indeed absent, i.e. has not yet had the time to form or was destroyed, or is just invisible because it is composed of dark remnants or because the tracer stars (giants) have been rendered invisible through collisions. Future measurements of stellar dynamics and sensitive, high angular resolution observations with 30m-class telescopes may help us to better understand the mystery of the missing cusp.

Although we can still not reconstruct the Milky Way's NSC's formation history, it appears to be clear that the majority of its stars formed many gigayears ago with a standard IMF. Recent star formation occurred close to Sgr A* a few Myr ago. The stars probably formed in a dense gas disc around the black hole with a top-heavy IMF.

Acknowledgements

The research leading to these results has received funding from the European Research Council under the European Union's Seventh Framework Programme (FP/2007-2013) /

ERC Grant Agreement n. 614922, and by grants AYA2010-17631 and AYA2012-38491-CO2-02, cofunded with FEDER funds, of the Spanish Ministry of Economy and Competitiveness.

References

Amaro-Seoane, P. & Chen, X. 2014, *ApJ* (Letters), 781, L18

Alexander, T. & Hopman, C. 2009, *ApJ*, 697, 1861

Bartko, H., Martins, F., Fritz, T. K., Genzel, R., Levin, Y., Perets, H. B., Paumard, T., Nayakshin, S., Gerhard, O., Alexander, T., Dodds-Eden, K., Eisenhauer, F., Gillessen, S., Mascetti, L., Ott, T., Perrin, G., Pfuhl, O., Reid, M. J., Rouan, D., Sternberg, A., & Trippe, S. 2009, *ApJ*, 697, 1741

Bartko, H., Martins, F., Trippe, S., Fritz, T. K., Genzel, R., Ott, T., Eisenhauer, F., Gillessen, S., Paumard, T., Alexander, T., Dodds-Eden, K., Gerhard, O., Levin, Y., Mascetti, L., Nayakshin, S., Perets, H. B., Perrin, G., Pfuhl, O., Reid, M. J., Rouan, D., Zilka, M., & Sternberg, A. 2010, *ApJ*, 708, 834

Böker, T. 2014, *Proceedings of the International Astronomical Union*, IAU Symposium 266, 58

Bonnell, I. A. & Rice, W. K. M. 2008, *Science*, 321, 1060

Buchholz, R. M., Schödel, R., & Eckart, A. 2009, *A&A*, 499, 483

Chatzopoulos, S., Fritz, T. K., Gerhard, O.,Gillessen, S., Wegg, C., Genzel, R., & Pfuhl, O. 2015, *MNRAS*, 447, 948

Dale, J. E., Davies, M. B., Church, R. P., & Freitag, M. 2009, *MNRAS*, 393, 1016

Do, T., Ghez, A. M.,Morris, M. R., Lu, J. R., Matthews, K., Yelda, S., & Larkin, J. 2009, *ApJ*, 703, 1323

Do, T.,Martinez, G. D.,Yelda, S.,Ghez, A.,Bullock, J., Kaplinghat, M., Lu, J. R., Peter, A. H. G., & Phifer, K. 2013, *ApJ* (Letters), 779, L6

Eckart, A. & Genzel, R. 1996, *Nature*, 383, 415

Feldmeier, A., Neumayer, N., Seth, A.,Schödel, R., Lützgendorf, N., de Zeeuw, P. T., Kissler-Patig, M., Nishiyama, S., & Walcher, C. J. 2014, *A&A*, 570, id.A2

Fritz, T. K., Gillessen, S., Dodds-Eden, K., Lutz, D., Genzel, R., Raab, W., Ott, T., Pfuhl, O., Eisenhauer, F., & Yusef-Zadeh, F. 2011, *ApJ*, 737, 73

Fritz, T. K., Chatzopoulos, S., Gerhard, O., Gillessen, S., Genzel, R.,Pfuhl, O.,Tacchella, S.,Eisenhauer, F., & Ott, T. 2014, *ArXiv e-prints*, 1406.7568F

Genzel, R., Eisenhauer, F., & Gillessen, S. 2010, *Rev. Mod. Physics*, 82, 3121

Ghez, A. M., Klein, B. L., Morris, M., & Becklin, E. E. 1998, *ApJ*, 509, 678

Ghez, A. M., Duchêne, G., Matthews, K., Hornstein, S. D., Tanner, A., Larkin, J., Morris, M., Becklin, E. E., Salim, S., Kremenek, T., Thompson, D., Soifer, B. T., Neugebauer, G., & McLean, I. 2003, *ApJ* (Letters), 586, L127

Ghez, A. M., Salim, S., Weinberg, N. N., Lu, J. R., Do, T., Dunn, J. K., Matthews, K., Morris, M. R., Yelda, S., Becklin, E. E., Kremenek, T., Milosavljevic, M., & Naiman, J. 2008, *ApJ*, 689, 1044

Gillessen, S., Eisenhauer, F.,Trippe, S.,Alexander, T.,Genzel, R., Martins, F., & Ott, T. 2009, *ApJ*, 692, 1075

Gültekin, K., Richstone, D. O., Gebhardt, K., Lauer, T. R.,Tremaine, S., Aller, M. C., Bender, R., Dressler, A., Faber, S. M., Filippenko, A. V., Green, R., Ho, L. C., Kormendy, J., Magorrian, J., Pinkney, J., & Siopis, C. 2009, *ApJ*, 698, 198

Launhardt, R., Zylka, R., & Mezger, P. G. 2002, *A&A*, 384, 112

Levin, Y. & Beloborodov, A. M. 2003, *ApJ* (Letters), 590, L33

Löckmann, U., Baumgardt, H., & Kroupa, P. 2010, *MNRAS*, 402, 519

Lu, J. R., Ghez, A. M., Hornstein, S. D., Morris, M. R., Becklin, E. E., & Matthews, K. 2009, *ApJ*, 690, 1463

Lu, J. R., Do, T., Ghez, A. M., Morris, M. R., Yelda, S., & Matthews, K. 2013, *ApJ*, 764, 155

&Merritt, D. and Szell, A. 2006, *ApJ*, 648, 890

Merritt, D. 2010, *ApJ*, 718, 739

Meyer, L., Ghez, A. M., Schödel, R., Yelda, S., Boehle, A., Lu, J. R., Do, T., Morris, M. R., Becklin, E. E., & Matthews, K. 2012, *Science*, 338, 84

Nayakshin, S. & Sunyaev, R. 2005, *ApJ*, 364, L23

Nishiyama, S., Nagata, T., Tamura, M., Kandori, R., Hatano, H., Sato, S., & Sugitani, K. 2008, *ApJ*, 680, 1174

Paumard, T., Genzel, R., Martins, F., Nayakshin, S., Beloborodov, A. M., Levin, Y., Trippe, S., Eisenhauer, F., Ott, T., Gillessen, S., Abuter, R., Cuadra, J., Alexander, T., & Sternberg, A. 2006, *ApJ*, 643, 1011

Perets, H. B., Gualandris, A., Kupi, G., Merritt, D., & Alexander, T. 2009, *ApJ*, 702, 884

Pfuhl, O., Fritz, T. K., Zilka, M., Maness, H., Eisenhauer, F., Genzel, R., Gillessen, S., Ott, T., Dodds-Eden, K., & Sternberg, A. 2011, *ApJ*, 741, 108

Preto, M. & Amaro-Seoane, P. 2010, *ApJ* (Letters), 708, L42

Schödel, R., Ott, T., Genzel, R., Eckart, A., Mouawad, N., & Alexander, T. 2003, *ApJ*, 596, 1015

Schödel, R.,Eckart, A., Alexander, T.,Merritt, D.,Genzel, R., Sternberg, A., Meyer, L., Kul, F., Moultaka, J., Ott, T., & Straubmeier, C. 2007, *A&A*, 469, 125

Schödel, R.,Merritt, D., & Eckart, A. 2009, *A&A*, 502, 91

Schödel, R., Najarro, F., Muzic, K., & Eckart, A. 2010, *A&A*, 511, id.A18

Schödel, R., Feldmeier, A., Kunneriath, D., Stolovy, S., Neumayer, N., Amaro-Seoane, P., & Nishiyama, S. 2014, *A&A*, 566, id.A47

Schödel, R., Feldmeier, A., Neumayer, N., Meyer, L., & Yelda, S. 2014, *Classical and Quantum Gravity*, 31, id.244007

Seth, A. C., van den Bosch, R., Mieske, S., Baumgardt, H., Brok, M. D., Strader, J., Neumayer, N., Chilingarian, I., Hilker, M., McDermid, R., Spitler, L., Brodie, J., Frank, M. J., & Walsh, J. L. 2014, *Nature*, 513, 398

Weinberg, N. N., Milosavljević, M., & Ghez, A. M. 2005, *ApJ*, 622, 878

Star clusters and black holes in galaxies across cosmic time
Proceedings IAU Symposium No. 312, 2014
Y. Meiron, S. Li, F.-K. Liu & R. Spurzem, eds.

© International Astronomical Union 2016
doi:10.1017/S1743921315008030

Star-formation in nuclear clusters and the origin of the Galactic center apparent core distribution

Danor Aharon and Hagai B. Perets

Physics Department, Technion – Israel Institute of Technology, Haifa 3200003, Israel
email: `danor@tx.technion.ac.il`

Abstract. Nuclear stellar clusters (NSCs) are known to exist around massive black holes (MBHs) in galactic nuclei. Two formation scenarios were suggested for their origin: build-up of NSCs and Continuous in-situ star-formation. Here we study the effects of star formation on the build-up of NSCs and its implications for their long term evolution and their resulting structure. We show that continuous star-formation can lead to the build-up of an NSC with properties similar to those of the Milky-way NSC. We also find that the general structure of the old stellar population in the NSC with in-situ star-formation could be very similar to the steady-state Bahcall-Wolf cuspy structure. However, its younger stellar population does not yet achieve a steady state. In particular, formed/evolved NSCs with in-situ star-formation contain differential age-segregated stellar populations which are not yet fully mixed. Younger stellar populations formed in the outer regions of the NSC have a cuspy structure towards the NSC outskirts, while showing a core-like distribution inwards; with younger populations having larger core sizes.

1. Introduction

Nuclear stellar clusters (NSCs), hosting massive black holes (MBHs) are thought to exist in a significant fraction of all galactic nuclei. Their origin is still not well understood. Two main scenarios were suggested for their origin: (1) The cluster infall scenario, in which stellar clusters inspiral to the galactic nucleus are disrupted, and thereby build up the nuclear cluster (Tremaine *et al.* (1975); Antonini (2013)). (2) The nuclear star formation (SF) scenario, in which gas infalls into the nucleus and then transforms into stars through star formation processes (Loose *et al.* (1982)). Here we focus on the latter process, and study the long term effects of SF on the formation and evolution of NSCs.

The structure, evolution and dynamics of NSCs have been extensively studied in recent years. These studies explored the general dynamics of NSCs, and in particular NSCs similar to the well-observed NSC in the Milky Way Galactic Center (GC). The presence of a young stellar disk in the central pc of GC, as well as a dense concentration of HII regions and young stars throughout the central 100 pc of the Milky-way (Figer *et al.* (2004)) provide evidence for a continuous star-formation in this region (Genzel *et al.* (2010)). Evidence for star formation exists in other extagalactic NSCs (e.g. Seth *et al.* (2006)). Walcher *et al.* (2005) argued that NSCs are protobulges that grow by repeated accretion of gas and subsequent star formation, McLaughlin *et al.* (2006) suggested a NSC in-situ star formation model regulated by momentum feedback.

These various studies provide further motivation and suggest that star-formation has an important role in shaping NSCs and their evolution. Here, we summarize our results of the role of in-situ SF in NSCs, and explore its implication both for the build-up of the NSC, as well as the long term evolution and structure of NSCs (detailed information can be found in Aharon & Perets (2014)). Our work makes use of the Fokker-Planck (FP) diffusion equation, first used by Bahcall & Wolf (1976) in this context, to describe the dynamics of stellar populations in dense clusters around MBHs.

2. Evolution of NSCs around MBHs: by Fokker-Planck analysis

NSCs are complex interacting systems. Their evolution and dynamics are mainly affected by two-body relaxation. Here we follow the evolution of NSCs by numerically solving the FP equation following the approach first used by Bahcall & Wolf (BW; 1976, 1977) in this context. However, we supplement the basic equation, for the first time, with a source term accounting for SF, as well as use a large number of distinct stellar populations to account for different SF epochs.

In our model we simulate the star formation in the GC through adding an extra source term component to the FP equation. Its value and range are determined according to the number of new stars added in the appropriate region. We simulate multiple stellar populations forming at different epochs, and follow the evolution of their distribution. The modified FP equation with the addition of the source term has the form:

$$\frac{\partial f(E,t)}{\partial t} = -AE^{-\frac{5}{2}} \frac{\partial F}{\partial E} - F_{LC}(E,t) + F_{SF} \tag{2.1}$$

where the $F_{LC}(E,t)$ term corresponds to the empty loss-cone term (see Lightman & Shapiro (1977); Young (1977); Perets *et al.* (2007)).

The source term added to the FP equation is:

$$F_{SF} = \frac{\partial}{\partial t} \left(\Pi(E) E_0 E^{\alpha} \right) \tag{2.2}$$

where $\Pi(E)$ is a rectangular function, which boundaries correspond to the region where new stars are assumed to from; E_0 is the source term amplitude; and F_{SF} is a power-law function with a slope α, defining the SF distribution in phase space . We simulated a number of NSC evolutionary scenarios, taking different models for the SF function (rate and spatial structure) and for the background population. The chosen slope of the SF function was motivated by the observed power-law (Do *et al.* (2009)) distribution of young stars observed in the young stellar disk in the GC.

3. Results

We have followed the evolution of multiple stellar populations formed at different epochs. The build-up and structure evolution of an NSC which grows through a continuous long-term in-situ star formation. The final configuration of this NSC is very similar to that of a steady state BW cusp, and the number densities are comparable to those observed in the GC. The final structures of these NSCs after 10 Gyr of evolution for the two main scenarios (2b and 7) are summarized in Fig. 1. The other scenarios can be found in Aharon & Perets (2014). These models show the existence of a core-like structure for the young stellar populations, where the cores vary in size, and are systematically bigger for younger populations.

4. Discussion

The build up of nuclear stellar clusters through in-situ star formation. We explored two evolutionary scenarios of NSCs: (1) pre-existing NSCs with a BW-like structure that experience later SF and (2) NSCs built-up completely from in-situ SF. Both type of models

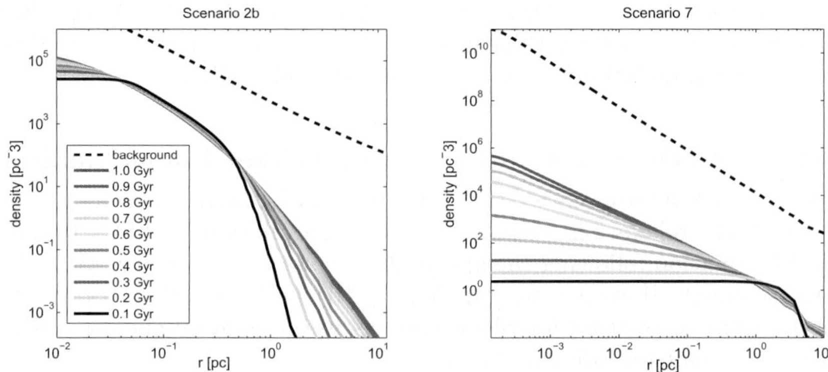

Figure 1. The number density profile of a 10 Gyr evolved NSC with SF inside the central 2 pc (scenario 2b) and outside (Scenario 7). The number of gigayears presents the age of the new population. The difference in the SF range affects the final distribution of an evolved NSC. The old background stellar population (black dashed lines) correspond either to stars produced through in-situ SF (over the first 9 Gyr; right) or to a pre-existing BW cusp population (left). The structures of these old populations show an almost BW-like steady state behavior in both of the models, while the young stellar populations formed in the last Gyr are not yet relaxed, and show large cores ranging in size. The young populations arise from in-situ star formation at a rate of 10^{-4} stars yr^{-1}, during the last Gyr, where each SF episode continues for 100 Myr.

assume that several epochs of gas-infall into the nuclear region triggered SF, transforming the infalling gas into newly born stars. We show that NSCs built-up from in-situ SF give rise to NSCs dominated by the stellar population formed at earlier stages (first few gigayears). The structure of the older population is very similar to a steady-state BW-cusp, and the total number of stars is comparable to that inferred for the GC NSC. We note that when lower mass stellar populations are assumed (e.g. if different initial mass function are considered) the relaxation times become longer, as expected, and late-formed younger populations are far from achieving a steady state structure, producing larger core-like structures, as discussed in more details below.

Core-cusp structure. After ~ 10 Gyr of NSC evolution, the younger stellar populations in NSCs may evolve to a core-like distribution, while older population already become progressively cuspier. These results are of great interest in light of the recent findings about the structure of the GC NSC.

There are a number of models suggesting to explain the origin of the GC core, for example Merritt (2010) suggested that a binary merger, or a triaxial potential could deplete the inner regions of an NSC producing a large core, and have shown that the long relaxation times would not be sufficient to regrow a cusp. A similar behavior is seen in our models, where progressively younger populations of stars formed in the outer regions of the NSC do not relax and grow an inner cusp. Though in both cases slow relaxation explains the non-growth of the inner cusp, the origins of the initial core in both models differ, and the outcomes could significantly differ as well. In particular, the SF models studied here suggest that cores of different sizes could exist for different stellar populations, and in particular an NSC can have both a cusp distribution of old stars and a core distribution for young and intermediate age stars.

We note that younger, more massive red giants could be more luminous and more easily detected in observations (Pfuhl *et al.* (2011)). We therefore hypothesize that if such younger red giants (up to 2-3 Gyr old) are overly represented in observations then the observed core could be limited to these younger populations, while the underlying population of older stars might still have a cusp distribution. This can be well

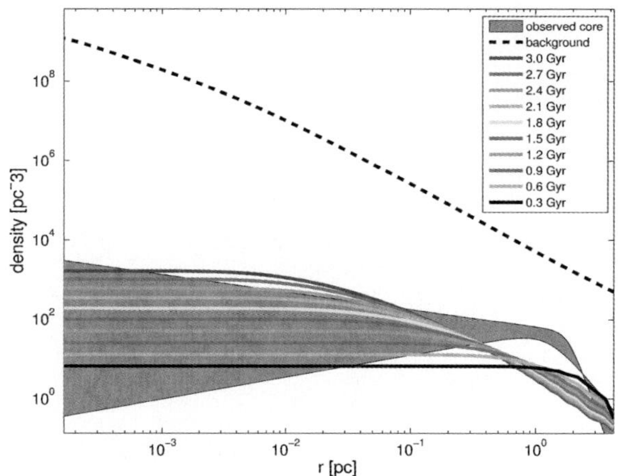

Figure 2. The number density profile of the GC nuclear cluster after 10 Gyr evolution (and total star formation epoch of 3 Gyr) compared to the inferred range of number density profiles from 3D modeling of GC observations Do *et al.* (2013). The number of Gyrs presents the age of the new population.

demonstrated both in Fig. 2 where the model results are compared with the 3D modeled number density profile determined by Do *et al.* (2013) based on observations. As can be seen, in some models a large, parsec-size core of up to a few gigayears old stellar populations can exist. Such a core might be consistent with the density profiles inferred from observations, while the old stellar population preserves a typical BW-like cusp profile.

References

Aharon, D. & Perets, H.B. 2014 *ArXiv e-prints*, 1409.5121

Antonini, F. 2013, *ApJ*, 763, 62

Bahcall, J. N. & Wolf, R. A. 1976, *ApJ*, 209, 214-232

Bahcall, J. N. & Wolf, R. A. 1977, *ApJ*, 216, 883-907

Do, T., Ghez, A. M., Morris, M. R., Lu, J. R., Matthews, K., Yelda, S., & Larkin, J. 2009, *ApJ*, 703, 1323-1337

Do, T., Martinez, G. D., Yelda, S., Ghez, A., Bullock, J., Kaplinghat, M., Lu, J. R., Peter, A. H. G.., & Phifer, K. 2013, *ApJL*, 779, L6

Figer, D. F., Rich, R. M., Kim, S. S., Morris, M., & Serabyn, E. 2004, *ApJ*, 601, 319-339

Genzel, R., Eisenhauer, F., & Gillessen, S. 2010, *Reviews of Modern Physics*, 82, 3121-3195

Lightman, A.P. & Shapiro, S.L. 1977 *ApJ*, 211, 244-262

Loose, H. H., Kruegel, E., & Tutukov, A. 1982, *AAP*, 105, 342-350

McLaughlin, D. E., King, A. R., & Nayakshin, S. 2006, *ApJL*, 650, L37-L40

Merritt, D 2010, *ApJ*, 718, 739-761

Perets, H. B., Hopman, C., & Alexander, T. 2007, *ApJ*, 656, 709-720

Pfuhl, O., Fritz, T. K., Zilka, M., Maness, H., Eisenhauer, F., Genzel, R., Gillessen, S., Ott, T., Dodds-Eden, K., & Sternberg, A. 2011, *ApJ*, 741, 108

Seth, A. C., Dalcanton, J. J., Hodge, P. W., & Debattista, V. P. 2006, *AJ*, 132, 2539-2555

Tremaine, S.D., Ostriker, J.P., & Spitzer, L. Jr. 1975 *ApJ*, 196, 407-411

Walcher, C. J., Van der Marel, R. P., McLaughlin, D., Rix, H. W., Böker, T., Häring, N., Ho, L. C., Sarzi, M., & Shields, J. C. 2005, *ApJ*, 618, 237-246

Young, P. J. 1977, *ApJ*, 215, 36-52

PART THREE
Gravitational wave emission, observations,
and the link to astrophysics

Star clusters and black holes in galaxies across cosmic time
Proceedings IAU Symposium No. 312, 2014
Y. Meiron, S. Li, F.-K. Liu & R. Spurzem, eds.

© International Astronomical Union 2016
doi:10.1017/S1743921315008042

White dwarf binaries and the gravitational wave foreground

Matthew Benacquista

Center for Gravitational Wave Astronomy, University of Texas at Brownsville,
One West University Blvd, Brownsville, TX 78520, USA
email: `matthew.benacquista@utb.edu`

Abstract. Galactic white dwarf binaries will be an abundant source of gravitational waves in the mHz frequency band of space-based detectors such as eLISA. A few thousand to a few tens of thousands of these systems will be individually resolvable by eLISA, depending on the final detector configuration. The remaining tens of millions of close white dwarf binaries will create an unresolvable anisotropic foreground of gravitational waves that will be comparable to the instrument noise of eLISA at frequencies below about a mHz. Both the resolvable binaries and the foreground can be used to better understand this population. Careful choice of the initial orientation of eLISA can mitigate this foreground in searches for other sources.

Keywords. Gravitational Radiation, Ultracompact Binaries

1. Introduction

Ultracompact binaries with orbital periods less than a day will be the dominant source of continuous gravitational radiation in the millihertz band of the spectrum. Most of the proposed designs for space-based gravitational wave observatories are sensitive in this frequency band and the Galactic population of ultracompact binaries will produce a confusion-limited foreground at or above the designed instrumental noise. This acts as a source of noise for observations of extragalactic and cosmological gravitational wave sources, such as supermassive binary black hole inspirals, extreme mass ratio inspirals, or cosmological strings. However, the Galactic signal also contains a wealth of information about close binary evolution, the star formation history of the Galaxy, and the production of binaries through dynamical interactions in dense stellar systems. In this review, we will provide a brief introduction to the physics of gravitational radiation, a review of the current designs for eLISA, a history of population synthesis of close white dwarf binaries in the Galaxy, and conclude with a discussion of the potential of eLISA to observe this population.

2. Gravitational radiation from binary systems

Gravitational radiation is a propagating perturbation in the curvature of spacetime, which manifests itself as a time-varying perturbation of the metric. The most common coordinate (or gauge) choice used to describe a gravitational wave is the transverse-traceless gauge, which highlights the fact that gravitational radiation is a transverse wave. In the far zone, where we expect to measure the gravitational wave, we can approximate the background metric as a flat spacetime metric ($\eta_{\mu\nu}$) and the gravitational wave as a perturbation ($h_{\mu\nu}$), so:

$$g_{\mu\nu} = \eta_{\mu\nu} + h_{\mu\nu}. \tag{2.1}$$

A passing gravitational wave is measured as a strain, h. Gravitational waves generated by binary systems produce a strain in a detector whose amplitude is proportional to the orbital frequency, f, the distance to the binary, d, and the masses of the components, M_1 and M_2:

$$h \propto \frac{M_1 M_2}{(M_1 + M_2)^{1/3}} \frac{f^{2/3}}{d}. \tag{2.2}$$

The combination of masses is referred to as the "chirp mass" with:

$$\mathcal{M}_c = \left(\frac{M_1 M_2}{(M_1 + M_2)^{1/3}} \right)^{3/5} = \mu^{3/5} M^{2/5}, \tag{2.3}$$

where μ is the reduced mass and M the total mass.

For circularized binaries, the gravitational wave signal is a nearly monochromatic wave with a frequency that is twice the orbital frequency. As the binary loses energy through gravitational wave emission, the orbital frequency increases as the components slowly spiral together. The rate of change of the frequency is known as the chirp and is related to the chirp mass by:

$$\frac{df}{dt} \propto \mathcal{M}_c^{5/3} f^{11/3}. \tag{2.4}$$

Since the chirp is a measure of the rate at which gravitational wave energy is emitted by the binary, it can be considered as providing the absolute magnitude of the binary in gravitational radiation. The observed strain amplitude provides the apparent magnitude of the binary. Both depend upon the masses by the chirp mass, so the distance can be determined for a chirping binary if f, $\dot{f}$, and h are measured:

$$d = \frac{5c\dot{f}}{384\pi^2 h f^3}. \tag{2.5}$$

3. Space-based detectors

Designs for space-based gravitational wave detectors have been around since the mid 1980's. The initial design was LAGOS (Faller *et al.* 1989). This had four spacecraft flying in formation, creating an interferometer with very large arms. This design evolved to become LISA in the mid-1990s as an ESA cornerstone mission (Folkner, Bender, & Stebbins 1998). In the early 2000s, LISA was a proposed joint NASA/ESA mission with well defined characteristics (Bender 1998). In 2011, NASA withdrew from the partnership. Over the past few years, the original LISA mission has evolved into an ESA led mission, known as eLISA (Amaro Seoane *et al.* 2013). It is currently selected as the 3rd Large mission (L3) with an expected launch of 2034. Much of the design of eLISA retains features of the original LISA. The detector itself is a constellation of three spacecraft in orbit about the sun. The constellation forms a nearly equilateral triangle that tumbles and precesses as the spacecraft follow their orbits, as shown in Figure 1. Each spacecraft is maintained in a geodesic orbit by tracking an internal proof mass that is shielded from external forces. The separation between pairs of spacecraft are measured through laser links. The measurement of a passing gravitational wave appears as a strain, $h = \Delta L / L$, where the armlength, L, is the separation between spacecraft. The major sources of noise are spurious forces, position sensing, and shot noise. The spurious forces produce an acceleration noise that generates a variation in ΔL which is proportional to f^{-2}. It

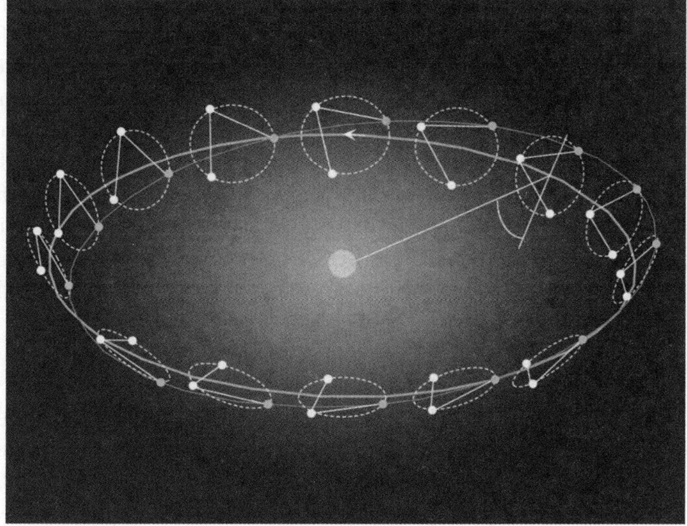

Figure 1. Orbit of the eLISA constellation about the sun.

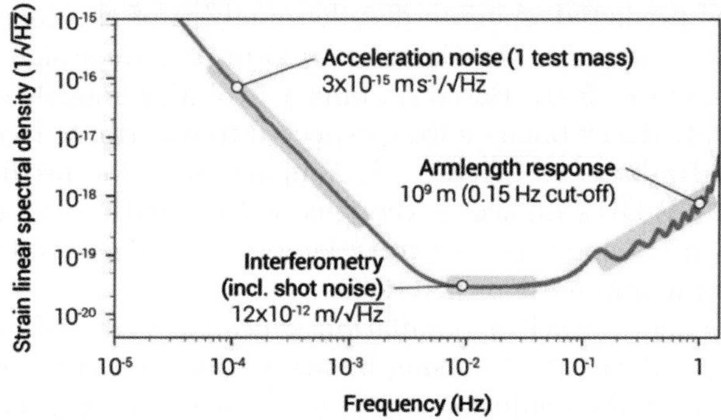

Figure 2. Typical eLISA sensitivity curve showing the contributions from acceleration, position and shot noise, and the armlength response.

dominates the instrument noise at low frequencies. Both the position and shot noise are assumed to be flat in frequency and form the floor of the detector sensitivity.

Current design choices for eLISA involve choosing armlengths that are shorter than the 5 million km of the initial LISA design. The effect of shortening the arms is to increase the acceleration and position noises inversely proportional to the change in armlength. Shortening the armlength also has the effect of decreasing the shot noise by increasing the received laser power at each spacecraft. If the shot noise drops significantly below the position noise, then additional savings can be found by lowering the laser power or decreasing the size of the optics.

Armlength also has an effect on the response of the detector for gravitational waves with wavelengths that are comparable to or shorter than the length of each arm. In this case, the effective armlength is proportional to the wavelength. Thus, the response of the detector decreases with increasing frequency beyond a threshold frequency set by the armlength of the detector, resulting in a rise in the sensitivity curve at high frequencies. Shortening the armlength then has the effect of increasing the upper frequency of the sensitivity band. See Figure 2, for a typical eLISA sensitivity curve.

The initial LISA design had three laser links, joining all three pairs of spacecraft in the constellation. This allowed for the synthesis of two linearly independent interferometers, and provided a coverage of the gravitational wave sky that was uniform in ecliptic longitude. The current eLISA designs call for only two pairs of spacecraft to be joined by laser links. As a result, there are four directions relative to the constellation to which the detector is insensitive to gravitational waves. These null directions trace out figure-eights in the sky as eLISA orbits the sun. The effect of this is to introduce four regions of reduced sensitivity in ecliptic longitude. The position of these regions in the sky depend upon the initial orientation of eLISA (Jani, Finn & Benacquista 2013). Judicious choice of the initial orientation of eLISA can then be used to suppress gravitational wave signals from certain regions in the sky while enhancing the response in other regions.

4. Population syntheses

Close white dwarf binaries (CWDBs) are compact enough to have short orbital periods, yet massive enough to have chirp masses that are large enough to produce meaningful strain amplitudes at typical Galactic distances. Thus, they are the most likely sources of gravitational radiation. Early estimates of the detectability of close white dwarf binaries were based on the rates of Type Ia supernovae, assuming a double degenerate progenitor scenario (Evans, Iben & Smarr 1987). Although highly uncertain, a Galactic foreground level was predicted. A more detailed analysis of the contribution of several types of ultracompact binaries to the Galactic gravitational wave signal was done in 1990 (Hils, Bender, & Webbink 1990). Based on binary evolution models of Webbink (1984), roughly 10^8 close white dwarf binaries were expected to contribute to a confusion-limited foreground in the mHz band of LAGOS. A comparison of the predicted space density with the number of CWDB's known at the time led the authors to reduce the number of modeled CWDBs by a somewhat arbitrary factor of 10. The resulting foreground still dominated the instrumental noise of LAGOS.

Ten years later, a more detailed population synthesis was performed by Nelemans, Yungelson & Portegies Zwart (2001) using binary evolution models of Hurley. This produced a similar level for the confusion noise, but was based on a population of $\sim 10^8$ CWDBs. Thus, although their population was 10 times larger than the reduced number of Hils, Bender, & Webbink (1990), the overall levels of the confusion noise were comparable. This is because the improved binary evolution models result in an average chirp mass of nearly half the value found by Hils, Bender, & Webbink (1990). By this time, LISA was the expected gravitational wave observatory. After another decade, a new population synthesis by Ruiter et al. (2010) using the `StarTrack` population synthesis code produced results similar to Nelemans et al. This indicates that the predicted value of the confusion-limited foreground is fairly robust, given the uncertainties in the binary evolution assumptions.

Much of the calibration of the population synthesis codes has been based on a small number of observed systems. In recent years, dedicated searches for white dwarf binaries such as the extreme low mass (ELM) survey have discovered a growing number of low mass detached and mass transferring systems (Kilic, et al. 2010; Brown et al. 2010; Kilic, et al. 2012). In addition, a number of AM CVn systems have been found in the Palomar Transient Factory (PTF) data and Sloan Digital Sky Survey data (Rau et al. 2010). The number of AM CVn systems found is lower than that predicted by population synthesis methods (Roelofs, Nelemans, & Groot 2007). There are two proposed solutions to this discrepancy (Nissanke et al. 2012). Once could be a broader spatial distribution of CWDBs, leading to an unchanged total number of CWDBs, but a lower local space density of the nearby and more easily observable AM CVn systems. Another could be

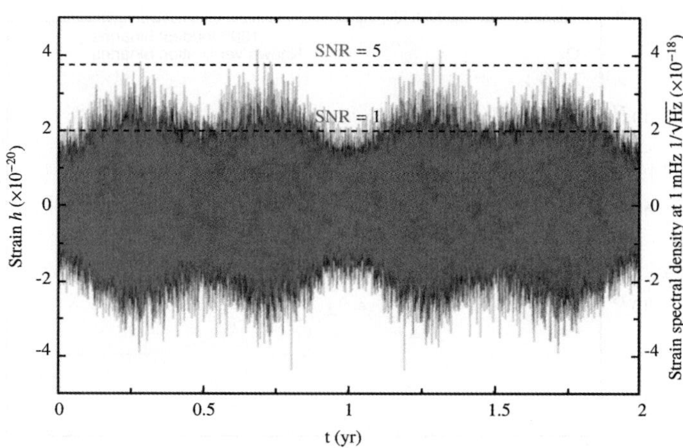

Figure 3. Level of the Galactic gravitational wave signal as a function of time. Black is the total signal, the red after removal of the resolved binaries. The yearly variation of the Galactic foreground is clearly seen. Based on the Ruiter *et al.* (2010) Galactic model.

that the conditions for stable mass transfer are more restrictive than current assumptions. This would lead to fewer AM CVns surviving after the onset of mass transfer. In this case the local space density of CWDBs would remain the same, but the number of observable AM CVns would be reduced.

5. Gravitational wave observations

The expected gravitational wave signal from the Galactic population of close white dwarf binaries can be separated into two classes—the unresolved confusion-limited foreground, and individually resolved systems. The confusion-limited foreground dominates at frequencies below a few mHz. In this region, there are multiple binaries within each resolvable frequency bin. If these overlapping signals cannot be disentangled, then they appear as a stochastic noise. The distribution of CWDBs in the sky is anisotropic with the majority of them located in the direction of the Galactic center. Consequently, the level of the foreground noise varies as the sensitivity pattern of eLISA passes over the Galactic center, as shown in Figure 3. Since the confusion foreground dominates in the frequency band where acceleration noise is the main contributor to the instrument noise, a reduction in the armlength tends to reduce the contribution of the foreground signal to the overall noise at low frequencies. See Figure 4, for an illustration of the confusion noise relative to different eLISA armlength choices.

The modulation of the confusion foreground may be used to identify and separate the foreground from the instrument noise. If the spectral slope of the signal can be determined, it may also be possible to constrain star formation histories of the Milky Way (Yu & Jeffery 2013).

The resolvable signals generally come from high-mass or high-frequency systems. These systems are very bright and can be detected from anywhere in the Galaxy. Thus, the resolvable systems will provide a complete census of double degenerate progenitors to type Ia supernovae (Stroeer, Benacquista, & Ceballos 2013). Additionally, some of the resolvable systems will have measurable chirps and so their distances can also be measured. This may provide a better understanding of the spatial distribution of the massive, high-frequency population of CWDBs in the Galaxy.

If a binary can be resolved, its sky location can be determined through modulations in

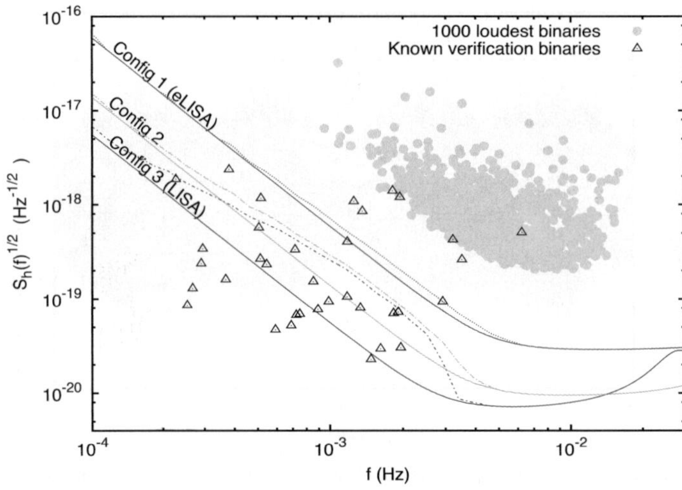

Figure 4. Sensitivity curves for 5, 2, and 1 million km armlengths. The solid lines show the sensitivity set by the measurement noise while the dashed curves include an estimate of the confusion-limited foreground. Known binaries are indicated with black triangles and resolved binaries from a population synthesis are grey circles. Figure taken from Littenberg, *et al.* (2013).

the amplitude, frequency, and polarization of the observed gravitational wave as eLISA executes its tumbling motion during its orbit about the sun. Typical angular resolutions for these systems will be below a few square degrees. This angular resolution will be sufficient to identify the compact binary population of individual globular clusters.

The gravitational wave strain amplitude produced by close white dwarf binaries is too small for any extragalactic systems to be observed. However, stellar mass binary black holes will have substantially larger chirp masses and may be detectable at the distance of the Virgo cluster galaxies. The field population of stellar mass black holes is small enough that fewer than a couple of Galactic systems may exist in the eLISA band. The increased number of galaxies gained by looking out to the Virgo cluster may bring this number up to 10 or so, if extragalactic systems are included (Benacquista *et al.* 2014).

The potential for detection of extragalactic sources within the Virgo cluster provides incentive to take advantage of the ability to suppress eLISA sensitivity to regions of the sky. The bulk of the foreground signal will come from the large population of CWDBs in the direction of the Galactic center. Choosing the initial orientation of eLISA so that one of the nulls passes over the Galactic center reduces the sensitivity of eLISA to the strongest source of the foreground. This orientation also reduces the sensitivity of eLISA to signals from the direction of the Virgo cluster, but the two sensitivity nulls that cause these reductions are out of phase with each other. Thus, the sensitivity to the Virgo cluster is at its highest during those times that the sensitivity to the Galactic center is at its lowest (Jani, Finn & Benacquista 2013).

6. Conclusions

Space-based interferometry will be the most likely detector for observations of gravitational waves in the mHz band. The most numerous sources for gravitational radiation in this band are the Galactic population of close white dwarf binaries. Decades of population synthesis work has produced a robust prediction of the nature of the gravitational wave signal from this population. There will be a confusion-limited foreground signal from millions of CWDBs at frequencies below 3 mHz. This signal will be comparable or larger than the instrument noise for several designs of eLISA. In addition to the foreground, a few thousand of Galactic CWDBs will be individually resolvable. These systems will be high mass/high frequency sources, and will be detectable from throughout the Galaxy. Because the distribution of CWDBs is concentrated toward the Galactic center, the foreground is anisotropic in the sky and it will be modulated as the eLISA sensitivity peak sweeps over the Galactic center during its orbit about the sun. Some black hole binaries produced in globular clusters may also be observable at extragalactic distances, with the majority of them coming from the Virgo cluster. By choosing the initial orientation of the eLISA constellation, it will be possible to suppress the foreground signal while enhancing the sensitivity to sources in the Virgo cluster.

References

Amaro Seoane, P., *et al.* (The eLISA Consortium) 2013, *arXiv*, 1305.5720

Benacquista, M., Hinojosa, J., Mata, A., & Belczynski, K. 2014, *arXiv* 1410.1177

Bender, P. L. 1998, *Bull. Amer. Astron. Soc.*, 30, 1326

Brown, W. R., Kilic, M., Allende Prieto, C., & Kenyon, S. J. 2011, *ApJ*, 723, 1072

Evans, C. R., Iben, I., & Smarr, L. 1987, *ApJ*, 323, 129

Faller, J. E., Bender, P. L., Hall, J. L., & Hils, D., Stebbins R. T. 1989, *Adv. Space Res.*, 9, 107

Folkner, W. M., Bender, P. L., & Stebbins, R. T. 1998, *Publication 97-16*, Jet Propulsion Laboratory, California Institute of Technology, Pasadena, CA.

Hils, D., Bender, P. L., & Webbink, R. F. 1990, *ApJ*, 360, 75

Jani, K. P., Finn, L. S., & Benacquista, M. J. 2013, *arXiv*, 1306.3253

Kilic, M., Brown, W. R., Allende Prieto, C., Kenyon, S. J., & Panei, S. J. 2010, *ApJ*, 716, 122

Kilic, M., Brown, W. R., Allende Prieto, C., Kenyon, S. J., Heinke, C. O., Agüeros, M. A., & Kleinman, S. J. 2012, *ApJ* 751, 141

Littenberg, T. B., Larson, S. L., Nelemans, G., & Cornish, N. J. 2013, *MNRAS*, 429, 2361

Nelemans, G., Yungelson, L. R., & Portegies Zwart, S. F. 2001, *A&A*, 375, 890

Nissanke, S., Vallisneri, M., Nelemans, G., & Prince, T. A. 2012, *ApJ*, 758, 131

Rau, A., Roelofs, G. H. A.., Groot, P. J., Marsh, T. R., Nelemans, G, Steeghs, D., Salvato, M., & Kasliwal, M. M. 2010, *ApJ*, 708, 456

Roelofs, G. H. A.., Nelemans, G., & Groot, P. J. 2007, *MNRAS* 382, 685

Ruiter, A. J., Belczynski, K., Benacquista, M., Larson, S., & Williams, G. 2010, *ApJ*, 717, 1006

Stroeer, A., Benacquista, M., & Ceballos, F. 2013, in *Proc. IAU Symposium 281: Binary Paths to Type Ia Supernovea Explosions*, ed. R. Di Stefano, M. Orio & M. Moe (Cambridge University Press), 217

Webbink, R. F. 1984, *ApJ*, 277, 355

Yu, S. & Jeffery, S. 2013, *MNRAS*, 429, 1602

Star clusters and black holes in galaxies across cosmic time
Proceedings IAU Symposium No. 312, 2014
Y. Meiron, S. Li, F.-K. Liu & R. Spurzem, eds.

© International Astronomical Union 2016
doi:10.1017/S1743921315008054

The gravitational wave signal from close galaxy pairs

Jinzhong Liu and Yu Zhang

National Astronomical Observatory/Xinjiang Observatory, Chinese Academy of Sciences, 150
Science 1-Street, Urumqi, Xinjiang 830011, China
email: liujinzh@xao.ac.cn

Abstract. The early phase of coalescence of supermassive black hole binaries (SMBHBs) from
their host galaxies provides a guaranteed source of low-frequency gravitational wave (GW)
radiation by pulsar timing observations. Nowadays, SMBHBs are ubiquitous in the nuclei of
galaxies. A latest sample of close galaxy pairs has been released from the Sloan Digital Sky
Survey (SDSS) Data. A binary population synthesis (BPS) approach has been applied to study
the characteristics of clusters and galaxies. Here we report how BPS, using SDSS results, can
be used to determine the GW radiation from SMBHBs. In this study we show numerical results
under the assumption that SMBHBs formed through the merger of two galaxies and give the
waveform evolution using post-Newtonian approximation methods. Based on the sensitivity of
the International Pulsar Timing Array (IPTA) and Square Kilometer Array (SKA) detectors,
we show that the value of strain amplitude h can be changed from about 10^{-14} to 10^{-15} during
the observation of 20 years, which can be considered as a precise evolution.

Keywords. gravitational wave, black hole, galaxy.

1. Introduction

The various frequency ranges of the GW detectors can respectively observe different
GW sources (Jaffe & Backer 2003; Belczynski, Kalogera & Bulik 2002; Liu 2009; Liu
et al. 2010a; Liu *et al.* 2010b; Liu *et al.* 2012, Liu *et al.* 2014). The supermassive black
hole binaries (SMBHBs) are thought to be the LISA verification binary GW sources.
This paper provides new detailed specifics and characteristics of the SMBHBs with the
latest findings of Liu *et al.* (2012).

2. Computations

Few studies can accurately determine galaxy pairs internal compositions and kinematics equations. The BL Lacertae object OJ287, will allow us to investigate a precise orbital
evolution due to GW radiation in our BPS simulations, where GW radiation energy loss
is appeared in the galaxy pairs. To systematically investigate the GW radiation of galaxy
pair systems, we perform a Monte Carlo simulation where we follow the evolution of a
sample of 1 million binaries. The properties of the overall galaxy pair sample have been
displayed by the released SDSS data, including projected and angular separation, relative
velocities, stellar mass ratio and the redshift. We present polynomial curve fitting for-
mulae, which is to construct the distribution function of galaxy pairs, that approximate
the evolution of SMBHBs for a wide range of mass and stellar environment. And the star
formation rate is taken to be constant in the simulation. According to the derivation of
black hole mass function at low redshift, a simple polynomial curve fitting approximation
of SBHMF is also used. In addition to all aspects of SDSS data, the accretion disk model

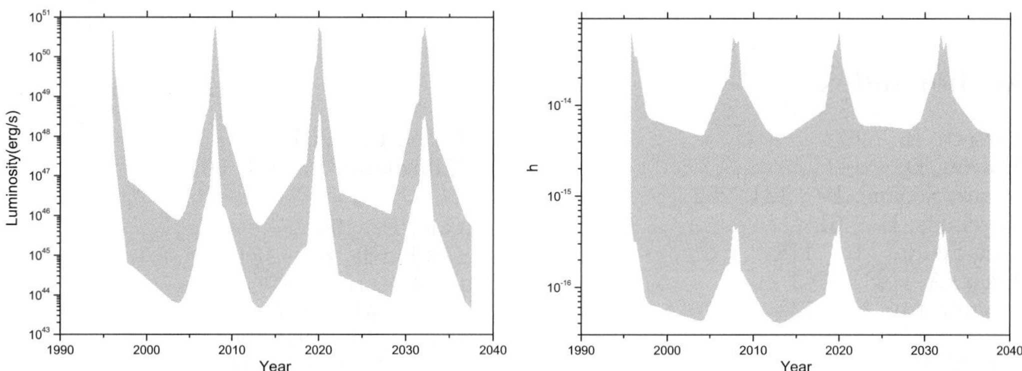

Figure 1. The strain amplitude h and luminosity L of GW radiation variation with time.

of primary galaxy is assumed by the analytical fitting function. And for the eccentricity of the MBHB samples, we obtain it from the random number.

3. Results and discussion

Figure 1 shows that the range of strain amplitude and luminosity can be covered by GW radiation from MBHBs in our Monte Carlo simulations. From these two Figures we find that there is major double-peaked GW radiation outburst at approximately 12 year intervals, which is consistent with the light curves of OJ 287. This is because that we adopt a precise orbital evolution of OJ 287 as our orbital evolutionary model to investigate the orbital track changes due to GW radiation in the BPS simulations. In other words, according to the proposal of accretion disk model, the maximum outbursts of GW radiation and the optical outburst are happening at the same time when the secondary BH crashes into the disk of the primary BH. From the right panel, we see that the strain amplitude of GW radiation will be detected by the SKA detector. From the left panel, we can see that our calculated MBHB samples can radiate a high luminosity due to GWs, which possibly means the influence of GW dominates the total energy loss of the galaxy pairs in the phase of coalescence. Meanwhile this variation of strain amplitude or luminosity with time maybe provides an indirect evidence for GWs existence. Especially using 20 years of observational time the pulsar timing measurements will confirm a period variation of 12 yr in residual data. Finally, we can find some irregular "sawtooth" appearance at the maxima, which indicates that the precession of orbital period due to GW radiation and the accretion disk can influence the evolution of galaxy pairs.

Acknowledgements

This work is supported by the program of the light in China's Western Region (LCWR) (No. XBBS201221) and Natural Science Foundation (No. 11303080).

References

Belczynski, K., Kalogera, V., & Bulik, T. 2002, *ApJ*, 572, 407

Jaffe, A. H. & Backer, D. C. 2003, *ApJ*, 583, 616

Liu, J. Z. 2009, *MNRAS*, 400, 1850

Liu, J. Z., Han, Z., Zhang, F., & Zhang, Y. 2010a, *ApJ*, 719, 1546

Liu, J. Z., Zhang, Y., Han, Z., & Zhang, F. 2010b, *Ap&SS*, 329, 297

Liu, J. Z., Zhang, Y., Zhang, H., Sun, Y., & Wang, N. 2012, *A&A*, 540, 67

Liu, J. Z. & Zhang, Y. 2014, *PASP*, 126, 211

Author index